Miniaturized Analytical Devices

Miniaturized Analytical Devices

Materials and Technology

Edited by Suresh Kumar Kailasa and Chaudhery Mustansar Hussain

WILEY-VCH

Editors

Prof. Suresh Kumar Kailasa
Department of Chemistry
Sardar Vallabhbhai National Institute of Technology (SVNIT)
Ichchhanath Surat-Dumas Road
395007 Surat
Gujarat
India

Prof. Chaudhery Mustansar Hussain
New Jersey Institute of Technology
Department of Chemistry & Environmental Sciences
Newark
New Jersey
United States

Cover Image: © gan chaonan/Getty Images

Library of Congress Card No.: applied for

British Library Cataloguing-in-Publication Data
A catalogue record for this book is available from the British Library.

Bibliographic information published by the Deutsche Nationalbibliothek The Deutsche Nationalbibliothek lists this publication in the Deutsche Nationalbibliografie; detailed bibliographic data are available on the Internet at <http://dnb.d-nb.de>.

Print ISBN: 978-3-527-34758-2
ePDF ISBN: 978-3-527-82720-6
ePub ISBN: 978-3-527-82722-0
oBook ISBN: 978-3-527-82721-3

Contents

Section 1

Miniaturized Devices in Analytical and Bioanalytical Sciences

1

Miniaturized Capillary Electrophoresis for the Separation and Identification of Biomolecules

Suresh K. Kailasa[1], Vaibhavkumar N. Mehta[2], and Jigneshkumar V. Rohit[3]

[1] *Department of Chemistry, Sardar Vallabhbhai National Institute of Technology, Surat-395007, Gujarat, India*
[2] *ASPEE Shakilam Biotechnology Institute, Navsari Agricultural University, Surat-395007, Gujarat, India*
[3] *Department of Chemistry, National Institute of Technology Srinagar, Hazratbal Kashmir 190006, Kashmir, India*

1.1 Introduction

Microchip capillary electrophoresis (MCE) is one of the efficient bioanalytical tools for rapid separation and detection of bioactive molecules with high separation resolution [1–3]. It has proven to be a prominent tool for identification of nucleic acids and proteins in food and clinical microbiology [2]. Separation of biomolecules is a key platform for quantitative and qualitative analysis of target biomolecules in biological matrices. For the first time, Manz's group integrated a simple analytical procedure on a small glass chip for the separation and detection of target chemical species [4], collectively referred to as "lab-on-a-chip" or micro total analysis systems (μTAS). In this concept, MCE is included due to the separation mechanism in a microchip with very short channels. As a result, target molecules from the mixture are effectively separated by using high electric field strengths. MCE has been widely applied as a rapid separation tool in various fields of science, i.e. proteomics, genomics, biomarkers, and forensics [5–9]. These reviews reported that MCE has shown better performance for the separation of target analytes compared to traditional capillary electrophoresis. The MCE has successfully separated >30 000 proteins from a single cell [10].

In this chapter, we summarize the recent developments of MCE for separation and identification of nucleic acids and proteins from clinical and food samples. We briefly describe the history and role of MCE in clinical and food microbial research. A section is devoted to applications of MCE for separation and identification of nucleic acids, proteins, and biomarkers from clinical and food samples. The analytical features of MCE for rapid separation and detection of biomolecules are tabulated, which provides significant information to scientists to know potential advancements of MCE in molecular biology.

Miniaturized Analytical Devices: Materials and Technology, First Edition.
Edited by Suresh Kumar Kailasa and Chaudhery Mustansar Hussain.

1.2 Brief Summary of MCE

Generally, MCE consists of four core parts: microfluidic chip, electric field, separation, and detector. The electric field is applied for sample concentration and separation. Figure 1.1 displays a T-shaped microfluidic chip. The microfluidic chip contains few reservoirs such as sample and buffer reservoirs. These reservoirs should be filled with a background solution, and sieving gels and pipetting and syringe pumps are used for fluidic control. Once the microfluidic chip is set up with these parts, a high electric field is applied to the reservoirs (sample) to separate the target analytes. The detector is placed at the end of the separation channel, which results in registering the zones for separation and transmitting the data for signal processing unit, which generates an electropherogram. In this section, an overview of fabrication of microfluidic chips, sample preparation (on-microfluidic chip), separation, and analyte detection is given.

1.2.1 Fabrication of Microfluidic Chips

So far, several methods have been adopted for the fabrication of microfluidic chips, such as reactive ion etching, wet etching, photolithography, conventional

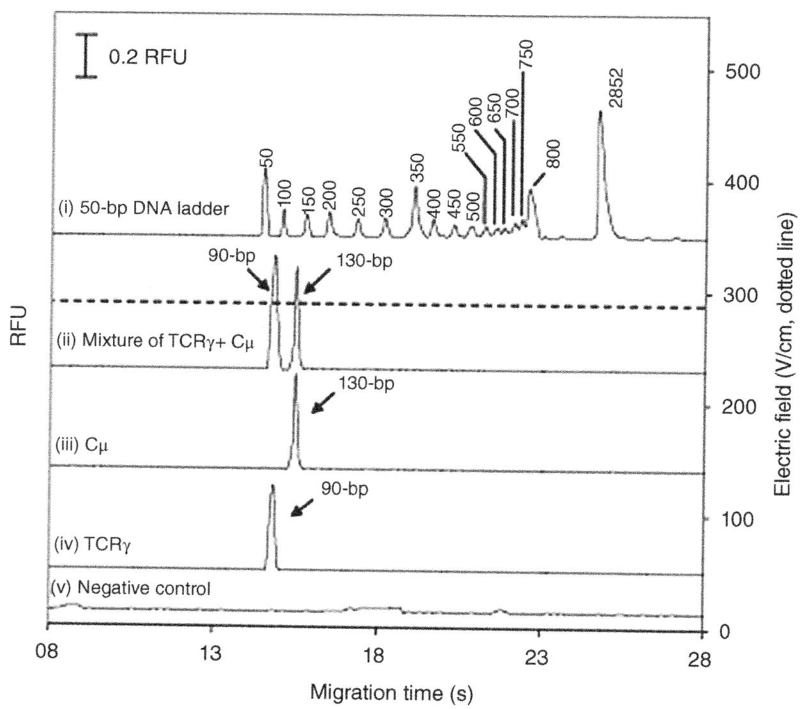

Figure 1.1 Separation and identification of amplified-PCR products of T-cell lymphoma. 50-bp DNA ladder (i); mixture of TCRγ and Cμ (ii); positive control Cμ(130-bp) (iii); TCRγ (90-bp) (iv); and negative control (v). The dotted linesrepresent the applied electric field. Figure reprinted from Ref. [25] with permission.

machining, hot embossing, injection molding, soft lithography, in situ construction, laser ablation, and plasma etching [11, 12]. Silicon or glass is used as raw materials for the fabrication of microfluidic chips. Microfluidic electrophoresis chips consist of two reservoirs (sample and buffer) connected to the separation channel. A wide variety of materials including ceramics, glass, and polymers (poly(methyl methacrylate), cyclic olefin copolymers, polycarbonate, polystyrene, and fluorescent poly(*p*-xylylene) polymer (Parylene-C) have been used for preparation of microfluidic chips. Paper and fabric-based disposal chips have also received attention in MCE [7]. Electroosmotic flow (EOF) is generated in microfluidic chips as the reservoirs are filled with a background solution or electrolyte. Since EOF significantly obstructs separation, microfluidic chips are coated with various chemicals and hydrogels to suppress EOF.

1.2.2 Designing Microfluidic Channels

Crossed-channel and T-shaped microfluidic chips are widely used in MCE, and the microchip channel is connected perpendicularly to other channels (sample and buffer). Microfluidic chips are also prepared with different designs. For example, a microfluidic chip (Agilent Bioanalyzer™ chip) is prepared with 16 reservoirs, of which 12 are for sample reservoirs and four for references and reagents. Accordingly, MCE has been successfully applied for the analysis of various bioactive molecules with high precision and accuracy. Microfluidic chip channels with a width of 10–100 μm and a depth of 15–40 μm are considered the best design for separation of analytes. Also, separation channel area is designed to be 165 mm and $8 \times 8\ mm^2$ and the number of channels of microchips is increased to 12–384 for separation of nucleic acids and multiple genotyping (384) molecules with reduced time and increased accuracy [3]. Furthermore, microchips are designed with 8, 12, 16, 48, and 384 parallel channels for rapid and efficient separation of a wide variety of analytes.

The electric field (voltage application) and hydrodynamic pressure are applied for sample injection in MCE. On-chip peristaltic pump is used for hydrodynamic injection [13]. Inkjet and array techniques are droplet injection systems, which provide high throughput and the sample injection volume ranges from nano- to picoliters [14]. As microfluidic chips are compact in size, electrokinetic injection is the preferred method of sample injection, where the injection volume of the sample is strongly dependent on the applied voltage and injection time [15]. Further, hydrodynamic injection system requires a pump or pipette, which limits its use in MCE. It is usually carried out by variations in pressure, vacuum, reservoir (sample waste), and fluid levels of sample.

Electrokinetic injection system contains several injection modes, i.e. floating, gated, dynamic, and pinched. In the pinched injection mode, a voltage is applied at various channels including sample, buffer waste, and buffer reservoirs, and, as a result, the sample is injected into the channel junctions, and further enters the separation channels. Although the pinched injection mode is well illustrated and low volume of analyte plugs, it decreases the sensitivity due to sample plug. In the floating injection mode, which is similar to the pinched design, the potential is not required in the buffer and buffer waste reservoirs, increasing the sample load due to diffusion of the sample into the separation channel. Voltage can be applied at the sample and buffer reservoirs, and two waste reservoirs are grounded in this mode. The sample is injected into the separation channel by switching off the voltage at the reservoir (buffer), loading several amounts of the sample, which could help to improve the sensitivity of MCE. Then, the voltage is again switched on at the buffer reservoir. In dynamic injection mode, electroosmosis is required to inject the sample into separation channels, which can also improve the sample load.

1.2.3 Electrophoretic Separation

Target analytes are effectively separated in separation channels via electrophoretic separation by applying an electric field. Electrophoretic migration of analytes occurs due to the electric field and flow of liquid, collectively known as EOF. Various electrophoretic separations including electrokinetic chromatography, gel electrophoresis, and zone electrophoresis have been described in MCE [7]. Dielectrophoresis (DEP), gel electrophoresis, and zone electrophoresis modes are generally used in MCE. The DEP allows the separation of particles and cells by applying irregular electric fields (nonuniform) [7]. PC-3 human prostate cancer cells and polystyrene microbeads were separated by ionic liquid electrodes [16], and target analytes and particles were successfully separated via the on-chip procedure from human plasma [17]. The charged analytes migrate using the background solution (BGS) and are separated through electrophoretic mobility of target analytes, which leads to detection of analytes on the basis of descending mobility. As a result, small organic molecules including metabolites and drugs are effectively separated by the zone electrophoresis mode.

Electrophoretic mobility of analytes takes place when an electric field is applied along a sieving matrix. The analytes migrate through the matrix, and the degree of migration is dependent on the size/weight of the analytes. Porous gels are used for the preparation of the sieving matrix, and the size of the gel mesh affects separation efficiency of MCE. Generally, separation efficiency of MCE will be greater with the increasing concentration of cross-linked gels. Viscosity of gels also increases with increasing concentration of the sieving matrix, making it difficult to load the gels into microfluidic channels. Common gels are prepared by using starch, agarose, cellulose, and polyacrylamide. Large biomolecules such as nucleic acids (DNA and RNA), proteins, and biomarkers can be successfully separated by using gel electrophoresis based on size-dependent separation. Polymerase chain reaction (PCR) is often used for the amplification of RNA and DNA to concentrate nucleic acids prior to their separation by MCE. Importantly, MCE has ability to separate base pairs with reduced time and improved resolution, exhibiting better analytical features compared to slab gel electrophoresis, which requires <10 min for the analysis of base pairs.

1.2.4 Detectors

Target analytes can be detected by various detectors, once analytes are separated in separation channels of microfluidic chips. In MCE, various detectors including absorbance, electrochemical, chemiluminescence, and laser-induced fluorescence (LIF) are widely used to detect a wide variety of analytes. Among them, LIF is a prominent detector for the detection of trace-level target analytes due to its simplicity and sensitivity. It has been widely applied for the analysis of nucleic acids from various sample matrices [7].

In the LIF detection, the analyte (fluorescent) is excited by a laser, and light is emitted when the excited molecule returns to the ground state, which can be useful for detection signal. To eliminate autofluorescence, fluorescent derivatization is done in LIF for effective detection of target analytes. Fluorescent derivatization is done in channel before sample injection [19], which can also extend for parallel detection of multiple analytes. For parallel excitation of molecules, two lasers at visible and near-infrared wavelengths are used for the excitation of three metabolites [20]. In LIF, laser and light emitting diodes have proven to be promising and inexpensive portable devices for the analysis of various target analytes. Although LIF is a sensitive detection device, microfluidic chips (narrow structures) and small volumes of injected samples decrease its analytical sensitivity. Furthermore, MCE was successfully used for the detection of interferon-gamma at the picomolar level by using the LIF detection mode [21].

In chemiluminescence, the target analyte is excited by absorbing energy generated from chemical reactions. In this detection mode, molecules are effectively detected, even target analytes in femto- and atto-molar concentrations. Oxidation of luminol is an example of chemiluminescence reaction where luminol oxidizes in a basic medium in the presence of hydrogen peroxide (oxidizer) and enzyme as a biocatalyst (horseradish peroxidase, HRP). To enhance the chemiluminescence signal, various chemicals are used as modifiers in the luminol reaction. Amperometric and capacitively are coupled with contactless conductivity for the detection of analytes by the electrochemical detection mode in MCE. Similarly, amperometric detection mechanism is based on measuring changes in electric current when the analyte reaches the detector. In recent years, electrochemical MCE approaches are considered promising biotools for point-of-care diagnostics in various clinical applications [22].

1.2.4.1 Capability of Microchip Electrophoresis for the Separation and Identification of Biomolecules

In this section, an overview of the potential applications of microchip electrophoresis for the separation of biomolecules is provided. One of the simplest ways of detecting DNA molecules is an assay based on MCE. The MCE is a promising miniaturized analytical device for the separation of target molecules in various fields of science including clinical, pharmaceuticals, forensic, biology, genomic, and biomedical sciences. The rapidity of ME was improved by applying an electric field with a programmed field strength gradient (PFSG), facilitating separation of DNA with high speed [23–26]. A novel rapid MCE approach was developed for the separation and detection of *Mycoplasma haemofelis* and *Candidatus M. Haemominutum* in Korean feral cats [24]. The developed method was able to separate and differentiate two polymerase chain reaction (PCR) products in <11 seconds. Another report illustrated the rapidity of MCE with PFSG for the detection of T-cell lymphoma in dogs [25]. In this method, polyethylene oxide (PEO, Mr 8 000 000, 0.5%) was used as a sieving matrix for the separation of amplified-PCR products from canine T-cell lymphoma. This method successfully separated the PCR products of T-cell lymphoma within 7.0 s using PFSG (Figure 1.1).

1.2.4.2 Detection of Cancer Biomarkers

Furthermore, MCE is applied to investigate cancer via the identification of various biomolecules (peptides, proteins, DNA, and RNA). Telomerase activity, carcinoembryonic antigen, and DNA mutations in p53 oncogenes are recognized as cancer biomarkers. A chip-based analytical approach was established for the analysis of genetic codes to assess the telomerase activity, separating the amplicons within 120 seconds [27]. An aptamer-based MCE with laser-induced fluorescence (LIF) was developed for the detection of thymidine kinase 1 [28] and carcinoembryonic antigen [29, 30]. To suppress the adsorption of biomolecules on the microchip channel, 30 mM sodium dodecyl sulfate (SDS) was added and target analytes were detected within 4.0 min, offering an impressive limit of detection (LOD) at 45.7 pg/ml. The p53 gene consists of more than 20 000 bp, 11 exons, and 10 introns, and p53 gene mutations are investigated by single-strand conformational polymorphism (SSCP) and heteroduplex analysis (HA) combined with MCE–LIF [31]. This method successfully analyzed 106 tissue samples, and p53 gene mutations were correctly identified with high accuracy. Similarly, MCE–LIF was applied to detect gene mutations in exons 19 and 21 of patients with non–small-cell lung cancer [32]. MCE with chemiluminescence resonance energy transfer mechanism was developed for the detection of neuron-specific enolase [33]. The developed method was able to detect neuron-specific enolase events at 4.5 pM, which confirms the potentiality of MCE for separation and detection of cancer biomarkers.

Low-abundance KRAS mutation detection, glycomic serum analysis, and microsatellite instability were used for the investigation of colorectal cancer. MCE method was developed for the detection of mutations in the KRAS gene in patients with colorectal cancer [34]. Figure 1.2 represents the use of MCE for the detection of the KRAS gene. This method involved the amplication of 107-bp fragment of KRAS from DNA templates extracted from cancer cells with mismatched primer PCR, generating two peaks for 77 and 30 bp fragments. Prostate cancer was examined by identifying the circulating cell-free DNA in patients with hormone-refractory prostate cancer [35]. The concentration of cell-free DNA was estimated during chemotherapy. The concentration of cell-free DNA increased from 13.3 to 46.8 ng/ml after treatment of the first cycle of chemotherapy. DNA methylation analysis and MCE–LIF were used for the identification of skin cancer in patients with oral squamous cell carcinoma [36]. Eight genes (transmembrane protein with epidermal growth factor-like and 2 follistatin-like domains 2 [TMEFF2], e-cadherin [ECAD], *O*-6 methyl-guanine DNA methyltransferase [MGMT], retinoic acid receptor beta [RARβ], fragile histidine triad gene, cyclin-dependent kinase inhibitor 2A, WNT inhibitory factor 1, and death-associated protein kinase 1) were used as signals for the estimation of elevated levels of DNA methylation. Among these, four hypermethylated genes (RARβ, TMEFF2, ECAD, and MGMT) were

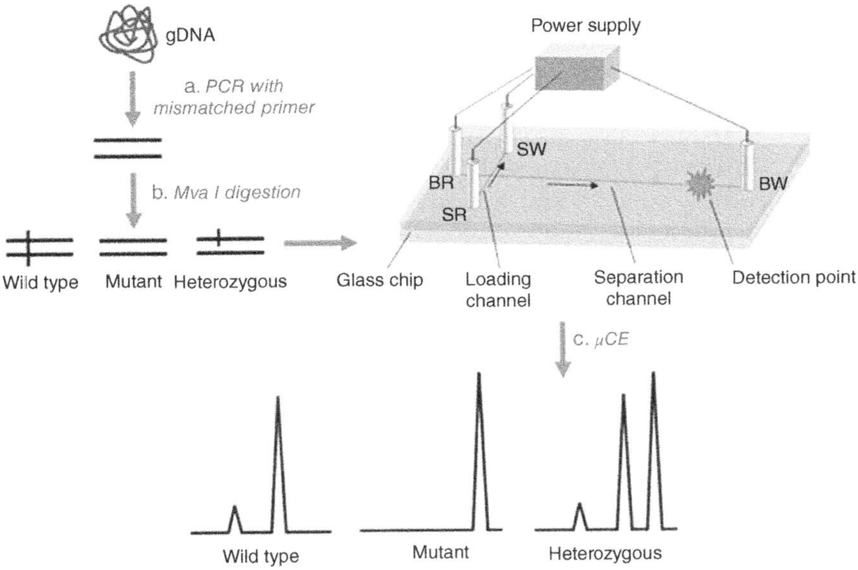

Figure 1.2 Schematic representation for the detection of low-abundance KRAS mutations in colorectalcancer. Figure reprinted from Ref. [2] with permission.

selected for the diagnosis of oral squamous cell carcinoma with 87.5% specificity and 100% sensitivity. Further, α-fetoprotein acts as an important biomolecule for the detection of small tumors at the early stage of cancer growth, which could be used for the treatment of cancer. MCE with fluorescence detection and lectin-affinity chip electrophoresis were successfully applied to detect α-fetoprotein in various tumor cells [37, 38]. Lectin-affinity electrophoresis exhibited the limit of detection at 0.3 ng/ml in serum, whereas the transient isotachophoresis approach exhibited a detection limit of 5 pM. Both analytical techniques exhibited a much lower level of α-fetoprotein (280 pM). This immunoassay was also applied to detect lens culinaris agglutinin-reactive α-fetoprotein, des-γ-carboxy prothrombin, and total α-fetoprotein [39]. Three biomarkers fucosylated fraction of α-fetoprotein, α-fetoprotein, and prothrombin induced by vitamin K absence-II were used for the accurate diagnosis of liver cancer, allowing detection of small-sized tumor, early-stage hepatocellular carcinoma, and single tumor with an accuracy of 86.7%, 81.8%, and 91.7%, respectively [40].

The profiling of *N*-glycans in patient serum [41], human papillomavirus [42] and determination of β-subunits of human chorionic gonadotropin [43] can be used for the investigation of ovarian and cervical cancers. The glycans were extracted by solid-phase extraction using activated carbon columns followed by fluorescence labeling with APTS (8-aminopyrene-1,3,6-trisulfonic acid). The electropherogram provides the differentiation between the before and after drug treatment of patients. The assay of β-subunits of the hormone human chorionic gonadotropin was used as a tumor biomarker for choriocarcinoma. A noncompetitive immunoassay was developed for the detection of β-subunits [44]. The practical application of the method was demonstrated for the detection of β-subunits in serum samples from healthy subjects and patients with ovarian cancer. The concentrations of β-subunits were found in the range of 9.5–15.7 and 160.9–210.4 mIU/ml in healthy and cancer samples, respectively. Several biomarkers such as lactate dehydrogenase B, galectin-1, and bladder cancer–specific genes (FGFR3, HRAS, TERT, RALL3, ALX4, MT1A, and RUNX3) were widely used for the detection of bladder cancer. In this connection, the cell lysate of bladder cancer cell lines RT4 and T24 were investigated for the protein expression of galectin-1 and lactate dehydrogenase B by microfluidic immunoassays [45, 46]. With this approach, the bladder cancer cells were rapidly extracted from human urine [46]. The amplicons were identified by MCE–LIF within 2 minutes.

1.2.4.3 Assays of Immune Disorders and Microbial Diseases by MCE

A reusable glass capillary array electrophoresis (CAE) chip was developed for the detection of standard K562 DNA [47]. In situ PCR was performed in the fabricated chip. The CAE chip consists of two DEA chips that are reversibly attached to polydimethylsiloxane (PDMS) blocks, thereby facilitating simple prewashing of

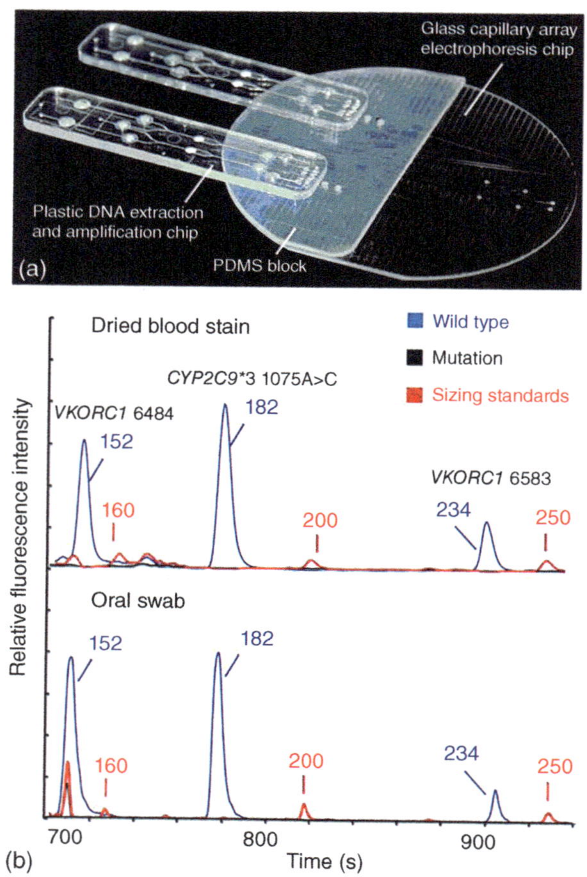

Figure 1.3 (a) Photographic image of microfluidic chip for the rapid pharmacogenetic typing of multiple warfarin-related single-nucleotide polymorphisms. (b) Electropherogram of pharmacogenetic analyses carried out using a dried blood stain and buccal swab. Reprinted with permission from Ref. [2].

PDMS surface (Figure 1.3a). As a result, effective electrophoretic separation was achieved (Figure 1.3b), suggesting that this method was successfully applied to detect target analytes directly from blood, buccal swabs, and dried blood spots. A new portable microfluidic chip was developed for the detection of uric acid in urine of gout patients [48]. MCE was successfully developed for the identification of proteins for different patients [49]. This method successfully differentiated several proteins including albumins and α-, β-, and γ-globulins from cerebrospinal fluid and serum samples of patients and healthy persons. A novel and rapid immunoassay was developed for the detection of phenobarbital in blood samples of epileptic patients [50, 51]. In this approach, liquid–liquid extraction was performed for the extraction of phenobarbital, and effectively analyzed target analytes within the therapeutic range from 43 to 172 μM.

Recently, MCE has also been widely applied to detect various pathogenic microorganisms such as viruses (influenza and hepatitis B) and bacterial species. MCE was used to detect influenza A (H1N1) virus [52]. A high-throughput separation was achieved for hemagglutinin (HA) and nucleocapsid protein (NP) gene PCR products (116 and 195 bp, respectively) by multichannel MCE. A spatially multiplexed microdevice was applied to detect influenza A and influenza B in nasopharyngeal and throat swabs of humans [53]. This method enables to complete the analysis within 40 min, offering an LOD of 10 RNA copies per reaction. Similarly, MCE–LIF method evaluated alanine aminotransferase (ALT) enzyme kinetics [54]. This method exhibited LOD of 4.0×10^{-7} and 2.0×10^{-7} M for l-Glu and l-Ala, respectively. This method was effectively applied to evaluate ALT enzyme kinetics in serum samples.

A novel paper-based CE microdevice was fabricated for point-of-care (POC) pathogen diagnostics [55]. This newly designed paper-based CE was successfully applied to differentiate single-stranded DNA (ssDNA) with 4 bp resolution. PCR amplicons of two target genes of *Escherichia coli O157:H7* (rrsH gene, 121 bp) and *Staphylococcus aureus* (glnA gene, 225 bp) were successfully separated and identified within 3 minutes. Figure 1.4 shows the design of a paper-based CE microdevice. Figure 1.5 shows the electropherogram of bracket ladders of genomic DNA (from 0.5 ng to 0.5 pg). The developed paper-based CE achieved impressive LOD of 9.3×10^1 and 1.6×10^2 copies for *E. coli O157:H7* and *S. aureus*, respectively. A dual-channel PDMS-based microchip was fabricated for the detection of o-aminophenol, *p*-aminophenol, and *m*-aminophenol using a modified indium tin oxide (ITO) microelectrode detector [56]. This method exhibited LOD of 0.41, 0.24, and 0.42 μM for *o*-, *p*-, and *m*-aminophenols, respectively. A portable MCE combined with on-chip contactless conductivity detector was developed for the analysis of vancomycin in human plasma [57]. This method was applied to detect vancomycin within 5.0 min, offering an LOD of 1.2 μg/ml. This method exhibited good recovery ranges of 99.00–99.20% in plasma samples.

1.2.4.4 Assays of Biomarkers by MCE

Assay of biomarkers plays a key role in diagnosing various disorders by MCE. Multichannel chips were fabricated using poly(methyl methacrylate) (PMMA)

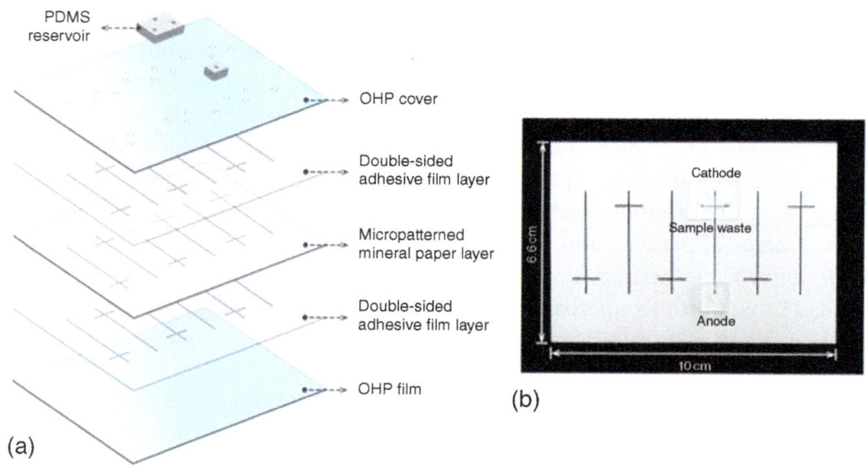

Figure 1.4 (a) An view of paper-based capillary electrophoresis that consists six layers. (b) Photographic image of paper-based CE. Reprinted with permission from Ref. [55].

and applied for the separation and identification of free bilirubin in biofluids [58]. The developed method exhibited a wider linear range from 10 to 200 μM with an LOD of 9.0 μM. This method was successfully applied to detect total bilirubin (118.3–119.4 μM) from the samples of jaundiced infants. An MCE with fluorescence detector was established for the detection of creatinine in urine [59]. The creatinine was quantified by using exciting and emission wavelengths at 490 and 523 nm, respectively. Authors obtained an impressive LOD of 2.87 μM and good recovery in the range of 96.0–107.0%. MCE has been used as a promising analytical tool for the rapid and selective detection of urine proteins in clinical samples [60]. This method exhibited a wider linear range in the concentration of 1.0–15.0 and 1.0–10.0 g/l for human albumin and human transferrin with an LOD of 0.4 g/l. This method has the ability to detect urine proteins in various clinical samples, i.e. 105 nonselective proteinuria, 60 selective proteinuria, and 6 overflow proteinuria. Thyroxine (T4) was successfully quantified by MCE method [61]. The detection principle was based on the immunoreaction between T4 and HRP-labeled T4 (HRP-T4) with anti-T4 mouse monoclonal antibody (Ab), which favored the formation of HRP–T4 and the HRP-T4–Ab complex, which can be easily separated and identified by MCE with the chemiluminescence technique. The MCE separated the target analytes within 60 seconds, exhibiting an LOD of 2.2 nM. Furthermore, Au/chitosan/Fe_3O_4 nanocomposites were synthesized and inserted into the microfluidic device [62]. The Au/chitosan/Fe_3O_4 nanocomposites inserted microfluidic device was connected to Au microelectrode for immunoassay of hemoglobin A1c (HbA1c). In this work, Au/chitosan/Fe_3O_4 nanocomposites were functionalized with anti-human hemoglobin-A1c antibody (HbA1c mAb), enabling rapid immunoreactions between the sequence of HbA1c, Hb mAb, and secondary alkaline phosphatase-conjugated antibodies. This method exhibited an LOD of 0.025 μg/ml and successfully evaluated assay7 of HbA1c in blood.

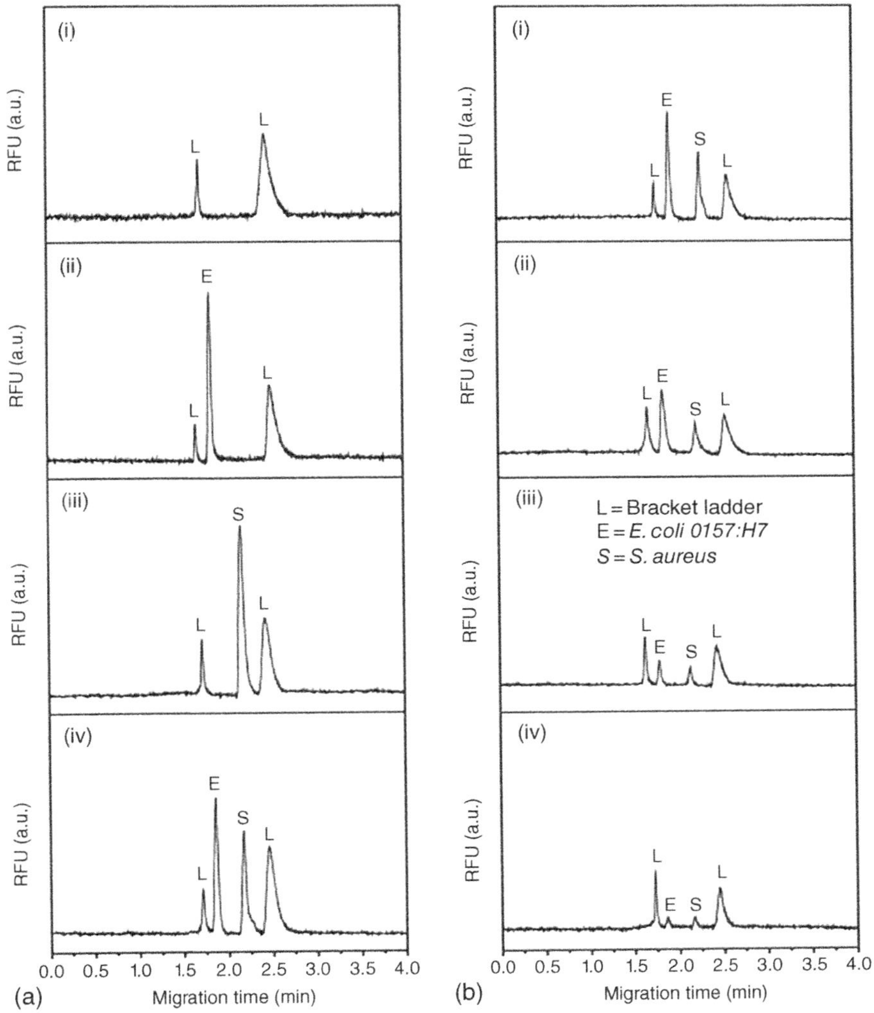

Figure 1.5 Separation and identification of pathogens (i) bracket ladders, (ii) *E. coli O157:H7* with bracket ladders, (iii) *S. aureus* with bracket ladders, and (iv) *E. coli O157:H7* and *S. aureus* with bracket ladders by paper-based CE. (b) Electropherogram of *E. coli O157:H7* and *S. aureus* at different DNA concentrations (i) 0.5 ng, (ii) 0.05 ng, (iii) 5 pg, and (iv) 0.5 pg. Reprinted with permission from Ref. [55].

Four urinary proteins, namely, immunoglobulin G (IgG), transferrin, human serum albumin (HSA), and β2-microglobulin were successfully separated and identified by isotachophoresis/capillary zone electrophoresis [63]. Using this approach, sample desalting/preconcentration was carried out on-chip, achieving low LOD with high selectivity. Testosterone was sensitively detected using MCE with chemiluminescence detectors [64]. The detection mechanism was based on the immunoreaction between testosterone and *N*-(4-aminobutyl)-*N*-ethylisoluminol-labeled T

(ABEI-T) with antibody (Ab). The formed products ABEI-T–Ab complex and free ABEI-T were efficiently separated and detected. This method displayed an LOD of 1.0 nM in human serum. An electrokinetic-based microfluidic device was developed for the detection of preterm birth biomarkers [65]. In this approach, solid-phase extraction was used for the extraction of preterm birth biomarkers. The developed method required ~10 μl sample volume and exhibited 30-fold faster time for the extraction and preconcentration of preterm birth biomarkers (lactoferrin and ferritin) in the clinical samples. A microfluidic device was developed for the extraction and detection of preterm birth biomarkers using pH-mediated solid-phase extraction [66]. In this method, the preterm birth biomarker peptide was effectively separated using nM of biomarkers, demonstrating the detection of preterm birth biomarkers even at 400 pg (0.2 pM). These methods have illustrated that MCE can be successfully applied to separate and detect trace-level target analytes from complex samples.

1.3 Summary

In summary, MCE has significantly proven to be a miniaturized analytical device for separation and identification of trace-level organic molecules and biomolecules from various sample matrices. Microfluidic devices enable on-line sample preparation procedures where trace-level analytes are isolated and enriched, thereby achieving low LOD in microvolume samples of various biomarkers. To enhance sensitivity and selectivity of MCE, nanocomposites have been inserted into microfluidic devices, facilitating immunoreactions in microfluidic channels, which enables detection of target biomarkers with high selectivity and rapidity. Overall, microfluidic capillary electrophoresis has proven to be a miniaturized analytical tool for the rapid analysis of various target molecules with small sample volumes.

Acknowledgments

The authors gratefully acknowledge the financial support (EMR/2016/002621/IPC) for this work by the Department of Science and Technology (DST), Government of India . The authors thank the Director, SVNIT, Surat, for providing necessary facilities to carry out this work.

References

1 Dolnik, V. and Liu, S. (2005). Application of capillary electrophoresis on microchip. *J. Sep. Sci.* 28: 1994–2009.
2 Wuethrich, A. and Quirino, J.P. (2019). A decade of microchip electrophoresis for clinical diagnostics – a review of 2008–2017. *Anal. Chim. Acta* 1045: 42–66.

3 Kailasa, S.K. and Kang, S.H. (2009). Microchip-based capillary electrophoresis for DNA Analysis in modern biotechnology: a review. *Sep. Purif. Rev.* 38: 242–288.
4 Manz, A., Graber, N., and Widmer, H.M. (1990). Miniaturized total chemical analysis systems: a novel concept for chemical sensing. *Sens. Actuators B Chem.* 1: 244–248.
5 Štěpánová, S. and Kašička, V. (2017). Analysis of proteins and peptides by electromigration methods in microchips. *J. Separ. Sci.* 40: 228–250.
6 Kašička, V. (2016). Recent developments in capillary and microchip electroseparations of peptides (2013–middle 2015). *Electrophoresis* 37: 162–188.
7 Sonker, M., Sahore, V., and Woolley, A.T. (2017). Recent advances in microfluidic sample preparation and separation techniques for molecular biomarker analysis: a critical review. *Anal. Chim. Acta* 986: 1–11.
8 Han, J.P., Sun, J., Zhuang, B. et al. (2015). The electrophoresis on microfluidic chips and its application in forensic science. *Prog. Biochem. Biophys.* 42: 1128–1135.
9 Kim, Y.T., Heo, H.Y., Oh, S.H. et al. (2015). Microchip-based forensic short tandem repeat genotyping. *Electrophoresis* 36: 1728–1737.
10 Kang, C.C., Lin, J.M.G., Xu, Z. et al. (2014). Single-cell western blotting after whole-cell imaging to assess cancer chemotherapeutic response. *Anal. Chem.* 86: 10429–10436.
11 Fiorini, G.S. and Chiu, D.T. (2005). Disposable microfluidic devices: fabrication, function, and application. *Biotechniques* 38: 429–446.
12 Becker, H. and Gärtner, C. (2012). Microfluidics and the life sciences. *Sci. Prog.* 95: 175–198.
13 Sahore, V., Kumar, S., Rogers, C.I. et al. (2016). Pressure-actuated microfluidic devices for electrophoretic separation of pre-term birth biomarkers. *Anal. Bioanal. Chem.* 408: 599–607.
14 Hassan, S.U., Morgan, H., Zhang, X., and Niu, X. (2015). Droplet interfaced parallel and quantitative microfluidic-based separations. *Anal. Chem.* 87: 3895–3901.
15 Huang, X., Gordon, M.J., and Zare, R.N. (1988). Bias in quantitative capillary zone electrophoresis caused by electrokinetic sample injection. *Anal. Chem.* 60: 375–377.
16 Sun, M., Agarwal, P., Zhao, S. et al. (2016). Continuous on-chip cell separation based on conductivity-induced dielectrophoresis with 3D self-assembled ionic liquid electrodes. *Anal. Chem.* 88: 8264–8271.
17 Szydzik, C., Khoshmanesh, K., Mitchell, A., and Karnutsch, C. (2015). Microfluidic platform for separation and extraction of plasma from whole blood using dielectrophoresis. *Biomicrofluidics* 9: 1–16.
18 Hsieh, J.-F., Sinchaikul, S., Ou, B.-M., and Chen, S.-T. (2010). Studies on the analysis of human serum proteins by the gel-based microchip and SDS-PAGE. *J. Food Drug Anal.* 18: 434–439.
19 Wuethrich, A. and Quirino, J.P. (2017). Derivatisation for separation and detection in capillary electrophoresis (2015–2017). *Electrophoresis* 39: 82–96.

20 Li, Q., Chen, P., Fan, Y. et al. (2016). Multicolor fluorescence detection-based microfluidic device for single-cell metabolomics: simultaneous quantitation of multiple small molecules in primary liver cells. *Anal. Chem.* 88: 8610–8616.

21 Qin, Y., Zhang, L., Li, S. et al. (2017). A microchip electrophoresis-based fluorescence signal amplification strategy for highly sensitive detection of biomolecules. *Chem. Commun.* 53: 455–458.

22 Wang, J. (2006). Electrochemical biosensors: towards point-of-care cancer diagnostics. *Biosens. Bioelectron.* 21: 1887–1892.

23 Kailasa, S.K. and Kang, S.H. (2007). Ultra-fast simultaneous analysis of GMOs in maize by microchip electrophoresis with laser-induced fluorescence detector. *Electrophoresis* 28: 4247–4254.

24 Kailasa, S.K., Lee, H.G., Yoo, D.J., and Kang, S.H. (2008). Ultra-fast detection and differentiation of Mycoplasma haemofelis and Candidatus M. Haemominutum in Korean feral cats by microchip electrophoresis with programmed field strength gradients. *Bull. Korean Chem. Soc.* 29: 153–158.

25 Kailasa, S.K., Lee, M.J., Park, J., and Kang, S.H. (2008). Microchip gel electrophoresis with programmed field strength gradients for ultra-fast detection of canine T-cell lymphoma in dogs. *Talanta* 75: 49–55.

26 Lee, H.G., Kailasa, S.K., Soh, J.R. et al. (2008). Ultra-fast simultaneous detection of obesity-related coenzymes in mice using microchip electrophoresis with a LIF detector. *Anal. Chim. Acta* 619: 94–100.

27 Karasawa, K. and Arakawa, H. (2015). Detection of telomerase activity using microchip electrophoresis. *J. Chromatogr. B: Anal. Technol. Biomed. Life Sci.* 993–994: 14–19.

28 Pagaduan, J.V., Ramsden, M., O'Neill, K., and Woolley, A.T. (2015). Microchip immunoaffinity electrophoresis of antibody-thymidine kinase 1 complex. *Electrophoresis* 36: 813–817.

29 Pan, L., Zhao, J., Huang, Y. et al. (2015). Aptamer-based microchip electrophoresis assays for amplification detection of carcinoembryonic antigen. *Clin. Chim. Acta* 450: 304–309.

30 Ye, F., Shi, M., Huang, Y., and Zhao, S. (2010). Noncompetitive immunoassay for carcinoembryonic antigen in human serum by microchip electrophoresis for cancer diagnosis. *Clin. Chim. Acta* 411: 1058–1062.

31 Hestekin, C.N., Lin, J.S., Senderowicz, L. et al. (2014). Blinded study determination of high sensitivity and specificity microchip electrophoresis-SSCP/HA to detect mutations in the p53 gene. *Electrophoresis* 32: 2921–2929.

32 Malapelle, U., Russo, S., Pepe, F. et al. (2013). EGFR mutation detection by microfluidic technology: a validation study. *J. Clin. Pathol.* 66: 982–984.

33 Yang, T., Vdovenko, M., Jin, X. et al. (2014). Highly sensitive microfluidic competitive enzyme immunoassay based on chemiluminescence resonance energy transfer for the detection of neuron-specific enolase. *Electrophoresis* 35: 2022–2028.

34 Zhang, H., Song, J., Ren, H. et al. (2013). Detection of low-abundance KRAS mutations in colorectal cancer using microfluidic capillary electrophoresis-based restriction fragment length polymorphism method with optimized assay conditions. *PLoS One* 8.

35 Kwee, S., Song, M.A., Cheng, I. et al. (2012). Measurement of circulating cell-free DNA in relation to 18F-fluorocholine PET/CT imaging in chemotherapy-treated advanced prostate cancer. *Clin. Transl. Sci.* 5: 65–70.

36 Nagata, S., Hamada, T., Yamada, N. et al. (2012). Aberrant DNA methylation of tumor-related genes in oral rinse. *Cancer* 118: 4298–4308.

37 Kawabata, T., Wada, H.G., Watanabe, M., and Satomura, S. (2008). "Electrokinetic Analyte Transport Assay" for α-fetoprotein immunoassay integrates mixing, reaction and separation on-chip. *Electrophoresis* 29: 1399–1406.

38 Kobayashi, M., Hosaka, T., Ikeda, K. et al. (2011). Highly sensitive AFP-L3% assay is useful for predicting recurrence of hepatocellular carcinoma after curative treatment pre- and postoperatively. *Hepatol. Res.* 41: 1036–1045.

39 Kanke, F., Kumada, T., Toyoda, H., and Satomura, S. (2012). Reference change values for lens culinaris agglutinin-reactive α-fetoprotein and des-γ-carboxy prothrombin in patients with chronic hepatitis C. *Clin. Chem. Lab. Med.* 50: 957–960.

40 Choi, J.Y., Jung, S.W., Kim, H.Y. et al. (2013). Diagnostic value of AFP-L3 and PIVKA-II in hepatocellular carcinoma according to total-AFP. *World J. Gastroenterol.* 19: 339–346.

41 Mitra, I., Alley, W.R., Goetz, J.A. et al. (2013). Comparative profiling of N-Glycans isolated from serum samples of ovarian cancer patients and analyzed by microchip electrophoresis. *J. Proteome Res.* 12: 4490–4496.

42 Liu, Q., Lin, X., Lin, L. et al. (2015). A comparative study of three different nucleic acid amplification techniques combined with microchip electrophoresis for HPV16 E6/E7 mRNA detection. *Analyst* 140: 6736–6741.

43 Li, S., Yang, T., Zhao, J. et al. (2017). Chemiluminescence noncompetitive immunoassay based on microchip electrophoresis for the determination of β-subunit of human chorionic gonadotropin. *J. Chromatogr. B: Anal. Technol. Biomed. Life Sci.* 1053: 42–47.

44 Li, S., Yang, T., Zhao, J. et al. (2017). Chemiluminescence noncompetitive immunoassay based on microchip electrophoresis for the determination of β-subunit of human chorionic gonadotropin. *J. Chromatogr. B: Anal. Technol. Biomed. Life Sci.* 1053: 42–47.

45 Chuang, C.-H., Wu, T.-F., Chen, C.-H. et al. (2015). Lab on a chip for multiplexed immunoassays to detect bladder cancer using multifunctional dielectrophoretic manipulations. *Lab Chip* 15: 3056–3064.

46 Deng, Y., Yi, L., Lin, X. et al. (2015). A non-invasive genomic diagnostic method for bladder cancer using size-based filtration and microchip electrophoresis. *Talanta* 144: 136–144.

47 Zhuang, B., Han, J., Xiang, G. et al. (2016). A fully integrated and automated microsystem for rapid pharmacogenetic typing of multiple warfarin-related single-nucleotide polymorphisms. *Lab Chip* 16: 86–95.

48 Fernández-la-Villa, A., Bertrand-Serrador, V., Pozo-Ayuso, D.F., and Castaño-Álvarez, M. (2013). Fast and reliable urine analysis using a portable

platform based on microfluidic electrophoresis chips with electrochemical detection. *Anal. Methods* 5: 1494.

49 Trenčevska, O., Aleksovski, V., and Stojanoski, K. (2009). Advanced techniques in clinical practice: use of lab-on-a-chip electrophoresis and other methods in protein profiling. *J. Med. Biochem.* 28: 274–278.

50 Huang, Y., Zhao, S., Shi, M. et al. (2011). Competitive immunoassay of phenobarbital by microchip electrophoresis with laser induced fluorescence detection. *Anal. Chim. Acta* 694: 162–166.

51 Huang, Y., Zhao, S., Shi, M., and Liang, H. (2011). One-way multiplexed immunoassay strategy for simultaneous determination of multi-analytes by microchip electrophoresis. *Analyst* 136: 2119.

52 Zhang, P., Nan, H., Lee, S., and Kang, S.H. (2015). Fast high-throughput screening of H1N1 virus by parallel detection with multichannel microchip electrophoresis. *Methods Mol. Biol.* 183–195.

53 Prakash, R., Pabbaraju, K., Wong, S. et al. (2014). Multiplex, quantitative, reverse transcription PCR detection of influenza viruses using droplet microfluidic technology. *Micromachines* 6: 63–79.

54 Mu, X., Qi, L., Qiao, J. et al. (2012). Study on alanine aminotransferase kinetics by microchip electrophoresis. *Anal. Biochem.* 421: 499–505.

55 Lee, J.W., Lee, D., Kim, Y.T. et al. (2017). Low-cost and facile fabrication of a paper-based capillary electrophoresis microdevice for pathogen detection. *Biosens. Bioelectron.* 91: 388–392.

56 Zhu, G., Song, Q., Liu, W. et al. (2017). A gold nanoparticles-modified indium tin oxide microelectrode for in-channel amperometric detection in dual-channel microchip electrophoresis. *Anal. Methods* 9: 4319–4326.

57 Chong, K.C., Thang, L.Y., Quirino, J.P., and See, H.H. (2017). Monitoring of vancomycin in human plasma via portable microchip electrophoresis with contactless conductivity detector and multi-stacking strategy. *J. Chromatogr. A* 1485: 142–146.

58 Nie, Z. and Fung, Y.S. (2008). Microchip capillary electrophoresis for frontal analysis of free bilirubin and study of its interaction with human serum albumin. *Electrophoresis* 29: 1924–1931.

59 Wang, S., Li, X., Yang, J. et al. (2012). Rapid determination of creatinine in human urine by microchip electrophoresis with LED induced fluorescence detection. *Chromatographia* 75: 1287–1293.

60 Song, H., Wang, H., Ju, S. et al. (2014). Simplified microchip electrophoresis for rapid separation of urine proteins. *J. Clin. Lab. Anal.* 28: 104–109.

61 Huang, Y., Zhao, S., Shi, M., and Liu, Y.M. (2010). Chemiluminescent immunoassay of thyroxine enhanced by microchip electrophoresis. *Anal. Biochem.* 399: 72–77.

62 Chen, S.P., Yu, X.D., Xu, J.J., and Chen, H.Y. (2011). Gold nanoparticles-coated magnetic microspheres as affinity matrix for detection of hemoglobin A1c in blood by microfluidic immunoassay. *Biosens. Bioelectron.* 26: 4779–4784.

63 Wu, R., Yeung, W.S.B., and Fung, Y.S. (2011). 2-D t-ITP/CZE determination of clinical urinary proteins using a microfluidic-chip capillary electrophoresis device. *Electrophoresis* 32: 3406–3414.

64 Huang, Y., Shi, M., Zhao, S., and Liang, H. (2011). A sensitive and rapid immunoassay for quantification of testosterone by microchip electrophoresis with enhanced chemiluminescence detection. *Electrophoresis* 32: 3196–3200.

65 Sonker, M., Yang, R., Sahore, V. et al. (2016). On-chip fluorescent labeling using reversed-phase monoliths and microchip electrophoretic separations of selected preterm birth biomarkers. *Anal. Methods* 8: 7739–7749.

66 Sonker, M., Knob, R., Sahore, V., and Woolley, A.T. (2017). Integrated electrokinetically driven microfluidic devices with pH-mediated solid-phase extraction coupled to microchip electrophoresis for preterm birth biomarkers. *Electrophoresis* 38: 1743–1754.

2

Portable Nanomaterials Impregnated Paper-Based Sensors for Detection of Chemical Substances

Khemchand Dewangan[1] *and Kamlesh Shrivas*[2]

[1] *Indira Gandhi National Tribal University, Department of Chemistry, Amarkantak-484887, Madhya Pradesh, India*
[2] *Pt. Ravishankar Shukla University, School of Study in Chemistry, Raipur-492010, Chhattisgarh, India*

2.1 Introduction

Today undesirable hazardous chemical substances in the environment have led to unsafe and unsanitary water sources, food contamination, and poor air quality, thereby cause a large number of hospitalizations and deaths worldwide [1]. While contaminated drinking water can increase water-borne diseases, the use of polluted and unsafe water for irrigation can contaminate the agricultural products, causing food-borne illnesses and affecting all living organisms [2, 3]. Metal ions (e.g., As, Hg, Pb, Cr, Cd, Al) present in the air, soil, and water are hazardous elements and known for their toxicity to human beings as well as animals, and some of them are cancerogenic due to their high toxicity, stability, and affinity of accumulation through food chains [4–6]. Therefore, serious actions are needed to tackle the pollution created by the use of toxic and harmful chemical substances.

In this context, rapid, precise, accurate recognition, and monitoring of toxic substances are a great challenge in preventing risks imposed by these toxic elements present in the environment. To overcome this, nanomaterials have great potential in terms of their high surface-to-volume ratio and remarkable electrical and optical characteristics. In addition, much attention needs to be focused on fabricating nanomaterial-based devices, which are easy to use, affordable, flexible, and scalable for a broad range of applications such as clinical diagnosis, food quality control, environmental monitoring [7, 8]. Also, nanomaterials allow design possibilities of devices with lightweight components and cost-effectiveness, so their demand is continuously increasing globally [9]. Therefore, this chapter mainly focuses on the general features of nanomaterials and their use in paper-based sensors in monitoring toxic chemical substances.

Miniaturized Analytical Devices: Materials and Technology, First Edition.
Edited by Suresh Kumar Kailasa and Chaudhery Mustansar Hussain.

2.2 General Aspects of Nanomaterials

Nanostructured materials are deliberately engineered materials at atomic/molecular levels having distinctive dimensions ranging from sub-nanometer to several hundred nanometers [10]. Nanomaterials exhibit remarkable and unique physio-chemical properties compared to their bulk counterparts; thus they are recognized in all fields of science and engineering [11–13] for potential and significant applications. Some of the topical fields that nanomaterials mainly affected are life science, information storage, magnetic refrigeration, catalysts, electrochemical energy storage, improved ceramics and insulators, environment along with various correlated areas [14–19]. Among these, some nanomaterial-based applications exhibit size tunable properties, which have already cemented a place as electronic devices in our daily life [12]. Also, nanomaterials have gained significant momentum in the past two decades to interplay with biology and medical sciences, leading to the emergence of a new interdisciplinary field called "nanomedicine" [15, 18].

Therefore, by interplaying with the arrangement of atoms/molecules in the order of few nanometers, it is possible to tune the properties of nanomaterials for desired applications and benefits to a great extent [20]. Three main size-dependent effects distinguish nanomaterials from their bulk counterparts:

(i) High surface-area-to-volume ratio: This means a greater portion of atoms/molecules are present at the surface of nanocrystalline materials. For example, a 3 nm spherical particle has 50% of its components on the surface, which is 20% for 10 nm and 5% for 30 nm particles. This is because the surface atoms/molecules experience forces that are totally different from those inside the crystallite. Thus, the arrangement of atoms/molecules on the surface happens in such a way that it leads to a minimum energy configuration.

(ii) Quantum confinement: In this phenomenon, electrons are just trapped by a potential well of nanometer-sized structures. Consequently, a discrete energy spectrum is observed and finally the bandgap becomes size dependent as shown in Figure 2.1. This may provide one of the most powerful resources to manipulate the electrical, optical, magnetic, catalytic, mechanical, and thermoelectric properties of a solid-state functional material [21].

(iii) High fraction of defects: A large fraction of defects results in reduced atomic density, a different local atomic arrangement, and a broad distribution of inter-atomic spacing. Thus, the overall structure of these materials deviates from the structure of the corresponding ideal crystals. It is known that the deviation from the perfect crystal structure leads to interesting properties. A large deviation from the perfect structure exists in nanocrystalline materials, resulting in unique and remarkable properties [22].

2.3 Synthesis of Nanomaterials

A variety of common nanosystems have widely been fabricated from various physical and chemical routes, such as metals (e.g., Au, Ag, Pt, Cu) [23], oxides

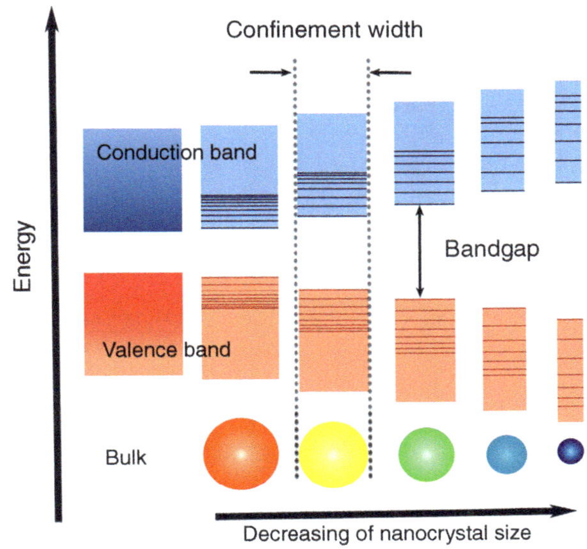

Figure 2.1 Quantum confinement.

(e.g., V_2O_5, SiO_2, ZnO, TiO_2, Cu_2O, Fe_3O_4, spinal and inverse spinal oxides) [24], nitrides (e.g., VN, TiN, Mo_2N, C_3N_4) [25, 26], carbides (e.g., VC, MoC, Ni_3C) [27], sulfides (e.g., ZnS, CdS, MoS_2, CuS) [28], phosphites (e.g., GaP, InP) [29], including nanocarbon compounds [carbon nanotubes (CNTs), graphene, C-dot, fullerene] [30–32]. Also, a number of different geometries of nanomaterials such as dot, particle, tube, rod, fiber, disk, plate, sheet, dendrimer, sphere, cube, and prism have been designed for various kinds of applications [11, 13].

In general, top-down, bottom-up, and their joint strategies are used for the fabrication of different shapes and sizes of required nanosystems (Figure 2.2) [11, 33–36]. Top-down techniques are based on cutting and milling of macroscopic-sized materials into the nanoscale size and providing a desired shape through solid-state physics strategies such as lithography, grinding, ball milling, and laser beam processing. In the other way of nanomanufacturing, bottom-up routes apply chemical or physical energy available at the atomic level to build up nanostructures by assembling fundamental units found in the reaction mixture. The most common bottom-up approaches are chemical synthesis, chemical vapor deposition, laser-induced assembly, self-assembly, colloidal aggregation, film deposition and growth, etc. Some of the very common wet chemical nanofabrication methods are summarized briefly in the following sections.

2.3.1 Solvothermal/Hydrothermal Technique

The solvothermal/hydrothermal process is also known as the green chemistry approach for nanofabrication due to the use of hydrothermal autoclave reactors to carry out the hydrothermal reaction at high pressures and high temperatures in a closed system [37]. Scheme 2.1 shows the general strategy of solvothermal process.

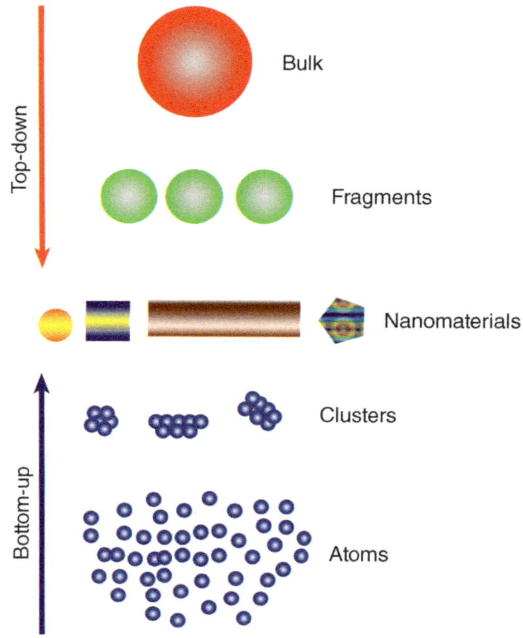

Figure 2.2 Top-down and bottom-up approaches of nanomaterials.

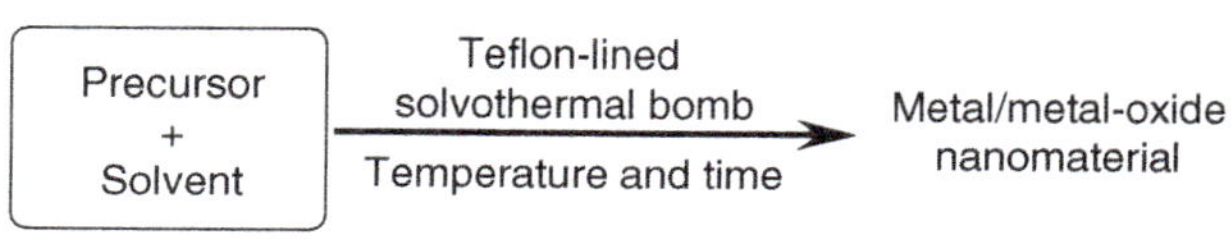

Scheme 2.1 General strategy of solvothermal technique.

Hydrothermal autoclave reactors consist two parts: (i) the outer jacket composed of high-quality stainless steel and (ii) the inner chamber made up of Teflon (polytetrafluoroethylene). In this process, metal salt/metal organic compounds act as starting materials, which are dispersed/dissolved along with an appropriate additive (if needed) in a hydrothermal solvent (water, alcohol, etc.). Then, this mixture is hydrothermally treated at the optimum reaction temperature and time for the preparation of desired nanomaterials. This technique is generally used for the preparation of metal [38] and metal oxide [39] nanosystems on the basis of the difference in solubility between the phase to be grown, which leads to nucleation, and the nutrient consisting of compounds that are thermodynamically unstable under the growth conditions. The conversion of product depends on the solubility of starting materials in the hydrothermal solvent and its properties such as stability, critical temperature, viscosity, density, and dielectric constant [40]. A major advantage of hydrothermal synthesis, in addition to being a green process, is that this method can be hybridized with other processes such as microwave [41], microemulsion [42], sonochemical [43], and mechanical rotational [44] to achieve advantages such as enhancement of reaction kinetics and increased ability to make new materials.

2.3.2 Reduction of Metal Salts

For the preparation of metal nanoparticles (NPs) (e.g. Au, Ag, Pt), a metal salt is reduced in the presence of a suitable stabilizer by using various reducing agents such as sodium citrate [45], ascorbate [46], sodium borohydride ($NaBH_4$) [47], tollens' reagent [48], and *N*,*N*-dimethylformamide [49]. Mostly, this synthesis strategy is carried out at room temperature, and it has advantages such as it allows assembly of particles with defined size, dimension, composition, and geometry.

2.3.3 Microemulsion Techniques

A microemulsion is a thermodynamically stable colloidal dispersion of two immiscible liquids in the presence of surface active agents (emulsifiers/surfactants). Microemulsions are classified into water-in-oil (w/o) microemulsions, where water is dispersed homogenously in an organic medium, and oil-in-water (o/w) microemulsions, where oil is dispersed in water. When the surfactant concentration dissolving in one phase exceeds the critical micelle concentration (CMC), the surfactant molecules spontaneously form aggregates, which are known as micelles (o/w) or reverse micelles (w/o). Surfactant packing parameter, which depends primarily on the volume of the polar head group and the length of the hydrocarbon chain, controls final geometry such as spherical, cylindrical, planar lamellar, onion-like lamellar of micelles/reverse micelles. Finally, the desired reactants are dissolved in this microemulsion, which serve as nanoreactor templates for the preparation of nanomaterials of metals, oxides, chalcogenides, etc. [50–52].

2.3.4 Sol–Gel

The sol–gel process refers to the fabrication of inorganic oxides using either metal alkoxide precursors or colloidal dispersions of hydrous oxides (sol). Components of metal alkoxides are dissolved together in a solvent (alcohol). The hydrolysis of alkoxides is referred to as "sol" and the "gel" is formed when alcohol is liberated. Sol–gel conversion is an irreversible process and often controls the size/shape of the powder product materials. Desiccation and calcination lead to xerogel and, finally, to the desired nanometer-sized metal [53] or metal oxide [54] product.

2.3.5 Polyol Processes

Polyol process (Scheme 2.2) is another versatile chemical approach for the synthesis of nanomaterials (metal [55] and oxides [56]) with well-defined shapes and controlled sizes. The polyol solvents, such as ethylene glycol (EG), triethylene glycol, diethylene glycol, tetraethylene glycol, and polyethyleneglycol (PEG), exhibit

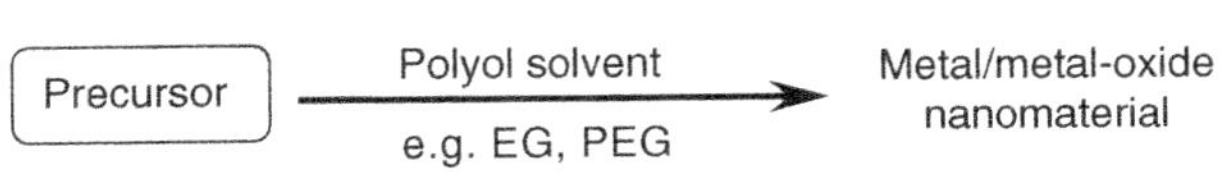

Scheme 2.2 General scheme of polyol technique.

interesting properties due to their high dielectric constants. They act as solvents and are capable of dissolving inorganic compounds. In addition, due to their relatively high boiling points, they offer a wide operating temperature range for preparing inorganic compounds. They can also serve as reducing agents as well as stabilizers to control particle growth and prevent interparticle aggregation. In comparison to the aqueous method, this approach offers several advantages. First, the surface of the prepared NPs is coated by hydrophilic polyol ligands in situ; therefore, the NPs can be easily dispersed in aqueous media and other polar solvents. Second, a relatively higher reaction temperature of this system favors particles with higher crystallinity, and, therefore, the size distribution of the NPs is much narrower than those particles produced by traditional methods.

2.3.6 Coprecipitation

Coprecipitation technique is a very easy, handy, and quick way of nanofabrication [57, 58]. In this technique, homogeneous precipitation is obtained by homogeneous aqueous solution of two or more cations by adding a base under ambient conditions or at higher temperatures:

$$xA^{y+}(aq) + yB^{x-}(aq) \rightarrow A_xB_y(s)$$

This reaction simultaneously involves nucleation and growth of nuclei, and then Ostwald ripening progression leads to agglomeration of nuclei. Typical coprecipitation routes are applied for the preparation of nanomaterials of metal oxides [57], metal hydroxides [59], and metal chalcogenides [60].

2.3.7 Thermal Decomposition of Metal–Organic Complex

In the solution chemistry, metal–organic complexes (metal–carbonyl, metal–acetylacetonate, metal–oleate, etc.) are used as precursors to produce metal/metal oxide nanosystems as shown in Scheme 2.3 [61].

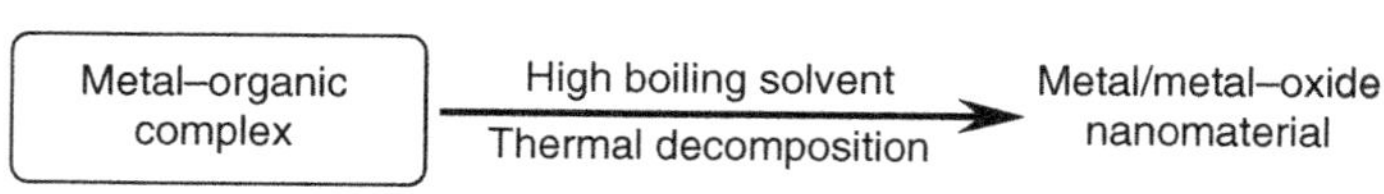

Scheme 2.3 General plan of thermal decomposition. Source: Modified from Park et al. [61].

Thermal decomposition is carried out in the solution phase (i.e., by nonaqueous chemical reactions). Basically, thermal decomposition of a metal–organic complex results in a short burst of metal nuclei, and then these nuclei grow at the same rate to produce monodispersed fine particles. Appropriate surfactants (e.g., oleic acid, oleylamine, polyvinyl pyrrolidone, trioctylphosphine) are used to cap the surface of these magnetic metal/metal–alloy/metal oxide NPs [62, 63].

2.3.8 Temperature-Programmed Reaction in the Presence of NH_3 Gas

Nitridation of nanosystems of inorganic oxides, hydrous oxides, and metal alloys will be carried out with ammonia gas at low temperatures compared to conventional methods. Nanocrystalline materials are formed through solid–gas reactions of an ultrafine precursor compound with ammonia gas [64, 65]. Decomposition of ammonia is effective at around 300 °C. The nanostructured oxide materials are reduced to the metal, and, consecutively, nitridation takes place very effectively because of very large surface area of nanomaterials. During nitridation, optimization of the flow rate of ammonia gas, temperature, and heating/cooling rate in ambient atmosphere plays an important role in controlling the formation of nanosized metal nitrides. The size and shape of the precursor nanomaterials can also be controlled by optimizing nitridation parameters [66].

2.3.9 Urea as Nitrogen Source

Using urea as a nitrogen source enables the synthesis of various metal nitride NPs. In this reaction, a metallic salt (usually chloride) is dissolved and solid urea is added to the alcoholic solution to obtain the required urea/metal precursor molar ratio. The mixture is stirred until the urea is completely dissolved to form a gel phase. The gel is placed in a cylindrical furnace and treated under a flow of N_2/NH_3 [27, 67]. In another method, urea is converted into graphitic carbon nitride (g-C_3N_4) by thermal poly-condensation process. Then, g-C_3N_4 will be used as a nitrogen source for the preparation of metal nitrides [68].

2.4 Characterization of Nanomaterials

Nanometrology is a subdivision of nanoscience and nanotechnology concerned with the characterization or measurement of physical and chemical properties of nanomaterials. A variety of instrumental techniques are available to characterize nanomaterials, and most of them are based on physical phenomena or forces exhibited at the nanoscale level. Table 2.1 summarizes some of the versatile and frequently used methods employed for the physicochemical characterization of nanomaterials.

Table 2.1 List of some measurement techniques.

Technique	Description	Purpose	References
X-ray diffraction (XRD)	Diffraction of X-ray by a crystalline structure	Crystal structure of solid, identification of unknown materials, orientation of single crystals, preferred orientation of polycrystals, defects, stresses	[69]
Scanning electron microscopy (SEM)	Energized focused electron beam interaction with bulk materials	Texture and surface topography	[70]

(Continued)

Table 2.1 (Continued)

Technique	Description	Purpose	References
Transmission electron microscopy (TEM)	Accelerated electrons are projected on ultra-thin specimens	High magnification images of grain boundary, dislocation, second phase particles, defects, selected area electron diffraction (SAED) for lattice parameters of crystalline materials and high-resolution TEM (HRTEM) provide atomic-level imaging for crystallographic structure information	[71]
Atomic force microscopy (AFM)	A precise cantilever mechanical probe is scanned over sample surface	Force measurement, topographic imaging, and manipulation	[72]
Energy dispersive X-ray (EDX) spectroscopy	Scattering power of individual elements as characteristic of the X-ray radiation	Qualitative and quantitative elemental analysis	[73]
Fourier transform infrared (FTIR) spectroscopy	Oscillations induced by infrared electromagnetic radiation	Atomic arrangements in molecules and concentration of the chemical bonds	[74]
Raman spectroscopy	Scattering of radiation by molecules that bring change in electrical polarizability	Structure determination of molecules	[75]
X-ray photoelectron spectroscopy (XPS)	Ionization and emission of core (inner shell) electrons	Elemental composition, empirical formula, chemical state, and electronic state of the elements	[76]
UV–Vis spectroscopy	The excitation of valence electron by UV–Vis radiation	Bonding between atoms into molecules	[77]
Superconducting quantum interference devices (SQUID)	Quantization of magnetic flux in a superconducting ring and Josephson junction	Magnetic properties of materials	[78]
Thermal gravimetric analysis (TGA)	Study of the mass loss of materials against temperature or time	Quantitative information of solvates/hydrates	[79]
Differential scanning calorimetry (DSC)	Recording the difference in heat flow between sample and reference against temperature or time	Phase transition, interference of recipient	[79]
Cyclic voltammetry (CV)	Electrochemical technique	Kinetics and thermodynamics details of electron transfer at the electrode/electrolyte interface	[80]

2.5 Paper Substrate and Functional Materials

The World Health Organization (WHO) has defined that target-specific diagnosing and monitoring devices should be designed with characteristics referred to as ASSURED: affordable, sensitive, specific, user-friendly, rapid and robust, equipment free, and deliverable to end-users [81]. Moreover, the fabrication of effective, point-of-need, and in-field testing analytical devices has been significantly attracted researchers to develop next-generation diagnostic systems especially for health and environmental perspectives. From this perspective, paper substrate has been realized as a flexible, less expensive, ubiquitous, and environmentally friendly material for various applications such as food safety, clinical test, environmental monitoring, and electronics [82–84].

2.5.1 Uniqueness of Paper Substrate

Paper is a thin sheet made of fibers and cellulose pulp obtained from trees and grasses. The suspension of raw pulp along with pigments, chemical premixes, color, etc. run through a series of physical and chemical processes such as pressing, drying, finishing, coating, calendaring to get suitable paper for a variety of applications. The composition of materials plays a key role in the characteristics of paper. For example, nanocellulose is used for the preparation of optically transparent paper [85].

The utility of paper was recognized 2000 years ago [86] and still its requirement is increasing progressively. The use of paper is not only limited to writing, printing, and drawing but it has been extensively employed in cleaning, packaging, decorating. Furthermore, development of paper-based techniques for environmental analysis, biomedical applications, and clinical assays will lead to several advantages compared to conventional materials. The following characteristics of paper make it advantageous over conventional materials:

- Abundant and less expensive.
- A yearly renewable and scalable substance than glass, plastic, and precious metal.
- Eco-friendly and biocompatible.
- Porous material having a high surface-to-volume ratio; consequently, functional reagents can be easily incorporated inside paper matrix.
- Contains high-aspect-ratio cellulose-rich fibers, which help to wick liquids by capillary action and self-pumping phenomenon.
- Lightweight and very flexible; it can be easily modified or bent to transform into two- and three-dimensional structures.
- Paper devices can be safely disposed of by simply ignition.

2.5.2 Functional Materials and Fabrication Methods

Generally, virgin paper cannot fulfill the requirements of paper-based devices; therefore, various functional materials are used to decorate paper. The conjunction between cellobiose units of paper and functional materials is very homogeny that allows paper to modify as per design requirements [87]. Active functional materials can be classified into conductors, semiconductors, and insulators. Also, they can be categorized as organic materials, inorganic materials, or their hybrid systems. Therefore, impregnation of paper with active reagents is an another important step, and for that, mostly well-known methods (painting, dip-coating, drop costing, drawing), including inkjet, screen, and laser printing, as well as roll-to-roll technologies, have been applied to fabricate sensors and electronic components. Each printing method requires different parameters such as resolution, accuracy, uniformity, compatibility, wettability to target substances, and throughput, which have been already extensively reviewed [88, 89].

The functionalized paper is generally categorized into (i) dipstick assays, (ii) lateral flow assays (LFAs), and (iii) microfluidic paper analytical devices (μPADs) [90, 91]. Dipstick paper assays are the easiest one of several existing measurement devices, and to execute chemical tests, they are just immersed into testing liquids, for example pH measurement by pH paper strip. Mostly, color changes of the assay confirm quantitatively or qualitatively the presence of toxic components. A major shortcoming of the dipstick method is to virtually designing sophisticated devices for accurate measurement and diagnostic purposes.

The LFAs are very similar to the dipstick method where active reagents are prestored in the strip, but they also combine with channels for sample flow, for example glucometer strips to measure blood glucose levels. The main advantage of flow canals is that they can be designed as different zones with different reagents for different functions; thus, they can also be used for multi-detection purposes [92]. However, quantitative analysis of LFAs is still sluggish due to large sample volume requirement and difficulties in optimization. Conversely, in μPADs, microchannels are engraved on paper by photolithography, inkjet printing, plasma etching, wax and laser printing, etc. [89]. Therefore, they overcome the challenges with LFAs, and microvolume of sample is sufficient for selective and quantitative analysis [91, 93].

Further, the use of nanomaterials as impregnated active functional components can improve sensitivity, selectivity, and quantitative/qualitative analysis of toxic elements as nanomaterials can introduce new active reaction sites and increase the contact area between active components in nanomaterials and target analytes [8, 94–99]. Moreover, nanomaterial-containing paper-based devices have the potential to surpass existing large size instruments and expensive techniques. Some of the commonly used nanomaterials for modifying paper substrate are summarized in Table 2.2.

2.6 Different Types of Detection Methods

Many detection methods can be applied for the aforementioned three different types of paper-based assays. For example, dipstick assays are suitable for optical (color change) detection, and mostly detection can be done with the naked eye without any microscope or other analytical devices such as pH strips. For LFAs, optical strip

Table 2.2 Commonly employed nanomaterials for functionalization of paper.

Nanomaterial	Detection methods	Applications
Au	Colorimetric, electrochemical, and SERS	Diagnosis of Alzheimer's disease [100], tuberculosis [101], forensic applications [102], and detection of foodborne pathogens [103]
Ag	Filtration, colorimetric	Water treatment [104, 105]
Au–Ag	SERS	Multiple explosive detection [106]
Graphene QD	Smartphone readout	Environmental monitoring [107, 108]
CNT	Charge transport	Humidity [109] and chemical [110] sensors
Ceria	Colorimetric	Glucose bioassay [111]
ZnO	Conductivity change and colorimetric	Ultraviolet sensor [112] and virus detection [113]
TiO_2	Photochemical	Carcinoembryonic Antigen [114]
Fe_3O_4-polydopamine	Fluorescence	Dengue detection [115]
Cobalt phthalocyanine colloids	Electrochemical	Determination of H_2O_2 and glucose [116]
Polyaniline	Color change	Acid concentration sensor [117]
$BaTiO_3$	Piezoelectric	Nanogenerator [118]
Hydroxyapatite	Absorption capacity	Water purification [119]

readers and electrochemical detectors have been used. For μPADs, simple optical detectors and more sophisticated devices and techniques such as microplate readers, luminescence, electrochemistry, piezoresistive sensors have been employed for various measurements [4, 7, 120, 121]. Some of the commonly employed detection techniques are briefly discussed in the following sections.

2.6.1 Colorimetric

Colorimetric sensing is a simple technique for detection of chemical substances and works on the basic principle of the color change of paper substrate after the addition of the sample containing the target analyte. The transition of color can be visually observed by the naked eye, while the quantitative determination can be done by colorimetry [122]. The pH strips are the example of paper-based sensors where the color change of paper, accordingly, after the addition of acidic, basic, and neutral sample solutions, can visually be observed with the naked eye qualitatively [123]. The obtained results are not accurate compared to analytical methods such as UV–Vis spectrophotometry, atomic absorption spectrometry. To improve the sensitivity and quantitative determination of chemical substances, digital images are captured by a smartphone, scanner, or digital color meter [124]. The color intensity of the image is calculated by transferring the data into imaging software such as Adobe Illustrator and ImageJ, where the pixel value is directly proportional to the concentration of chemical substances present in the sample [125, 126]. Klasner et al. reported the use of paper-based sensors for detection of ketones, glucose, and nitrites in artificial urine samples, where potassium iodide was used for the detection of glucose and *N*-(1-napthyl)ethylenediamine for nitrite detection [127]. Nanomaterials are also incorporated into paper substrate for colorimetric detection of a variety of chemical substances that are found to be very selective and sensitive. For example, Au NPs-decorated paper-based colorimetric immunosensor is demonstrated for the diagnosis of early Alzheimer's disease [100].

2.6.2 Electrochemical

In conventional electrochemical techniques such as cyclic voltammetry and potentiometry, the oxidation or the reduction reaction current on the surface of working electrode is monitored as a function of the applied voltage and time with respect to suitable counter and working electrodes [80]. The conventional electrodes are generally made up of platinum, carbon, gold, and mercury, which are very expensive and sophisticated metals [128, 129]. However, the paper-based nanomaterial-decorated strip is used as working, counter, and reference electrodes for detection of different electroactive compounds [88, 130–132]. Commercially available glucometers use paper-based strips for detection of glucose in blood samples of diabetic patients, where the conducting materials are screen printed on the paper strip and working on the principle of amperometry [133]. Jović et al. presented the Ag NP–based printed electrodes by inkjet printer for determination of chlorine in water using linear sweep voltammetry [134]. Thus, the use of paper-based electrochemical sensing devices will begin a new era of miniaturized, portable, rapid, sensitive, and cost-effective devices.

2.6.3 Fluorescence

In fluorescence sensing, the intensity of emitted radiation is measured after the target chemical substance is added into fluorescence materials such as quantum dots (QDs). Research in QD field has set up a landmark due to extensive significant and potential applications in electronics, biomedical, and optical fields. QDs can also be functionalized with various chemical groups to increase their affinity toward target compounds [135–138]. Huang group proposed a water-soluble CdSe/ZnS core–shell QDs for sensitive and quantitative detection of urea concentration in biological samples [2]. Chen et al. reported a turn-on photoluminescent sensor for detection of cysteine and homocysteine using CdTe/CdS QDs [139]. An alternative paper-based technique provides a miniaturization of devices and cost reduction for detection of a variety of chemical substances. Also, several paper-based fluorescence sensors have been reported for detection of proteins [140], DNA [141], cancer cells [142, 143], and bacteria [144].

2.6.4 Surface-Enhanced Raman Scattering (SERS)

Surface-enhanced Raman scattering (SERS) is a technique in which Raman scattered radiation is enhanced by absorption of chemical substances on metal surfaces such as Au, Ag, Pt nanomaterials [145–149]. For example, Au–Ag nanomaterial–impregnated paper-based SERS substrate was applied for detection of explosive materials [106] and thiram in soil [150]. Many works have demonstrated the detection of a variety of chemical substances using different substrates such as hollow Au nanospheres, Fe@Au core–shell structures, and SERS tags [106, 151]. With the advancement of engineering technology, there is a miniaturization of the portable Raman devices that can be low cost and rapid for the determination of clinical samples on SERS substrates [152]. Zeng et al. illustrated the use of nylon filter membrane paper deposited on Ag NPs combined with a smartphone and SERS for detection of rhodamine 6G, crystal violet, and malachite green, which yielded better sensitivity, stability, and reproducibility [153].

2.7 Applications of Nanomaterial-Based Paper Sensors

2.7.1 Environmental Aspects

Many nanomaterial-incorporated paper-based colorimetric, electrochemical, fluorescence, SERS techniques, and so on have been developed to monitor toxicants and pollutants in various environmental contaminants because of their easy-to-handle, portable, economical, and adaptable in-field testing. Environmental contaminants may be present in the air, land, and water originated from inorganic, organic, and/or biological toxicants [154, 155]. Various regulatory bodies such as US Environmental Protection Agency (EPA) and WHO have established guidelines of exposure to a large number of toxic substances for human beings and other living organisms.

Numerous paper-based sensors for analysis of metal ions such as Al^{3+}, As^{5+}, Cu^{2+}, Fe^{3+}, Cr^{6+}, Cd^{2+}, Hg^{2+}, Pb^{2+} have been studied that offer striking alternatives to more sophisticated and expensive techniques such as atomic absorption spectroscopy (AAS) and inductively coupled plasma mass spectrometry (ICP–MS). For example, recently, PADs are reported for determination of lead in water samples, where the polyvinylalcohol-capped Ag NPs fabricated on paper substrate was used and the color change of NPs was from yellow to reddish yellow. The change in color intensity on paper substrate was recorded using a smartphone and ImageJ software was used to determine the color intensity, which was found to be directly proportional to the concentration of the target analyte [105]. In another work, cetyltrimethyl ammonium bromide (CTAB)–stabilized Ag NPs fabricated on paper substrate was demonstrated for detection of Fe^{3+} in water samples using a smartphone. The mechanism for sensing of Fe^{3+} was based on discoloration of Ag NPs fabricated on paper substrate due to the electron transfer reaction taking place on the surface of NPs in the presence of CTAB. The methodology for determination of Fe^{3+} using paper-based sensors integrated with a smartphone is demonstrated in Figure 2.3 [156]. Smartphone-integrated paper device was exploited for detection of

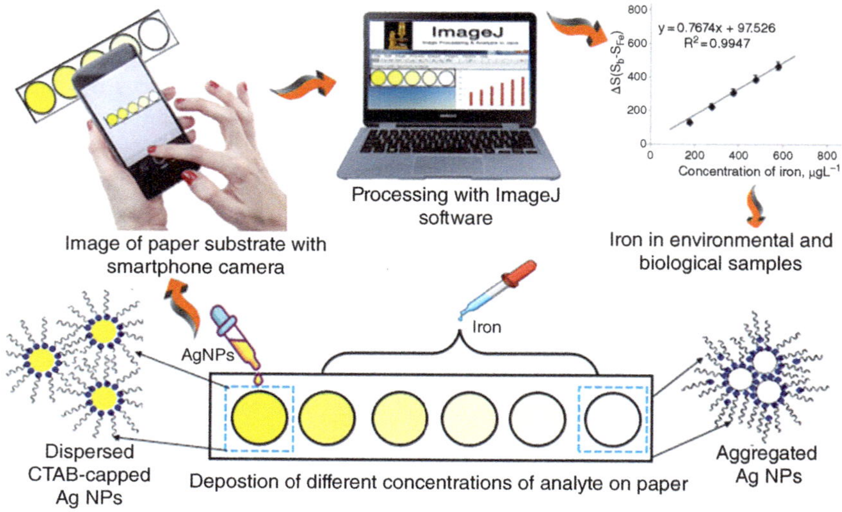

Figure 2.3 Schematic diagram demonstrating the use of smartphone–paper-based sensor followed by processing in ImageJ software to determine the calibration curve and total iron in unknown environmental water and biological samples. Source: Shrivas et al. [156]. Copyright (2020) Springer Nature.

As^{3+} using sucrose-modified gold NPs deposited on paper substrate, and the results were compared with colorimetric determination [156]. In another way, a paper electrode made from gold NP-modified boron-doped diamond was illustrated for anodic stripping voltammetric detection of arsenic in rice samples, where arsenic was quantified by reducing from As^{5+} to As^{3} using thiosulfate in the presence of HCl and depositing on the paper electrode. The obtained results showed better precision, working potential, and reproducibility [157]. Yakoh et al. illustrated application of paper-based colorimetric sensors fabricated with silver nanoprism for determination of chloride in environmental water samples. The detection was based on oxidative etching of silver nanoprism into smaller silver nanospheres, and the color of nanoprism changed from dark violet to red after the interaction with the chloride sample solution [158]. Lee and group designed a strip of tethered rhodamine carbon nanodots for selective detection of Al^{3+} using a Forster resonance energy transfer (FRET)–based ratiometric sensing mechanism [159].

In addition to metals ion recognition, nanomaterial-modified paper sensors have been applied to detect nonmetal inorganic and organic pollutants. Shrivas and group prepared a paper-based conductive track by simple direct writing with a pen using silver nano-ink for detection of H_2O_2 in water samples [160]. More recently, paper electrode has been prepared by fabricating silver nano-ink using office inkjet printer and exploited for determination of nitrate in water samples. The preparation steps of the paper electrode and then application in cyclic voltammetric analysis are shown in Figure 2.4 [161]. In another work, a graphene layer was directly transferred to paper substrate and employed as sensor for NO_2 gas [162]. Lee et al.

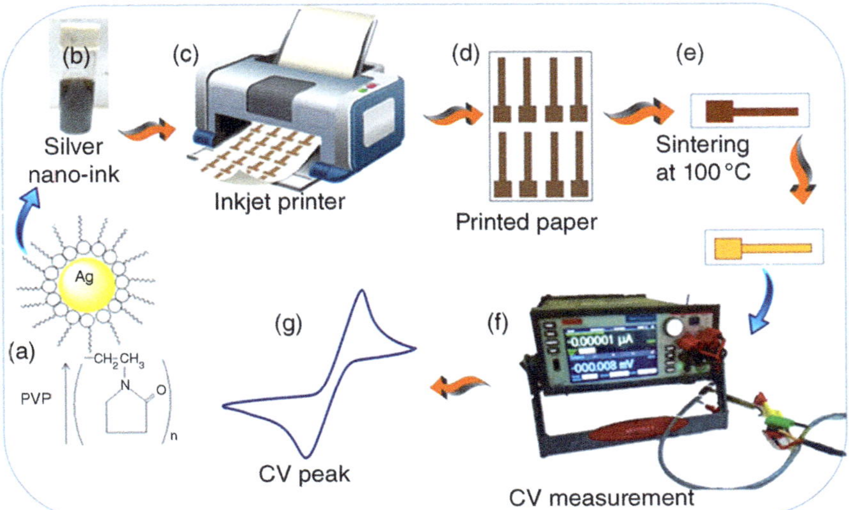

Figure 2.4 Schematic presentations: (a) synthesis of silver nano-ink, (b) formulation of silver nano-ink, (c) fabrication of nano-ink on paper substrate using inkjet printer, (d) printed electrodes, (e) sintering of electrode, (f) application of prepared electrode in CV measurements, and (g) CV responses. Source: Reproduced with permission from Kant et al. [161]. Copyright (2020) Elsevier.

designed a CdTe QDs–functionalized paper sensor coupled with a smartphone for detection of pictogram-level of trinitrophenol explosives [163]. Costa et al. demonstrated the use of inkjet printer for functionalizing the multiwalled CNTs on paper substrate where the working, reference, and counter electrodes were constructed. Then, mineral oil was spread on the substrate to form the hydrophobic barrier during the sensing of chemical substances from aqueous solution [164]. Nano-TiO_2-modified zeolite-based photocatalytic paper has been prepared for removal of volatile organic compounds under visible light [165]. Dye-encapsulated polymer NP-based ink printed on paper substrate was used as a colorimetric sensor for selective and specific discrimination of organic volatile amines [166], and a paper decorated with NPs (Ag, Au) can act as a sensor for the detection of volatile biogenic amines liberated from spoiled food [167]. CNT and MoS_2 integrated on the cellulose paper was demonstrated as a highly sensitive paper sensor for NO_2 gas [110]. A filter paper immobilized with Ag NPs and treated with alkyl ketene dimer generates enhanced SERS signals with the interaction of pesticides [168]. Moreover, inkjet-printed enzymatic–NP paper test strips generate colorimetric response that allows visual detection of bacterial pathogens in water [169].

2.7.2 Clinical Aspects

Monitoring blood glucose levels is a valuable routine for viewing, judgment, and continuing management of diabetes. In this regard, obtaining accurate and reliable glucose measurements from glucometers available on the market is challenging to consumers. Ornatska et al. developed paper-based sensor impregnated with ceria NPs for detection of glucose in biological samples. The sensing mechanism was based on the change of color of ceria NPs from white-yellowish to dark orange when H_2O_2 was liberated in the presence of glucose oxidase, which was proportional to the concentration of glucose present in the sample [111]. N-carbon dots/TiO_2–Pt modified paper was employed as a sensor for detection of carcinoembryonic antigen in a clinical serum sample by the photoelectrochemical sensing technique [114]. Another report used a TiO_2 nanotube–based rotatable paper-based photocontrolled cytosensor for ultrasensitive detection of cell-surface protein expressions of cancerous cells [170]. The folic acid in orange juice and urine samples was determined by a hybrid polydimethylsiloxane paper device functionalized with carbon dots as fluorophores by Schiff base [171]. Gold NPs immobilized on paper can be used as a nucleic acid probe by binding oligonucleotides. This is a highly sensitive and selective colorimetric probe working on the basis of separation of aptamer-assembled NP aggregates, and also cocaine in serum can be also detected [172].

2.7.3 Food Safety Aspects

Food quality and safety is now a major concern all over the world as the unsafe level of toxic chemicals affect the human health [108]. Also, it has been realized that the monitoring of contaminants in food samples should be done by nanomaterial-modified paper sensing devices for several advantages [108, 173]. For example, Quintela et al. illustrated the detection of nineteen strains of foodborne pathogen *Salmonella* spp. in food and environmental matrices by using oligonucleotide-functionalized gold NPs where the sensing was based on color change of the NPs from red to purple [174]. The cellulose-based device is modified with luminescent graphene QDs for detection of phenols and polyphenols in environmental and food samples. The graphene QD was synthesized from citric acid through pyrolysis and adsorbed on the nitrocellulose substrate for detection of target analytes from samples through the resonance energy transfer phenomenon [107]. The paper SERS substrate was coated with aluminum and a thin layer of Ag NPs, which enhances the signal intensity of tetracycline in milk samples. This work has the advantages of on-site detection of milk contaminants and the surveillance of food production and processing [175].

2.8 Conclusion and Future Prospects

Nanomaterial-incorporated paper sensors are versatile probes for the detection and monitoring of environmental toxicants present in air, water, and ground territories. Paper substrate is a biodegradable and scalable material, which makes these devices multipurpose and highly adoptable tools in terms of cost, easy handling, and portability. Also, impregnation of nanomaterials as active sensor components on paper substrate makes them suitable for miniaturization of devices with high sensitivity and selectivity toward analyte components. Fabrication of nanomaterials and decorating them on paper substrate may be challenging and provide opportunities to researchers to design new kinds of devices to overcome the limitations of existing devices.

This chapter briefly highlighted various wet chemical synthesis techniques of nanomaterials and potential applications of nanomaterial-combined paper substrate devices for a variety of applications such as environmental, biomedical, and food safety purposes.

Although there is considerable progress in this field, new trends are continuously emerging, and many more potential alternatives are evolving with modifications in paper and nanomaterials hybridization. For example, hybridization of more than one pristine nanomaterials and/or designing such a pattern on a paper substrate having different functionality and rationality of qualitative and quantitative analysis of multi-toxicants, simultaneously. Also, in future, theoretical approaches will lead to a better understanding of mechanisms and assist in instrumentation. We hope this chapter will contribute to future developments of paper-based devices and sensors.

References

1 Benedict, K.M., Reses, H., Vigar, M. et al. (2017). Surveillance for waterborne disease outbreaks associated with drinking water—United States, 2013–2014. *MMWR Morb. Mortal. Wkly Rep.* 66: 1216–1221.

2 Uyttendaele, M., Jaykus, L.-A., Amoah, P. et al. (2015). Microbial hazards in irrigation water: standards, norms, and testing to manage use of water in fresh produce primary production. *Compr. Rev. Food Sci. Food Saf.* 14: 336–356.

3 Viator, C., Blitstein, J., Brophy, J.E., and Fraser, A. (2015). Preventing and controlling foodborne disease in commercial and institutional food service settings: a systematic review of published intervention studies. *J. Food Prot.* 78: 446–456.

4 Aragay, G., Pons, J., and Merkoçi, A. (2011). Recent trends in macro-, micro-, and nanomaterial-based tools and strategies for heavy-metal detection. *Chem. Rev.* 111: 3433–3458.

5 Egorova, K.S. and Ananikov, V.P. (2017). Toxicity of metal compounds: knowledge and myths. *Organometallics* 36: 4071–4090.

6 Martin, B.R. (2006). Metal ion toxicity. *Encyclopedia Inorg. Chem.* https://doi.org/10.1002/0470862106.ia136.

7 Chen, J. and Park, B. (2016). Recent advancements in nanobioassays and nanobiosensors for foodborne pathogenic bacteria detection. *J. Food Prot.* 79: 1055–1069.

8 Ge, S., Zhang, L., Zhang, Y. et al. (2017). Nanomaterials-modified cellulose paper as a platform for biosensing applications. *Nanoscale* 9: 4366–4382.

9 Janković, N.Z. and Plata, D.L. (2019). Engineered nanomaterials in the context of global element cycles. *Environ. Sci. Nano* 6: 2697–2711.

10 Feynman, R.P. (1960). There's plenty of room at the bottom. *Eng. Sci.* 23: 22–36.

11 Cao, G. (2004). *Nanostructures and Nanomaterials Synthesis, Properties and Applications*. London: Imperial College Press.

12 Rahman, F. (2008). *Nanostructures in Electronics and Photonics*. Singapore: Pan Stanford.

13 Wang, Z.L. (2003). *Nanowires and Nanobelts, Materials, Properties and Devices*. Boston: Kluwer Academic.

14 Bell, A.T. (2003). The impact of nanoscience on heterogeneous catalysis. *Science* 299: 1688–1691.

15 De, M., Ghosh, P.S., and Rotello, V.M. (2008). Applications of nanoparticles in biology. *Adv. Mater.* 20: 4225–4241.

16 Han, X.-F., Shamaila, S., Sharif, R. et al. (2009). Structural and magnetic properties of various ferromagnetic nanotubes. *Adv. Mater.* 21: 4619–4624.

17 Kaushik, A., Kumar, R., Arya, S.K. et al. (2015). Organic–inorganic hybrid nanocomposite-based gas sensors for environmental monitoring. *Chem. Rev.* 115: 4571–4606.

18 Riehemann, K., Schneider, S.W., Luger, T.A. et al. (2009). Nanomedicine-challenge and perspectives. *Angew. Chem. Int. Ed.* 48: 872–897.

19 Simon, P. and Gogotsi, Y. (2008). Materials for electrochemical capacitors. *Nat. Mater.* 7: 845–854.

20 El-Sayed, M.A. (2001). Some interesting properties of metals confined in time and nanometer space of different shapes. *Acc. Chem. Res.* 34: 257–264.

21 Montalti, M., Cantelli, A., and Battistelli, G. (2015). Nanodiamonds and silicon quantum dots: ultrastable and biocompatible luminescent nanoprobes for long-term bioimaging. *Chem. Soc. Rev.* 44: 4853–4921.

22 Gleiter, H. (1989). Nanocrystalline materials. *Prog. Mater. Sci.* 33: 223–315.

23 Xia, Y., Xia, X., and Peng, H.-C. (2015). Shape-controlled synthesis of colloidal metal nanocrystals: thermodynamic versus kinetic products. *J. Am. Chem. Soc.* 137: 7947–7966.

24 Pinna, N. and Niederberger, M. (2008). Surfactant-free nonaqueous synthesis of metal oxide nanostructures. *Angew. Chem. Int. Ed.* 47: 5292–5304.

25 Hantanasirisakul, K. and Gogotsi, Y. (2018). Electronic and optical properties of 2D transition metal carbides and nitrides (MXenes). *Adv. Mater.* 30: 1804779. (1–30).

26 Yao, W., Makowski, P., Giordano, C., and Goettmann, F. (2009). Synthesis of early-transition-metal carbide and nitride nanoparticles through the urea route and their use as alkylation catalysts. *Chem. Eur. J.* 15: 11999–12004.

27 Giordano, C., Erpen, C., Yao, W. et al. (2009). Metal nitride and metal carbide nanoparticles by a soft urea pathway. *Chem. Mater.* 21: 5136–5144.

28 Wang, X., Zhuang, J., Peng, Q., and Li, Y. (2005). A general strategy for nanocrystal synthesis. *Nature* 437: 121–124.

29 Liu, Z., Sun, K., Jian, W.-B. et al. (2009). Soluble InP and GaP nanowires: self-seeded, solution–liquid–solid synthesis and electrical properties. *Chem. Eur. J.* 15: 4546–4552.

30 Jin, Q., Chen, H.-J., Li, X. et al. (2019). Reduced graphene oxide nanohybrid–assembled microneedles as mini-invasive electrodes for real-time transdermal biosensing. *Small* 15: 1804298.

31 Wang, Y. and Hu, A. (2014). Carbon quantum dots: synthesis, properties and applications. *J. Mater. Chem. C* 2: 6921–6939.

32 Yuan, T., Meng, T., He, P. et al. (2019). Carbon quantum dots: an emerging material for optoelectronic applications. *J. Mater. Chem. C* 7: 6820–6835.

33 Khare, H.S., Gosvami, N.N., Lahouij, I. et al. (2018). Nanotribological printing: a nanoscale additive manufacturing method. *Nano Lett.* 18: 6756–6763.

34 Liddle, J.A. and Gallatin, G.M. (2016). Nanomanufacturing: A perspective. *ACS Nano* 10: 2995–3014.

35 Spatz, J.P., Chan, V.Z.-H., Mößmer, S. et al. (2002). A combined top–down/bottom–up approach to the microscopic localization of metallic nanodots. *Adv. Mater.* 14: 1827–1832.

36 Stavis, S.M., Fagan, J.A., Stopa, M., and Liddle, J.A. (2018). Nanoparticle manufacturing – heterogeneity through processes to products. *ACS Appl. Nano Mater.* 1: 4358–4385.

37 Aleksandra, B.D., Yan, Y.X., Yan, F.H., and Wai, K.C. (2007). Hydrothermal synthesis of nanostructures. *Recent Pat. Nanotechnol.* 1: 121–128.

38 Huang, X., Zhang, H., Guo, C. et al. (2009). Simplifying the creation of hollow metallic nanostructures: one-pot synthesis of hollow palladium/platinum single-crystalline nanocubes. *Angew. Chem. Int. Ed.* 48: 4808–4812.

39 Dewangan, K., Sinha, N.N., Chavan, P.G. et al. (2012). Synthesis and characterization of self-assembled nanofiber-bundles of V_2O_5: their electrochemical and field emission properties. *Nanoscale* 4: 645–651.

40 Kim, M., Son, W.-S., Ahn, K.H. et al. (2014). Hydrothermal synthesis of metal nanoparticles using glycerol as a reducing agent. *J. Supercrit. Fluids* 90: 53–59.

41 Chung, C.-C., Chung, T.-W., and Yang, T.C.K. (2008). Rapid synthesis of titania nanowires by microwave-assisted hydrothermal treatments. *Ind. Eng. Chem. Res.* 47: 2301–2307.

42 Wu, M., Long, J., Huang, A. et al. (1999). Microemulsion-mediated hydrothermal synthesis and characterization of nanosize rutile and anatase particles. *Langmuir* 15: 8822–8825.

43 Jamshidi, M., Ghaedi, M., Dashtian, K. et al. (2016). Sonochemical assisted hydrothermal synthesis of ZnO: Cr nanoparticles loaded activated carbon for simultaneous ultrasound-assisted adsorption of ternary toxic organic dye: derivative spectrophotometric, optimization, kinetic and isotherm study. *Ultrason. Sonochem.* 32: 119–131.

44 Song, I., Lee, H., and Kim, D.H. (2018). Rotation-assisted hydrothermal synthesis of thermally stable multiwalled titanate nanotubes and their application to selective catalytic reduction of NO with NH_3. *ACS Appl. Mater. Interfaces* 10: 42249–42257.

45 Polte, J., Ahner, T.T., Delissen, F. et al. (2010). Mechanism of gold nanoparticle formation in the classical citrate synthesis method derived from coupled in situ XANES and SAXS evaluation. *J. Am. Chem. Soc.* 132: 1296–1301.

46 Malassis, L., Dreyfus, R., Murphy, R.J. et al. (2016). One-step green synthesis of gold and silver nanoparticles with ascorbic acid and their versatile surface post-functionalization. *RSC Adv.* 6: 33092–33100.

47 Chen, Y.-H., Hung, H.-H., and Huang, M.H. (2009). Seed-mediated synthesis of palladium nanorods and branched nanocrystals and their use as recyclable Suzuki coupling reaction catalysts. *J. Am. Chem. Soc.* 131: 9114–9121.

48 Yu, D. and Yam, V.W.-W. (2004). Controlled synthesis of monodisperse silver nanocubes in water. *J. Am. Chem. Soc.* 126: 13200–13201.

49 Pastoriza-Santos, I. and Liz-Marzán, L.M. (2009). N,N-dimethylformamide as a reaction medium for metal nanoparticle synthesis. *Adv. Funct. Mater.* 19: 679–688.

50 Ganguli, A.K., Ganguly, A., and Vaidya, S. (2010). Microemulsion-based synthesis of nanocrystalline materials. *Chem. Soc. Rev.* 39: 474–485.

51 Landfester, K. (2009). Miniemulsion polymerization and the structure of polymer and hybrid nanoparticles. *Angew. Chem. Int. Ed.* 48: 4488–4507.

52 Richard, B., Lemyre, J.-L., and Ritcey, A.M. (2017). Nanoparticle size control in microemulsion synthesis. *Langmuir* 33: 4748–4757.

53 Jiang, Y., Yang, S., Hua, Z., and Huang, H. (2009). Sol–gel autocombustion synthesis of metals and metal alloys. *Angew. Chem. Int. Ed.* 48: 8529–8531.

54 Niederberger, M. (2007). Nonaqueous sol–gel routes to metal oxide nanoparticles. *Acc. Chem. Res.* 40: 793–800.

55 Sun, Y. and Xia, Y. (2002). Shape-controlled synthesis of gold and silver nanoparticles. *Science* 298: 2176–2179.

56 Cao, A.-M., Hu, J.-S., Liang, H.-P., and Wan, L.-J. (2005). Self-assembled vanadium pentoxide (V_2O_5) hollow microspheres from nanorods and their application in lithium-ion batteries. *Angew. Chem. Int. Ed.* 44: 4391–4395.

57 Ahn, T., Kim, J.H., Yang, H.-M. et al. (2012). Formation pathways of magnetite nanoparticles by coprecipitation method. *J. Phys. Chem. C* 116: 6069–6076.

58 Dong, H. and Koenig, G.M. (2020). A review on synthesis and engineering of crystal precursors produced via coprecipitation for multicomponent lithium-ion battery cathode materials. *CrystEngComm* 22: 1514–1530. https://doi.org/10.1039/C9CE00679F.

59 Kim, Y. and Kim, D. (2012). Synthesis of high-density nickel cobalt aluminum hydroxide by continuous coprecipitation method. *ACS Appl. Mater. Interfaces* 4: 586–589.

60 Tsamouras, D., Dalas, E., Sakkopoulos, S., and Koutsoukos, P.G. (1999). Properties of Cu(II) and Ni(II) sulfides prepared by coprecipitation in aqueous solution. *Langmuir* 15: 7940–7946.

61 Park, J., An, K., Hwang, Y. et al. (2004). Ultra-large-scale syntheses of monodisperse nanocrystals. *Nat. Mater.* 3: 891–895.

62 Sun, S., Murray, C.B., Weller, D. et al. (2000). Monodisperse FePt nanoparticles and ferromagnetic FePt nanocrystal superlattices. *Science* 287: 1989–1992.

63 Xu, Z., Shen, C., Hou, Y. et al. (2009). Oleylamine as both reducing agent and stabilizer in a facile synthesis of magnetite nanoparticles. *Chem. Mater.* 21: 1778–1780.

64 Choi, D., Blomgren, G.E., and Kumta, P.N. (2006). Fast and reversible surface redox reaction in nanocrystalline vanadium nitride supercapacitors. *Adv. Mater.* 18: 1178–1182.

65 Krawiec, P., De Cola, P.L., Gläser, R. et al. (2006). Oxide foams for the synthesis of high-surface-area vanadium nitride catalysts. *Adv. Mater.* 18: 505–508.

66 Dewangan, K., Patil, S.S., Joag, D.S. et al. (2010). Topotactical nitridation of α-MoO_3 fibers to γ-Mo_2N fibers and its field emission properties. *J. Phys. Chem. C* 114: 14710–14715.

67 Buha, J., Djerdj, I., Antonietti, M., and Niederberger, M. (2007). Thermal transformation of metal oxide nanoparticles into nanocrystalline metal nitrides using cyanamide and urea as nitrogen source. *Chem. Mater.* 19: 3499–3505.

68 Fischer, A., Antonietti, M., and Thomas, A. (2007). Growth confined by the nitrogen source: synthesis of pure metal nitride nanoparticles in mesoporous graphitic carbon nitride. *Adv. Mater.* 19: 264–267.

69 Cullity, B.D. (1978). *Elements of X-Ray Diffraction.* California: Addison-Wesley Publishing Company Inc.

70 Reimer, L. (1998). *Scanning Electron Microscopy: Physics of Image Formation and Microanalysis.* New York: Springer-Verlag.

71 Williams, D.B. and Carter, C.B. (1996). *Transmission Electron Microscopy: A Textbook for Materials Science.* New York: Plenum Press.

72 Haugstad, G. (2012). *Atomic Force Microscopy: Understanding Basic Modes and Advanced Applications.* Hoboken, NJ: Wiley.

73 Goldstein, J.I., Newbury, D.E., Joy, D.C. et al. (2003). *Scanning Electron Microscopy and X-Ray Microanalysis.* Boston: Springer.

74 Brugel, W. (1962). *Introduction to Infrared Spectroscopy.* New York: Wiley.

75 Ferraro, J.R., Nakamoto, K., and Brown, C.W. (2003). *Introductory Raman Spectroscopy.* Elsevier.

76 Hüfner, S. (1995). *Photoelectron Spectroscopy: Principles and Applications.* Germany: Springer-Verlag.

77 De Caro, C. and Haller, C. (2015). *UV/VIS Spectrophotometry – Fundamentals and Applications.* Suiça: Mettler-Toledo.

78 Clarke, J. and Braginski, A.I. (2004). *The SQUID Handbook: Fundamentals and Technology of SQUIDs and SQUID Systems.* Wiley-VCH Verlag GmbH & Co.

79 Haines, P.J. (1994). *Thermal Methods of Analysis - Principle, Applications and Problems.* Blackie Academic and Professional.

80 Bard, A.J. and Faulkner, L.R. (1980). *Electrochemical Methods: Fundamentals and Applications.* New York: Wiley.

81 Mabey, D., Peeling, R.W., Ustianowski, A., and Perkins, M.D. (2004). Diagnostics for the developing world. *Nat. Rev. Microbiol.* 2: 231–240.

82 Parolo, C. and Merkoci, A. (2013). Paper-based nanobiosensors for diagnostics. *Chem. Soc. Rev.* 42: 450–457.

83 Vicente, A.T., Araújo, A., Mendes, M.J. et al. (2018). Multifunctional cellulose-paper for light harvesting and smart sensing applications. *J. Mater. Chem. C* 6: 3143–3181.

84 Zhao, W. and Berg, A.V.D. (2008). Lab on paper. *Lab Chip* 8: 1988–1991.

85 Nogi, M., Iwamoto, S., Nakagaito, A.N., and Yano, H. (2009). Optically transparent nanofiber paper. *Adv. Mater.* 21: 1595–1598.

86 Edwin, G. and Beal, J. (1988). *Paper and Printing. Vol. 5, Pt. 1 of Science and Civilisation in China* (eds. T. Tsuen-Hsuin and J. Needham). Cambridge: Chemistry and Chemical Rechnology.

87 Siegel, A.C., Phillips, S.T., Dickey, M.D. et al. (2010). Foldable printed circuit boards on paper substrates. *Adv. Funct. Mater.* 20: 28–35.

88 Ataide, V.N., Mendes, L.F., Gama, L.I.L.M. et al. (2020). Electrochemical paper-based analytical devices: ten years of development. *Anal. Methods* 12: 1030–1054.

89 Ghosh, R., Gopalakrishnan, S., Savitha, R. et al. (2019). Fabrication of laser printed microfluidic paper-based analytical devices (LP-μPADs) for point-of-care applications. *Sci. Rep.* 9: 7896.

90 Yang, Y., Noviana, E., Nguyen, M.P. et al. (2017). Paper-based microfluidic devices: emerging themes and applications. *Anal. Chem.* 89: 71–91.

91 Yetisen, A.K., Akram, M.S., and Lowe, C.R. (2013). Paper-based microfluidic point-of-care diagnostic devices. *Lab Chip* 13: 2210–2251.

92 Hui, C.Y., Liu, M., Li, Y., and Brennan, J.D. (2018). A paper sensor printed with multifunctional bio/nano materials. *Angew. Chem. Int. Ed.* 57: 4549–4553.

93 Lin, R., Skandarajah, A., Gerver, R.E. et al. (2015). A lateral electrophoretic flow diagnostic assay. *Lab Chip* 15: 1488–1496.

94 Alba-Patiño, A., Adrover-Jaume, C., and de la Rica, R. (2020). Nanoparticle reservoirs for paper-only immunosensors. *ACS Sens.* 5: 147–153.

95 Hayden, D.R., Mohan, S., Imhof, A., and Velikov, K.P. (2019). Fully biobased highly transparent nanopaper with UV-blocking functionality. *ACS Appl. Polym. Mater.* 1: 641–646.

96 Kumar, S., Pandey, C.M., Hatamie, A. et al. (2019). Nanomaterial-modified conducting paper: fabrication, properties, and emerging biomedical applications. *Global Chall.* 3: 1900041.

97 Ngo, Y.H., Li, D., Simon, G.P., and Garnier, G. (2011). Paper surfaces functionalized by nanoparticles. *Adv. Colloid Interf. Sci.* 163: 23–38.

98 Roig-Sanchez, S., Jungstedt, E., Anton-Sales, I. et al. (2019). Nanocellulose films with multiple functional nanoparticles in confined spatial distribution. *Nanoscale Horiz.* 4: 634–641.

99 Yao, S., Swetha, P., and Zhu, Y. (2018). Nanomaterial-enabled wearable sensors for healthcare. *Adv. Healthcare Mater.* 7: 1700889.

100 Brazaca, L.C., Moreto, J.R., Martín, A. et al. (2019). Colorimetric paper-based immunosensor for simultaneous determination of fetuin B and clusterin toward early Alzheimer's diagnosis. *ACS Nano* 13: 13325–13332.

101 Tsai, T.-T., Huang, C.-Y., Chen, C.-A. et al. (2017). Diagnosis of tuberculosis using colorimetric gold nanoparticles on a paper-based analytical device. *ACS Sens.* 2: 1345–1354.

102 Ameku, W.A., de Araujo, W.R., Rangel, C.J. et al. (2019). Gold nanoparticle paper-based dual-detection device for forensics applications. *ACS Appl. Nano Mater.* 2: 5460–5468.

103 Rodríguez-Lorenzo, L., Garrido-Maestu, A., Bhunia, A.K. et al. (2019). Gold nanostars for the detection of foodborne pathogens via surface-enhanced Raman scattering combined with microfluidics. *ACS Appl. Nano Mater.* 2: 6081–6086.

104 Dankovich, T.A. and Gray, D.G. (2011). Bactericidal paper impregnated with silver nanoparticles for point-of-use water treatment. *Environ. Sci. Technol.* 45: 1992–1998.

105 Shrivas, K., Sahu, B., Deb, M.K. et al. (2019). Colorimetric and paper-based detection of lead using PVA capped silver nanoparticles: experimental and theoretical approach. *Microchem. J.* 150: 104156.

106 Moram, S.S.B., Byram, C., Shibu, S.N. et al. (2018). Ag/Au nanoparticle-loaded paper-based versatile surface-enhanced Raman spectroscopy substrates for multiple explosives detection. *ACS Omega* 3: 8190–8201.

107 Álvarez-Diduk, R., Orozco, J., and Merkoçi, A. (2017). Paper strip-embedded graphene quantum dots: a screening device with a smartphone readout. *Sci. Rep.* 7: 976.

108 Weng, X. and Neethirajan, S. (2018). Paper-based microfluidic aptasensor for food safety. *J. Food Saf.* 38: e12412.

109 Han, J.-W., Kim, B., Li, J., and Meyyappan, M. (2012). Carbon nanotube based humidity sensor on cellulose paper. *J. Phys. Chem. C* 116: 22094–22097.

110 Lee, W.S. and Choi, J. (2019). Hybrid integration of carbon nanotubes and transition metal dichalcogenides on cellulose paper for highly sensitive and extremely deformable chemical sensor. *ACS Appl. Mater. Interfaces* 11: 19363–19371.

111 Ornatska, M., Sharpe, E., Andreescu, D., and Andreescu, S. (2011). Paper bioassay based on ceria nanoparticles as colorimetric probes. *Anal. Chem.* 83: 4273–4280.

112 Gimenez, A.J., Yáñez-Limón, J.M., and Seminario, J.M. (2011). ZnO–paper based photoconductive UV sensor. *J. Phys. Chem. C* 115: 282–287.

113 Xia, Y., Chen, Y., Tang, Y. et al. (2019). Smartphone-based point-of-care microfluidic platform fabricated with a ZnO nanorod template for colorimetric virus detection. *ACS Sens.* 4: 3298–3307.

114 Li, L., Wang, T., Zhang, Y. et al. (2018a). Editable TiO_2 nanomaterial-modified paper in situ for highly efficient detection of carcinoembryonic antigen by photoelectrochemical method. *ACS Appl. Mater. Interfaces* 10: 14594–14601.

115 Ortega, G.A., Pérez-Rodríguez, S., and Reguera, E. (2017). Magnetic paper-based ELISA for IgM-dengue detection. *RSC Adv.* 7: 4921–4932.

116 Sánchez-Calvo, A., Costa-García, A., and Blanco-López, M.C. (2020). Paper-based electrodes modified with cobalt phthalocyanine colloid for the determination of hydrogen peroxide and glucose. *Analyst* https://doi.org/10.1039/C9AN02413A.

117 Souza, F.G. Jr., Oliveira, G.E., Anzai, T. et al. (2009). A sensor for acid concentration based on cellulose paper sheets modified with polyaniline nanoparticles. *Macromol. Mater. Eng.* 294: 739–748.

118 Zhang, G., Liao, Q., Zhang, Z. et al. (2016). Novel piezoelectric paper-based flexible nanogenerators composed of $BaTiO_3$ nanoparticles and bacterial cellulose. *Adv. Sci.* 3: 1500257.

119 Zhang, Q.-Q., Zhu, Y.-J., Wu, J. et al. (2019). Ultralong hydroxyapatite nanowire-based filter paper for high-performance water purification. *ACS Appl. Mater. Interfaces* 11: 4288–4301.

120 Gong, M.M. and Sinton, D. (2017). Turning the page: advancing paper-based microfluidics for broad diagnostic application. *Chem. Rev.* 117: 8447–8480.

121 Yamada, K., Henares, T.G., Suzuki, K., and Citterio, D. (2015). Paper-based inkjet-printed microfluidic analytical devices. *Angew. Chem. Int. Ed.* 54: 5294–5310.

122 Martinez, A.W., Phillips, S.T., Whitesides, G.M., and Carrilho, E. (2010). Diagnostics for the developing world: microfluidic paper-based analytical devices. *Anal. Chem.* 82: 3–10.

123 Wang, Y., Liu, Y., Liu, W. et al. (2018). Quantification of combined color and shade changes in colorimetry and image analysis: water pH measurement as an example. *Anal. Methods* 10: 3059–3065.

124 Martinez, A.W., Phillips, S.T., Carrilho, E. et al. (2008). Simple telemedicine for developing regions: camera phones and paper-based microfluidic devices for real-time, off-site diagnosis. *Anal. Chem.* 80: 3699–3707.

125 Dungchai, W., Chailapakul, O., and Henry, C.S. (2011). A low-cost, simple, and rapid fabrication method for paper-based microfluidics using wax screen-printing. *Analyst* 136: 77–82.

126 Martinez, A.W., Phillips, S.T., Butte, M.J., and Whitesides, G.M. (2007). Patterned paper as a platform for inexpensive, low-volume, portable bioassays. *Angew. Chem. Int. Ed.* 46: 1318–1320.

127 Klasner, S.A., Price, A.K., Hoeman, K.W. et al. (2010). Paper-based microfluidic devices for analysis of clinically relevant analytes present in urine and saliva. *Anal. Bioanal. Chem.* 397: 1821–1829.

128 Rastogi, P.K., Ganesan, V., and Krishnamoorthi, S. (2014). A promising electrochemical sensing platform based on a silver nanoparticles decorated copolymer for sensitive nitrite determination. *J. Mater. Chem. A* 2: 933–943.

129 Zoski, C. (2006). *Handbook of Electrochemistry*. Elsevier Science.

130 Cinti, S., Cinotti, G., Parolo, C. et al. (2020). Experimental comparison in sensing breast cancer mutations by signal ON and signal OFF paper-based electroanalytical strips. *Anal. Chem.* 92: 1674–1679.

131 Mettakoonpitak, J., Boehle, K., Nantaphol, S. et al. (2016). Electrochemistry on paper-based analytical devices: a review. *Electroanalysis* 28: 1420–1436.

132 Oh, J.-M. and Chow, K.-F. (2015). Recent developments in electrochemical paper-based analytical devices. *Anal. Methods* 7: 7951–7960.

133 Yoo, E.-H. and Lee, S.-Y. (2010). Glucose biosensors: an overview of use in clinical practice. *Sensors* 10: 4558–4576.

134 Jović, M., Cortés-Salazar, F., Lesch, A. et al. (2015). Electrochemical detection of free chlorine at inkjet printed silver electrodes. *J. Electroanal. Chem.* 756: 171–178.

135 Harris, R.D., Bettis Homan, S., Kodaimati, M. et al. (2016). Electronic processes within quantum dot-molecule complexes. *Chem. Rev.* 116: 12865–12919.

136 Talapin, D.V., Lee, J.-S., Kovalenko, M.V., and Shevchenko, E.V. (2010). Prospects of colloidal nanocrystals for electronic and optoelectronic applications. *Chem. Rev.* 110: 389–458.

137 Xiao, L. and Sun, H. (2018). Novel properties and applications of carbon nanodots. *Nanoscale Horiz.* 3: 565–597.

138 Xu, G., Zeng, S., Zhang, B. et al. (2016). New generation cadmium-free quantum dots for biophotonics and nanomedicine. *Chem. Rev.* 116: 12234–12327.

139 Chen, S., Tian, J., Jiang, Y. et al. (2013). A one-step selective fluorescence turn-on detection of cysteine and homocysteine based on a facile CdTe/CdS quantum dots–phenanthroline system. *Anal. Chim. Acta* 787: 181–188.

140 Liu, H. and Crooks, R.M. (2011). Three-dimensional paper microfluidic devices assembled using the principles of origami. *J. Am. Chem. Soc.* 133: 17564–17566.

141 Ali, M.M., Aguirre, S.D., Xu, Y. et al. (2009). Detection of DNA using bioactive paper strips. *Chem. Commun.* 43: 6640–6642.

142 Abbasi Kajani, A., Bordbar, A.-K., Mehrgardi, M.A. et al. (2018). Green and facile synthesis of highly photoluminescent multicolor carbon nanocrystals for cancer therapy and imaging. *ACS Appl. Bio Mater.* 1: 1458–1467.

143 Liang, L., Su, M., Li, L. et al. (2016). Aptamer-based fluorescent and visual biosensor for multiplexed monitoring of cancer cells in microfluidic paper-based analytical devices. *Sensors Actuators B Chem.* 229: 347–354.

144 Shivaji, K., Mani, S., Ponmurugan, P. et al. (2018). Green-synthesis-derived CdS quantum dots using tea leaf extract: antimicrobial, bioimaging, and therapeutic applications in lung cancer cells. *ACS Appl. Nano Mater.* 1: 1683–1693.

145 Cialla-May, D., Zheng, X.S., Weber, K., and Popp, J. (2017). Recent progress in surface-enhanced Raman spectroscopy for biological and biomedical applications: from cells to clinics. *Chem. Soc. Rev.* 46: 3945–3961.

146 Fabris, L., Dante, M., Nguyen, T.-Q. et al. (2008). SERS aptatags: new responsive metallic nanostructures for heterogeneous protein detection by surface enhanced Raman spectroscopy. *Adv. Funct. Mater.* 18: 2518–2525.

147 Guo, H., He, L., and Xing, B. (2017). Applications of surface-enhanced Raman spectroscopy in the analysis of nanoparticles in the environment. *Environ. Sci. Nano* 4: 2093–2107.

148 Huang, Z., Zhang, A., Zhang, Q., and Cui, D. (2019). Nanomaterial-based SERS sensing technology for biomedical application. *J. Mater. Chem. B* 7: 3755–3774.

149 Witkowska, E., Niciński, K., Korsak, D. et al. (2020). Nanoplasmonic sensor for foodborne pathogens detection. Towards development of ISO-SERS methodology for taxonomic affiliation of *Campylobacter* spp. *J. Biophotonics* https://doi.org/10.1002/jbio.201960227.

150 Lin, X., Lin, S., Liu, Y. et al. (2019). Lab-on-paper surface-enhanced Raman spectroscopy platform based on self-assembled Au@Ag nanocube monolayer for on-site detection of thiram in soil. *J. Raman Specrosc.* 50: 916–925.

151 Lu, W., Singh, A.K., Khan, S.A. et al. (2010). Gold nano-popcorn-based targeted diagnosis, nanotherapy treatment, and in situ monitoring of photothermal therapy response of prostate cancer cells using surface-enhanced Raman spectroscopy. *J. Am. Chem. Soc.* 132: 18103–18114.

152 Realini, M., Conti, C., Botteon, A. et al. (2017). Development of a full micro-scale spatially offset Raman spectroscopy prototype as a portable analytical tool. *Analyst* 142: 351–355.

153 Zeng, F., Duan, W., Zhu, B. et al. (2019). Paper-based versatile surface-enhanced Raman spectroscopy chip with smartphone-based Raman analyzer for point-of-care application. *Anal. Chem.* 91: 1064–1070.

154 Anichini, C., Czepa, W., Pakulski, D. et al. (2018). Chemical sensing with 2D materials. *Chem. Soc. Rev.* 47: 4860–4908.

155 Nery, E.W. and Kubota, L.T. (2013). Sensing approaches on paper-based devices: a review. *Anal. Bioanal. Chem.* 405: 7573–7595.

156 Shrivas, K., Monisha, Kant, T. et al. (2020). Smartphone coupled with paper-based chemical sensor for on-site determination of iron(III) in environmental and biological samples. *Anal. Bioanal. Chem.* 412: 1573–1583.

157 Pungjunun, K., Chaiyo, S., Jantrahong, I. et al. (2018). Anodic stripping voltammetric determination of total arsenic using a gold nanoparticle-modified boron-doped diamond electrode on a paper-based device. *Microchim. Acta* 185: 324.

158 Yakoh, A., Rattanarat, P., Siangproh, W., and Chailapakul, O. (2018). Simple and selective paper-based colorimetric sensor for determination of chloride ion in environmental samples using label-free silver nanoprisms. *Talanta* 178: 134–140.

159 Kim, Y., Jang, G., and Lee, T.S. (2015). New fluorescent metal-ion detection using a paper-based sensor strip containing tethered rhodamine carbon nanodots. *ACS Appl. Mater. Interfaces* 7: 15649–15657.

160 Ghosale, A., Shrivas, K., Shankar, R., and Ganesan, V. (2017). Low-cost paper electrode fabricated by direct writing with silver nanoparticle-based ink for detection of hydrogen peroxide in wastewater. *Anal. Chem.* 89: 776–782.

161 Kant, T., Shrivas, K., Ganesan, V. et al. (2020). Flexible printed paper electrode with silver nano-ink for electrochemical applications. *Microchem. J.* 155: 104687.

162 Kumar, S., Kaushik, S., Pratap, R., and Raghavan, S. (2015). Graphene on paper: a simple, low-cost chemical sensing platform. *ACS Appl. Mater. Interfaces* 7: 2189–2194.

163 Tawfik, S.M., Sharipov, M., Kakhkhorov, S. et al. (2019). Multiple emitting amphiphilic conjugated polythiophenes-coated CdTe QDs for picogram detection of trinitrophenol explosive and application using chitosan film and paper-based sensor coupled with smartphone. *Adv. Sci.* 6: 1801467.

164 de Oliveira, T.R., Fonseca, W.T., de Oliveira Setti, G., and Faria, R.C. (2019). Fast and flexible strategy to produce electrochemical paper-based analytical devices using a craft cutter printer to create wax barrier and screen-printed electrodes. *Talanta* 195: 480–489.

165 Izadyar, S. and Fatemi, S. (2013). Fabrication of X zeolite based modified nano TiO_2 photocatalytic paper for removal of VOC pollutants under visible light. *Ind. Eng. Chem. Res.* 52: 10961–10968.

166 Soga, T., Jimbo, Y., Suzuki, K., and Citterio, D. (2013). Inkjet-printed paper-based colorimetric sensor array for the discrimination of volatile primary amines. *Anal. Chem.* 85: 8973–8978.

167 Tseng, S.-Y., Li, S.-Y., Yi, S.-Y. et al. (2017). Food quality monitor: paper-based plasmonic sensors prepared through reversal nanoimprinting for rapid detection of biogenic amine odorants. *ACS Appl. Mater. Interfaces* 9: 17306–17316.

168 Lee, M., Oh, K., Choi, H.-K. et al. (2018). Subnanomolar sensitivity of filter paper-based SERS sensor for pesticide detection by hydrophobicity change of paper surface. *ACS Sens.* 3: 151–159.

169 Creran, B., Li, X., Duncan, B. et al. (2014). Detection of bacteria using inkjet-printed enzymatic test strips. *ACS Appl. Mater. Interfaces* 6: 19525–19530.

170 Li, L., Zheng, X., Huang, Y. et al. (2018b). Addressable TiO_2 nanotubes functionalized paper-based cyto-sensor with photocontrollable switch for highly-efficient evaluating surface protein expressions of cancer cells. *Anal. Chem.* 90: 13882–13890.

171 Li, W., Zhang, X., Miao, C. et al. (2020). Fluorescent paper–based sensor based on carbon dots for detection of folic acid. *Anal. Bioanal. Chem.* https://doi.org/10.1007/s00216-020-02507-w.

172 Liu, J., Mazumdar, D., and Lu, Y. (2006). A simple and sensitive "dipstick" test in serum based on lateral flow separation of aptamer-linked nanostructures. *Angew. Chem. Int. Ed.* 45: 7955–7959.

173 Díaz-Amaya, S., Zhao, M., Lin, L.-K. et al. (2019). Inkjet printed nanopatterned aptamer-based sensors for improved optical detection of foodborne pathogens. *Small* 15: 1805342.

174 Quintela, I.A., de Los Reyes, B.G., Lin, C.-S., and Wu, V.C.H. (2019). Simultaneous colorimetric detection of a variety of *Salmonella* spp. in food and environmental samples by optical biosensing using oligonucleotide-gold nanoparticles. *Front. Microbiol.* 10: 1138.

175 Marques, A., Veigas, B., Araújo, A. et al. (2019). Paper-based SERS platform for one-step screening of tetracycline in milk. *Sci. Rep.* 9: 17922.

3

Miniaturized Analytical Technology in Agriculture

Vaibhavkumar N. Mehta[1], *Vimalkumar S. Prajapati*[1], *and Jigneshkumar V. Rohit*[2]

[1] *ASPEE SHAKILAM Biotechnology Institute, Navsari Agricultural University, Ghod-Dod Road, Surat-395007, Gujarat, India*
[2] *Department of Chemistry, National Institute of Technology, Hazratbal, Srinagar-190006, Jammu and Kashmir, India*

3.1 Introduction

Plant pathogens are among the major concern for agricultural losses around 3% of the whole costs of crop production due to environmental and biological stresses such as heat, drought, salinity, and flooding [1–3]. The early detection of plant pathogens is more important in terms of disease management during the various phases of plant growth to minimize the risk of infections mainly due to viruses, fungi, bacteria, and nematodes [4, 5]. Earlier, the physical symptoms of infected leaves, stems, and fruits were observed during the field surveillance to identify the diseases developed during the growth or else various biochemical-, physiological-, and pathogenicity-based laboratory tests were conducted using specific cultures of plant pathogens, which were time-consuming and also required subject expertise [6]. Plant disease diagnosis usually complies with the visual observation of the symptoms displayed on the leaves, stem, and fruits due to infections, and the top most various plant pathogens including bacteria, fungi, and viruses are listed in Table 3.1. Therefore, DNA-based immunoassays have been developed for the detection of plant pathogens by extracting the pathogen protein and nucleic acid from infected plants [7]. Furthermore, molecular diagnostic approaches such as antibody-based ELISAs (enzyme-linked immunosorbent assays) technique [8], PCR (polymerase chain reaction) [9, 10], lateral flow devices [11], immune-fluorescent test [12], tissue print ELISA or direct dot blot immunoassay (DTBIA) [13] have been excessively exploited for rapid, on-time, selective, and sensitive detection of plant pathogens. These molecular diagnostic approaches always require extensive sample preparation, expensive reagents, and sophisticated instrumentation, which restrict their use for on-field, cost-effective/affordable detection of plant pathogens. Moreover, these techniques are unable to detect low-level titers in seeds and insect vectors at the early stage of infection, and also false positive results are the major

Miniaturized Analytical Devices: Materials and Technology, First Edition.
Edited by Suresh Kumar Kailasa and Chaudhery Mustansar Hussain.

Table 3.1 Top 10 important plant pathogenic bacteria, fungi, and viruses published by Molecular Plant pathology.

Plant pathogen	Fungi	Bacteria	Viruses
1.	*Magnaporthe oryzae*	*Pseudomonas syringae*	*Tobacco mosaic virus*
2.	*Botrytis cinerea*	*Ralstonia solanacearum*	*Tomato spotted wilt*
3.	*Puccinia* spp.	*Agrobacterium tumefaciens*	*Tomato yellow leaf curl*
4.	*Fusarium graminearum*	*Xanthomonas oryzae*	*Cucumber mosaic*
5.	*Fusarium oxysporum*	*Xanthomonas campestris*	*Potato virus Y*
6.	*Blumeria graminis*	*Xanthomonas axonopodis*	*Cauliflower mosaic*
7.	*Mycosphaerella graminicola*	*Erwinia amylovora*	*African cassava mosaic*
8.	*Colletotrichum* spp.	*Xylella fastidiosa*	*Plum pox*
9.	*Ustilago maydis*	*Dickeya (dadantii and solani)*	*Brome mosaic*
10.	*Melampsora lini*	*Pectobacterium carotovorum*	*Potato virus X*

Source: Khater et al. [5].

concern due to cross-amplification of PCR-generated fragments of non-target DNA [14]. Therefore, the development of innovative technological approaches requires suitable plant disease management to increase the total crop productivity.

Recently, nanotechnology has been proved as a state-of-the-art technology with the integration of chemistry, physics, biology, and materials science to offer new opportunities to develop cost-effective nanomaterials and nanodevices such as nanopesticides, nanofertilizers, and nanobiosensors for boosting crop production and crop protection, which ultimately leads to the sustainable agriculture [15, 16]. Recently, nanobiosensors open up the new prospects for development of non-destructive, economical and easily handled probe for on-field detection of plant pathogens with more reliability, sensitivity, and specificity with enhanced limit of detection (LOD) due to their unique physico-chemical properties of nanomaterials [17, 18]. These unique properties of nanomaterials can be utilized for the fabrication of miniaturized nanobiosensor-based analytical technique by integration of bioreceptors or biomarkers on transducer *via* functionalization and immobilization to improve their analytical efficiency in complex plant structures. Furthermore, nanomaterial-based analytical techniques can convert plant chemical signals into the optical, electrochemical, or electric signals based on the deviations in biotic (fungus, viruses, bacteria, etc.) and abiotic stress (drought, temperature, salinity, etc.) for the real-time monitoring of plant health (Figure 3.1). In this connection, various nanomaterials such as metallic nanoparticles, metal oxides, magnetic nanoparticles, semiconductor quantum dots, carbon nanomaterials have already

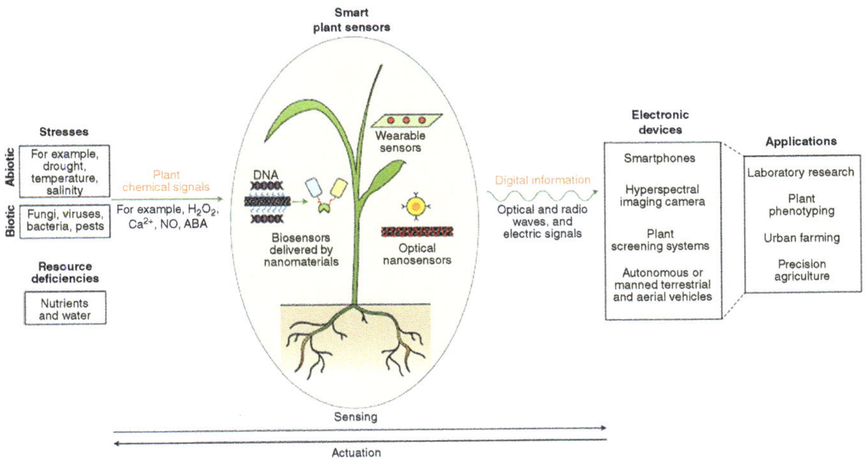

Figure 3.1 Nanobiotechnology approaches enable research and development of smart plant sensors that communicate plant chemical signals to agricultural and phenotyping equipment. Source: Giraldo et al. [3].

been used to develop miniaturized nanobiosensors for rapid, precise, and on-site detection of early stage infection in plant pathogens under field conditions [19].

This chapter focuses on recent developments of nanomaterial-based miniaturized analytical techniques for the selective and sensitive detection of plant pathogens such as fungi, bacteria, viruses, and nematodes based on the pathogen population genetics, gene transfer between the hosts and pathogens, and plant–microbe interactions. Moreover, the methods used to fabricate efficient nanobiosensors and mechanisms involved in detection such as optical and electrochemical are also highlighted, and issues and prospects in such applications are also briefly discussed.

3.2 Miniaturized Analytical Techniques for the Fungal Detection in Plants

Fungal diseases have the most adverse effects on the plant health as fungi are the most diverse group of plant pathogens responsible for 70–80% of plant diseases due to their symbiotic/saprophytic/pathogenic relation with plants [20]. The early identification of fungi is an essential prerequisite as plants may develop fungicide resistance because of their penetration into plant tissues *via* proliferation. Moreover, fungi may damage the entire crop of neighboring fields by spreading their spores through wind, water, soil, and animals [21]. Recently, nanomaterial-based analytical probes have gained more attention over the conventional molecular-based approaches due to their simplicity, miniaturization, and on-field detection of fungi plant pathogens. Lateral flow immunoassay-based biosensor integrated with gold nanoparticles was developed based on the universal primer-mediated asymmetric polymerase chain reaction (UP-APCR) by generating single-stranded DNA (ssDNA)

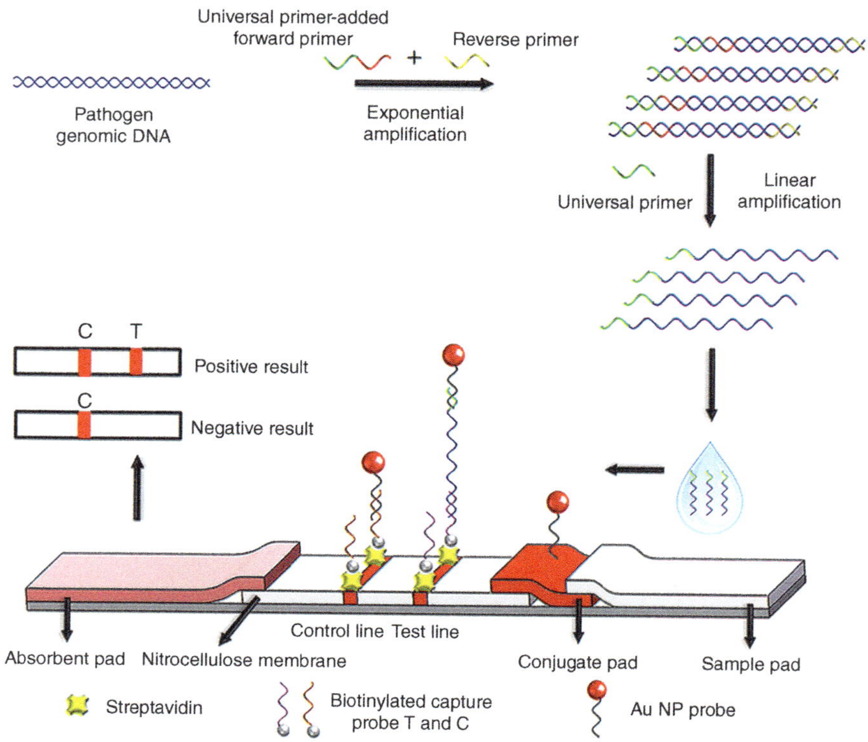

Figure 3.2 Schematic illustration of gold nanoparticle-based lateral flow biosensor for identification of *Phytophthora infestans* in potato plants. Source: Zhan et al. [22].

amplicons for the detection of *Phytophthora infestans* at the 0.1 pg/μl genomic DNA level in potato plants (Figure 3.2) [22]. Yüksel et al. fabricated surface-enhanced Raman scattering probe for the detection of *Phytophthora ramorum* in *rhododendron* leaves using bovine serum albumin-capped silver nanoparticles [23]. Sun's group developed electrochemical probes for the detection of salicylic acid infected with *Botrytis cinerea* in tomato leaves. Au@Fe_3O_4–CS nanocomposites were constructed with glassy carbon electrodes coated with chitosan to achieve a LOD up to 0.10 μM concentration of salicylic acid with a linear range of 1.0 μM to 1.2 mM in the Britton–Robinson buffer solution at pH 7.0 based on differential pulse voltammetry [24]. A gold electrode functionalized with copper nanoparticles was also developed for the detection of salicylic acid with the LOD 0.1 μM in oilseed rape infected with the fungal pathogen *Sclerotinia sclerotiorum* based on the electrocatalytic oxidation of salicylic acid at pH 13.3. The sensitive detection of *Sclerotinia sclerotiorum* was supported based on the changes measured in the levels of salicylic acid *via* anodic oxidation of salicylic acid at a electrode. The levels of salicylic acid sharply increased after 10 hours of inoculation with *Sclerotinia sclerotiorum*, which might be attributed to the stimulation of the octadecanoid pathway, and signal transduction and scavenging mechanisms in the cells might lead to the low concentration of salicylic acid to avoid potential toxic effects [25]. Ramasamy and coworkers developed an electrochemical probe for the detection of finger print compound *p*-ethylguaiacol, which is mostly present in fruits and plants infected with the pathogenic fungus *Phytophthora cactorum* by using the TiO_2 and SnO_2 nanoparticle–based screen-printed carbon electrode with lower detection limits in the range of 35–62 nM. The detection mechanism was based on the formation of phenoxy radical intermediates due to irreversible oxidation of *p*-ethylguaiacol, which further leads to the formation of benzoic acid derivatives and H_3O^+ after the reaction with phthalate anions in the electrolyte [26]. A highly sensitive surface plasmon resonance–based immunoassay was developed to detect *Pseudocercospora fijiensis,* which infects banana plants by resulting in Black Sigatoka. *P. fijiensis* was detected from the leaf extracts using the polyclonal antibody immobilized on the gold-coated self-assembled monolayer of alkanethiols with a detection limit of 11.7 μg/ml [27]. The detection of plant pathogenic fungi using nanomaterial-based miniaturized analytical devices is illustrated in Table 3.2.

3.3 Miniaturized Analytical Techniques for the Virus Detection in Plants

Viruses are the second most important plant pathogens after fungi as the infection of viruses resists the plant growth by shortening the life of the host *via* dominating over the allocation resources of plants produced through photosynthesis. The symptomatic viral infections of the plants such as chlorosis, necrosis, mottled, mosaics, yellowing, deformation of leaves and fruits mainly rely on the virus strain, variety of host plants, time of infection, and region, and early on-field detection of plant viruses during the initial replication of viral DNA or protein is required to avoid crop production loss [28]. The nanoparticle-based miniaturized analytical devices integrated with the classical techniques has been widely used over the conventional techniques such as serological, biological indexing, immunoassay techniques and PCR-based techniques to achieve the high sensitivity, specificity, and on-field detection of plant viruses [29]. Bare gold nanoparticles were used as visual optical probes for the selective and sensitive detection of *Maize chlorotic mottle virus* with 100% specificity among 15 sample viruses, and the LOD was found to be 30 pg/μl [30]. Wang et al developed an asymmetric polymerase chain reaction with unmodified gold nanoparticles coated with ssDNA for the visual detection of *Maize chlorotic mottle virus* with 23.9 pg/μl detection limit [31]. The same group successfully developed a colorimetric probe based on the unmodified gold nanoparticles for the detection of *Cucumber green mottle mosaic virus* (CGMMV) RNA with a detection limit of 30 pg/μl in the presence of NaCl [32]. Chartuprayoon et al. fabricated a chemiresistive immunosensor based on covalently conjugated antibody-functionalized polypyrrole (PPy) nanoribbon via *N*-(3-dimethylaminopropyl)-*N*′-ethylcarbodiimide (EDC)/*N*-hydrosuccinimide (NHS) cross-linking for the detection of *cucumber mosaic virus.* The sensitivity of the immunosensor was improved by decreasing the thickness of the nanoribbon, and the lower and upper limits of detection were found to be 10 ng/ml and 100 μg/ml, respectively [33]. Tabatabaei and coworkers detected the *Citrus tristeza virus* (CTV) in field samples by using Au NPs–CP/CdTe QDs–Ab

Table 3.2 Nanoparticle-based miniaturized analytical techniques for the plant fungi detection.

Sr. No.	Nanoparticle- based probe	Fungus	Method	Limit of detection	References
1.	Au NPs	*Phytophthora infestans*	Lateral flow immunosensor	0.1 pg/μl	[22]
2.	BSA–Ag NPs	*Phytophthora ramorum*	Surface enhanced Raman scattering	—	[23]
3.	Au@Fe_3O_4–CS	*Botrytis cinerea*	Differential pulse voltammetry	0.10 μM	[24]
4.	Cu NPs–GE	*Sclerotinia sclerotiorum*	Electrochemical detection	0.1 μM	[25]
5.	TiO_2–SP	*Phytophthora cactorum*	Cyclic voltammetry	82 nM	[26]
			Differential pulse voltammetry	62 nM	
6.	SnO_2–SP	*Phytophthora cactorum*	Cyclic voltammetry	126 nM	[26]
			Differential pulse voltammetry	35 nM	
7.	Au NPs–self-assembled monolayers	*Pseudocercospora fijiensis*	Lateral flow immunoassay	11.7 μg/ml	[27]

as a probe. In this work, cadmium telluride quantum dots were conjugated with polyclonal antibodies against the coat protein, which was immobilized on the surface of gold nanoparticles, resulting in the formation of the immune complex. The fluorescence resonance energy transfer (FRET) between the donor (QDs) and acceptor (Au NPs) led to a decrease in fluorescence intensity, whereas the fluorescence intensity was recovered as Au NPs–coat protein was substituted by free coat protein in the detection stage. The LOD for the *CTV* was found to be 0.13 μg/ml with 93% sensitivity and 94% specificity over ELISA [34]. The fluorometric immunoassay was also developed by the same group for the detection of *CTV* using carbon nanoparticles coated with the quenchers (coat protein) and antibodies labeled with CdTe quantum dots and 220 ng/ml LOD was attained [35]. Haji-Hashemi et al. fabricated a label-free electrochemical immunosensor for efficient, rapid, and sensitive detection of *Citrus tristeza* at 0.27 nM LOD. The electrochemical detection was carried out using the 11-mercaptoundecanoic acid and 3-mercaptopropionic acid modified gold electrode immobilized with a specific antibody against the coat protein of *Citrus tristeza via* the carbodiimide coupling reaction [36]. Furthermore, a simple, disposable, and rapid microfluidic electrochemical device was developed for the ultrasensitive detection of *CTV* in citrus sample. In this work, amperometry-based detection was carried out by using the electrodes modified with monoclonal antibody anti-CP-CTV *via* the catalytic redox reaction of hydroquinone/H_2O_2 due to the

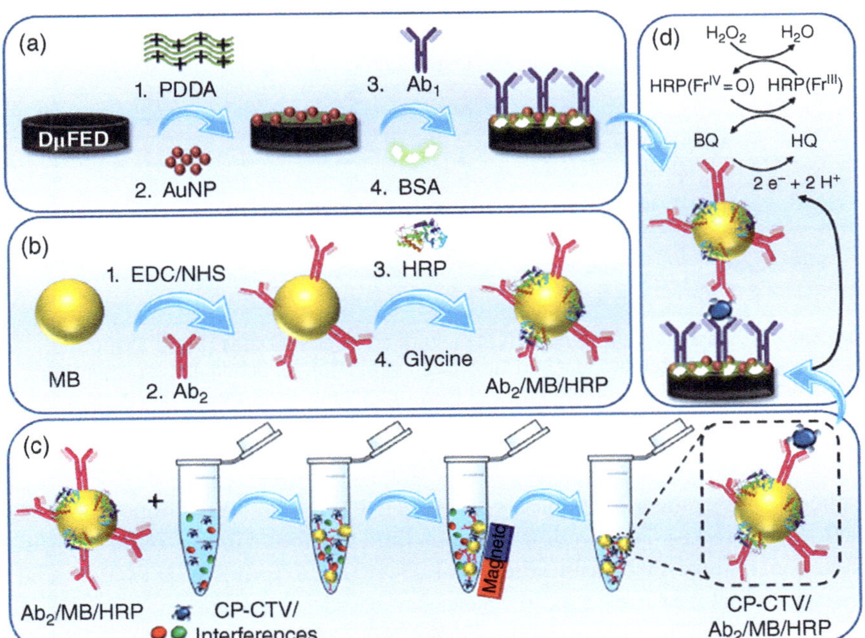

Figure 3.3 Fabrication of microfluidic device. (a) Surface modification of eight working electrodes with PDDA, AuNP, and monoclonal antibodies (Ab_1); (b) immunoconjugate preparation containing magnetic beads decorated with polyclonal antibodies and horseradish peroxidase enzymes; (c) coat protein *Citrus tristeza virus* capture using immunoconjugate; and (d) sandwich immunoassay formation and electrochemical analysis. Source: Freitas et al. [37].

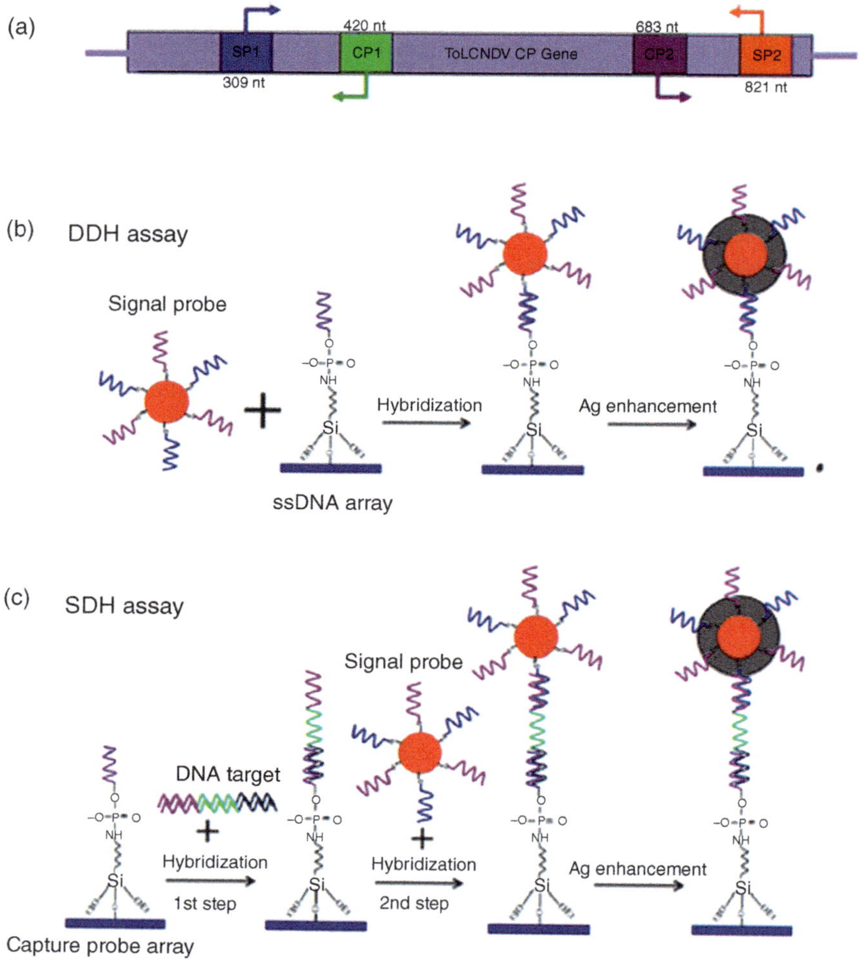

Figure 3.4 (a) Graphical representation of capture and signal probes (1 and 2) specific sites at the CP gene of tomato leaf curl New Delhi viral DNA. (b) Schematic representation of scanometric detection of tomato leaf curl New Delhi viral DNA through direct (single-step) DNA hybridization method and (c) sandwich (double-step) DNA hybridization methods with silver enhancement. Source: Dharanivasan et al. [47].

presence of horseradish peroxidase enzyme (HRP) label in an immune-sandwich structure (Figure 3.3). The ultralow detection limit was found to be around 0.3 fg/ml with a linear range of concentration of $1.95–10.0\times10^3$ fg/ml [37]. Khater et al. proposed a label-free impedimetric biosensor based on a screen-printed carbon electrode modified with gold nanoparticles immobilized with thiolated ssDNA probes for the selective and sensitive detection of nucleic acid of *CTV*. The LOD was around 100 nM [38]. The use of electrochemical impedance spectroscopy technique was also explored by the same group *via in-situ* solid-phase isothermal recombinase polymerase amplification/detection approach for *CTV* detection by using gold nanoparticle (Au NP)-modified substrates. The *Citrus tristeza* was detected at the level of 1000 fg/μl within a linear range from 1 to 500 pg/μl based on the alterations in electrochemical performance, which results from immobilization of the sensing layer and the amplified duplex DNA before and after *in-situ* amplification [39]. A novel self-dimerized DNA–silver nanocluster probe was developed by Rahaie's group for the ultrasensitive detection of *CTV* RNA based on the fluorescence enhancement of signals due to the structure conversion between dimeric and non-dimeric states upon DNA/RNA hybridization, and the LOD was 2.5 nM in the linearity range of 5–210 nM for *Citrus tristeza* RNA [40]. Majumder and Johari developed a novel dot immunobinding assay for rapid, selective, and sensitive detection of *Banana bunchy top* virus by using gold nanoparticles. The gold nanoparticles were conjugated with primary antibody (GCPab) and detected the virus at sap dilution up to 10^{-2} [41]. A simple and sensitive visual assay was also developed using ssDNA-thiolated (CR1) functionalized gold nanoparticles for the detection of *Banana bunchy top* virus with a detection limit of 1 pg/l of the plasmid (pTZBBTri 4, BBTV DNA) mixed with healthy banana DNA [42]. Xing et al. fabricated a miniaturized lateral flow device and gold nanoparticle colorimetric probes for the detection of *Banana bunchy top* virus. The developed lateral flow provided the quantitative and qualitative detection of amplified DNA based on the hybridization-mediated target capture within minutes with high specificity and without sophisticated instrumentation. The LOD was around 0.13 aM of the gene segment with confirmed amplified products, which is 10 times higher than that of electrophoresis [43]. Gold nanoparticle–based colorimetric probe was developed for visual optical detection of unamplified *Tomato yellow leaf curl virus* (TYLCV) genome in infected plants through localized surface plasmon resonance phenomenon [44]. The detection of virus was carried out by visual color change as well as measuring the changes in localized surface plasmon resonance peak due to the aggregation of nanoparticles in the presence of the virus by simple UV-Visible spectrometry. The developed colorimetric probe was able to detect the presence of TYLCV genome in 5 ng of the extracted DNA from infected plants without amplification. Wang and Yang also demonstrated the visual DNA detection of TYLCV within 20 minutes by integrating recombinase polymerase amplification and artificial cDNA-modified gold nanoparticles as a colorimetric probe without using any sophisticated instrumentation [45]. The LOD was found up to 1 copy/μl of the virus by amplifying the DNA with recombinase polymerase amplification at 25 °C for 5 minutes. Dharanivasan et al. prepared a bifunctional oligonucleotide probe conjugated with gold nanoparticles for the colorimetric detection of plant viral DNA of *Tomato leaf curl virus* at a detection limit of around 7.2 ng. In this work, the AuNP-bifunctional oligonucleotide probe detected the both forward and reverse strands of dsDNA target with a reduced concentration of DNA target for the detection [46]. Furthermore, scanometric detection based on the silver enhancement was also developed by the same research group using AuNP-conjugated mono- and bifunctional oligo probes through direct DNA hybridization assay (DDH assay) and sandwich DNA hybridization assay (SDH assay) for the detection of tomato leaf curl New Delhi viral DNA (Figure 3.4). In the DDH assay, the dark black/gray colors developed on the glass slide was due to the hybridization of signal probes with ssDNA array of positive control, and *tomato leaf curl New Delhi viral DNA* samples and signals were amplified with the silver enhancement method, which could be easily visualized, discriminated, and interpreted by the naked eyes. The SDH assay was performed through two rounds of hybridization between capture probes and target DNA; target DNA and signal probes followed by silver enhancement. The LOD was optimized between the 100 zM to 100 pM concentration for the tomato leaf curl New Delhi viral DNA by the scanometric method [47]. Panferov et al. fabricated the lateral flow immunoassay (LFIA) for the detection of potato virus X based on the enlargement of the size of gold nanoparticles *via* catalytic reduction of gold chloride in the presence of hydrogen peroxide (H_2O_2). The authors achieved the LOD as low as 17 pg/ml within 12 minutes without any false positive or false negative results in potato leave samples [48]. Razo et al. developed a double-enhanced LFIA using magnetic nanoparticles and gold nanoparticles for the selective and sensitive detection of potato virus X in potato leaves. The enhancement of LFIA was achieved by increasing the concentration of analytes in the samples using conjugates of magnetic nanoparticles with specific antibodies and increasing the visibility of magnetic nanoparticle aggregation due to gold nanoparticles. The double-enhanced LFIA was able to detect the potato virus X at 0.25 ng/ml without any sophisticated instrumentation [49]. The detection of potato virus X RNA was also proposed based on the nucleic acid lateral flow assay (LFA) coupled with reverse transcription recombinase polymerase amplification (RT–RPA) with a detection limit of 0.14 ng potato virus X per gram potato leaves in 30 minutes [50]. Furthermore, Masoudi et al. developed a simple and rapid colorimetric probe based on the change in the surface plasmon resonance of specific antibody–coated gold nanoparticles for the detection of *Potato virus S* with visual detection up to 0.05 μg/ml [51]. Tahir et al. detected the begomovirus using carbon nanotube–based copper nanoparticle composite probes at a 0.01 ng/μl DNA concentration detection limit. The electrical features of copper nanoparticles combined with the high conductivity of carbon nanotubes enhanced specificity toward the virus strain through electrostatically binding and resulted in the synergistic effects of carbon nanotubes and copper nanoparticles [52]. A selective and sensitive lateral flow immunochromatographic assay based on the sandwich immunoassay protocol was developed for the *Grapevine leafroll-associated virus 3* (GLRaV-3) by using antibody-labeled gold nanoparticles, which visually read out the virus within 10 minutes in grape leaf extracts [53]. The authors developed a simple one-step method for the detection of virus where control, test, and sample zones existing in the method would provide detection without any sophisticated instrumentation. In a similar work, a sandwich LFIA was developed by using the polyvalent antigen and gold nanoparticle–labeled antibodies for the detection of potato virus Y with the LOD of around 5.4 ng/ml, which was comparable with ELISA and surface plasmon resonance with a correlation coefficient of 0.891 [54]. Table 3.3 enlists the various miniaturized analytical devices based on the nanomaterials for the selective and sensitive detection of plant pathogenic viruses.

Table 3.3 Nanoparticle-based miniaturized analytical techniques for the plant virus detection.

Sr. No.	Nanoparticle-based probe	Viruses	Method	Limit of detection	References
1.	Au NPs	*Maize chlorotic mottle*	SPR-based colorimetric sensor	30 pg/μl	[30]
2.	Au NPs	*Maize chlorotic mottle*	Asymmetric PCR	23.9 pg/μl	[31]
3.	Au NPs	*Cucumber green mottle mosaic*	SPR-based colorimetric sensor	30 pg/μl	[32]
4.	Polypyrrole nanoribbon	*Cucumber mosaic*	Immunosensor	10 ng/μl	[33]
5.	Au NPs–coat protein/CdTe QDs–antibody	*Citrus tristeza*	FRET-based sensor	0.13 μg/ml	[34]
6.	Carbon NPs–coat protein/CdTe QDs–antibody	*Citrus tristeza*	Fluorometric immunoassay	220 ng/ml	[35]
7.	Anti-CTV/MUA–MPA/Au electrode	*Citrus tristeza*	Electrochemical immunoassay	0.25 nM	[36]
8.	DμFED/Ab_1/CP/Ab_2/MB/HRP	*Citrus tristeza*	Electrochemical immunoassay	0.3 fg/ml	[37]
9.	Au NPs–SPCE	*Citrus tristeza*	Electrochemical detection	100 nM	[38]
10.	Au NPs–SPCE	*Citrus tristeza*	Electrochemical detection	1000 fg/μl	[39]
11.	DNA–silver nanocluster	*Citrus tristeza*	Fluorescence detection	2.5 nM	[40]
12.	Au NPs–GCPab	*Banana bunchy top*	Dot immunobinding assay	—	[41]

13.	ssDNA-thiolated (CR1)–Au NPs	*Banana bunchy top*	Colorimetric probe	1 pg/l	[42]
14.	Au NPs–ssDNA	*Banana bunchy top*	Lateral flow immunoassay	0.13 aM	[43]
15.	Au NPs	*Tomato yellow leaf curl*	Colorimetric probe	5 ng/μl	[44]
16.	Artificial cDNA–Au NPs	*Tomato yellow leaf curl*	Recombinase polymerase amplification and colorimetric probe	1 copy/μl	[45]
17.	Au NPs–dsDNA	*Tomato leaf curl*	Colorimetric probe	7.2 ng	[46]
18.	Au NPs–oligonucleotide	*tomato leaf curl* New Delhi viral DNA	Scanometric probe	100 zM to 100 pM	[47]
19.	Au NPs–antibody	*Potato X*	Lateral flow immunoassay	17 pg/ml	[48]
20.	Magnetic and gold nanoparticle conjugates	*Potato X*	Lateral flow immunoassay	0.25 ng/ml	[49]
21.	Au NPs–antibody	*Potato virus S*	Colorimetric probe	0.05 μg/ml	[51]
22.	MWCNTs–Cu NPs	begomovirus	Electrochemical detection	0.01 ng/μl	[52]
23.	Au NPs–antibody	*Grapevine leafroll-associated virus 3*	Lateral flow immunochromatographic assay	—	[53]
24.	Au NPs–antibody	*Potato Virus Y*	Sandwich lateral flow immunoassay	5.4 ng/ml	[54]

3.4 Miniaturized Analytical Techniques for the Bacterial Detection in Plants

Numerous methods such as polymerase chain reaction and molecular immunoassay have been reported for the detection of plant pathogenic bacteria and are demonstrated to be accurate and specific enough to detect pathogenic bacteria at certain limits. Apart from these advantages, these methods have certain limitations to detect pathogens at low titers specifically at the early stage of infection. In recent years, to overcome such limitations, innovative, robust, and portable nanomaterial-based biosensors (quantum dots, nanostructured platforms, nanoimaging, and nanopore DNA sequencing) have gained much more focus mainly because they help, to a great extent, to cut down the cost of detection and achieve higher efficiency of testing and because of their rapidity over the conventional diagnostic techniques, which facilitate high-throughput analysis, and their ability to be used for selective and sensitive detection of pathogenic bacteria. Table 3.4 depicts certain examples of nanomaterial-integrated antibody-based and DNA-based bacterial pathogen detection methods employing optical and electrochemical techniques, which were previously reported. Rubpy-doped silica nanoparticles are used for the rapid detection of plant pathogens such as *Xanthomonas axonopodis* pv. *Vesicatoria (Xav)*, a causative agent of the bacterial spot disease in *Solanaceae* plant [55]. Scientists have adopted the reverse microemulsion method for the fabrication of Rubpy-doped silica nanoparticles, and they were further conjugated with goat anti-rabbit IgG antibodies through chemical processes. The fluorescence immunoassay method was used to detect Xav *in-vitro* using Ab-FSNP as a probe. The bacterial spot disease is one of the most devastating diseases of tomato around the globe caused by four different types of copper-resistant *Xanthomonas* spp. Silver-based nanocomposites have been characterized as a potential alternative to copper, and a minimum concentration of nanocomposites (10 μg/ml) can be used to manage the bacterial spot disease [56]. Among the phytopathogenic bacteria, *Pseudomonas syringae* pathovars are claimed to be a significant one and lead to a variety of diseases in many agricultural crops. A number of approaches have been employed for the detection of *P. syringae* including the immunological techniques (immunoassays, immunomagnetic separations), methods based on nucleic acid probes such as DNA hybridization, mass spectroscopy, DNA microarray, etc., though the detection of *P. syringae* always remains a controversy, uncertain, and debatable. To overcome all the limitations, it is well documented to employ a sensor-based approach comprising optical, electrochemical, and mass sensitivity measurements to detect such bacterial plant pathogens. A versatile and novel biosensor based on the DNA–gold nanoparticle (DNA–Au NPs) was employed for the detection of *P. syringae* with an advantage of detecting as little as 15 ng/μl of a genomic template, which is much promising for its selectivity and sensitivity of DNA targets [57]. Nanoprobes prepared using thiol-functionalized DNA attached to gold NPs are more efficient to detect specific DNA sequences and have gained noteworthy attention. In comparison to the reported biochemical method for the detection of *Helicobacter pylori*, the methods employing gold NPs have 98% sensitivity and specificity, which is considerably more

Table 3.4 Nanoparticle-based miniaturized analytical techniques for the bacterial plant pathogen detection.

Sr. No.	Nanoparticle-based probe	Bacteria	Method	Limit of detection	References
1.	Silica NPs	*Xanthomonas axonopodis* pv. *Vesicatoria*	Fluorescence-linked immunosorbent assay	10^3 cfu/ml	[55]
2.	Thiol-linked DNA–Au NPs	*Pseudomonas syringae*	Colorimetric probe	15 ng/μl	[57]
3.	Au NPs	*Ralstonia solanacearum*	Lateral flow immune assay	2×10^2 cfu/ml	[58]
4.	Au NPs	*Ralstonia solanacearum*	Lateral flow immunoassay	3×10^4 cells/ml	[59]
5.	Au NPs	*Listeria monocytogenes*	Dynamic light scattering immunoassay	3.5×10^{-1} cfu/ml	[60]
6.	Au NPs	*Pantoea stewartii* subsp. *Stewartii*	Dual amplified electrochemical sandwich immunosensor	8×10^3 cfu/ml	[61]
7.	Au NPs	*Acidovorax avenae* subsp. *citrulli*	Au NP-based DNA dip strip sensors	0.48 nM	[62]
8.	Au NPs	*Dickeya species*	Lateral flow immunoassay	1×10^5 cfu/ml for leaves 4×10^5 cfu/ml for tubers	[63]
9.	Carbon NPs	*X. arboricola* pv. *pruni*	Lateral flow immunoassay	10^4 cfu/ml	[65]

significant. Panferov et al. have developed the LFIA for the detection of *Ralstonia solanacearum* in potato tubers [58]. This test was based on silver enhancement and was successfully implemented on field for the on-site detection of bacterial pathogens as well as used in poorly equipped laboratories. Silver enhancement is a robust method for the signal amplification as silver ions are reduced on the gold nanoparticles, which enlarge the particle size. The produced silver particles are black in contrast to the gold nanoparticles, which are red against the white background of the membrane. The chief intention behind the implementation of the investigation was the rapid diagnosis and control of *R. solanacearum* with a low LOD by employing LFIA with silver enhancement. LFIA with silver enhancement was found to be tenfold more sensitive (LOD: 2×10^2 cfu/ml; 20 minutes) compared to the previously employed common analysis (LOD: 2×10^3 cfu/ml; 10 minutes). The specificity of the assay with silver enhancement was confirmed using different strains of *R. solanacearum* (54 samples) as well with other widespread bacterial pathogens (18 samples). As far as the visible symptoms-based diagnosis is concerned, gold nanoparticles and specific antibodies immobilized LFIA strips against the pathogen and tetrachloroauric(III) anion reduction (amplification solution) were employed for consistent and accurate detection of *R. solanacearum*, having an LOD of 3×10^4 cells/ml in potato tuber extract with overall 15-minutes assay time including the sample preparation [59]. The overall mechanism of the LFIA before and after signal amplification is presented in Figure 3.5. Huang et al. described the

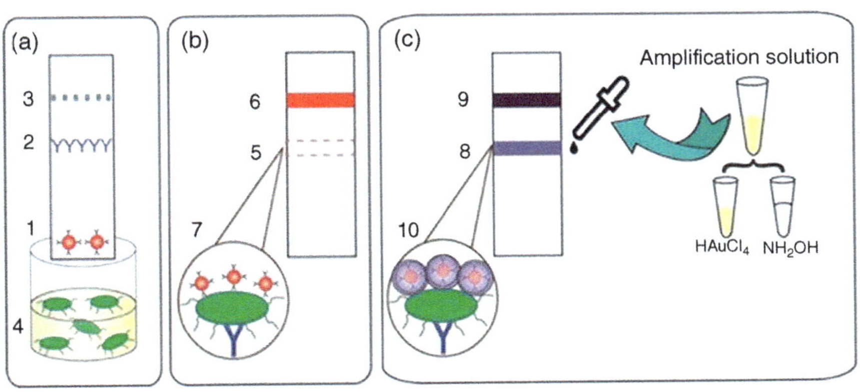

Figure 3.5 Lateral flow immunoassay (LFIA) scheme before and after signal amplification: (a) test strip dipped into a sample of potato tuber extract; (b) test strip after conventional LFIA, indicating a negative result because no color band exists in the test zone; (c) addition of amplification solution to the test strip and enlargement of the gold nanoparticle size. The numbers in the figure indicate (1) gold nanoparticle–antibody conjugate on the conjugate pad, (2) antibody specific to *R. solanacearum* immobilized in the test zone, (3) protein A immobilized in the control zone, (4) potato tuber extract containing *R. solanacearum*, (5) no visible color band in the test zone after analysis, (6) a clear red color band in the control zone, (7) formed immune complex (immobilized specific antibodies, *R. solanacearum*, gold nanoparticles conjugate with specific antibody) after conventional LFIA, (8) test zone after signal amplification, (9) control zone after signal amplification, and (10) enlarged gold nanoparticles particle size caused by the addition of amplification. Source: Razo et al. [59]. Licensed under CC-BY 4.0.

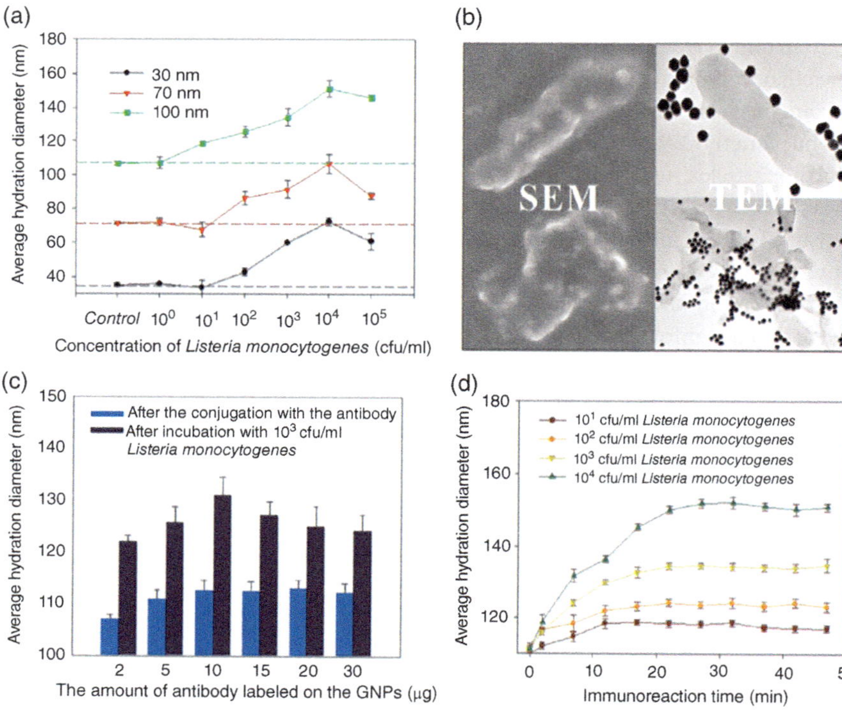

Figure 3.6 (a) Effect of gold nanoparticles size on the sensitivity of the assay. (b) SEM and TEM observations of gold nanoparticles incubated with 10^2 cfu/ml (upper) and 10^5 cfu/ml (lower) *Listeria monocytogenes*, respectively. (c) Effect of anti-*Listeria monocytogenes* mAbs concentration labeled on gold nanoparticles. (d) Immunoreaction time on DLS-based immunoassay. Source: Huang et al. [60].

use of dynamic light scattering-based immunoassay for ultrasensitive detection of *Listeria monocytogenes* in lettuces by using gold nanoparticles for the first time [60]. Gold nanoparticles are frequently used as the signal amplification probe for the engineering DLS-based biosensors because in comparison to biological samples light scattering ability of gold nanoparticles is much higher [64]. As far as the detection of *L. monocytogenes* is concerned, the major advantage of being larger isolate size, *L. monocytogenes* exhibit numerous antigen epitopes on their surface, which makes them highly suitable for the binding to gold nanoparticle-based probes and lead to the formation of the complex called "gold nanoparticle-coated bacteria" (Figure 3.6). To substantiate the presented investigation to a great extent, Huang et al. analyzed and optimized numerous parameters such as gold nanoparticles diameter and concentration, amount of labeled antibody, and immune reaction time. Using the aforementioned optimized conditions, the LOD for *L. Monocytogenes* was 3.5×10^{-1} cfu/ml in 10 mM phosphate buffer saline [60]. Zhao et al. have developed a dual amplified electrochemical sandwich immunosensor for highly sensitive detection of *Pantoea stewartii* subsp. *Stewartii (PSS)* – a causative agent of Stewart's vascular wilt in the maize crop, with an LOD of 0.8×10^3 cfu/ml [61].

This LOD was found to be 20 times lower than that of conventional ELISA. They prepared such an accurate immune sensor combining the Au NPs and horse radish peroxidase (HRP) for the implementation of dual amplification. The promising conductivity of Au NPs and its large surface area aid to retain the biological activity of antibodies. Bacterial fruit blotch (BFB) is one of the most damaging diseases in the production of cucurbits, including watermelon and melon, caused by *Acidovorax avenae* subsp. *citrulli* (AAC). Hence, looking into the economical perspective of import and export, the on-site detection seems to be a prerequisite. Zhao et al. developed a dipstick-type DNA biosensor using gold nanoparticle–labeled oligonucleotide probes for the first time to detect the AAC genomic DNA in the field, which leads to the assessment of pathogen's DNA both qualitatively and semiquantitatively [62]. The qualitative LOD of the Au NP–based DNA strip sensors was found to be 4.0 nM, while the LOD for the semiquantitative detection was calculated to be 0.48 nM. Potato blackleg is a serious potato disease caused by *Dickeya dianthicola* and *Dickeya solani* and its early diagnosis is really an essential prerequisite for the effective disease management. Safenkova et al. have developed the LFIA for rapid diagnosis of potato blackleg caused by *Dickeya* species [63]. The LIFA was developed with specific antibodies retrieved from rabbits for various *Dickeya* species and Au NPs. LFIA was rigorously optimized and the entire assay procedure was completed using samples of potato extract within 10 minutes, with visual LOD of 1×10^5 cfu/ml for leaves and 4×10^5 cfu/ml for tubers. Bacterial spot diseases of stone fruits and almond caused by *Xanthomonas arboricola* pv. *pruni* lead to a negative economic impact. Hence, reliable, specific, and prompt detection methods are required to mend the disease management. Soriano et al. developed a prototype of a LFIA for the on-site detection of *X. arboricola* pv. *pruni* in symptomatic field samples [65]. They efficaciously generated polyclonal antibody against *X. arboricola* pv. *pruni*, which were conjugated with carbon nanoparticles and assembled on nitrocellulose strips. Specificity and sensitivity of fabricated LFIA prototype were meticulously assayed against the 87 *X. arboricola* pv. *pruni* strains collected from different countries worldwide, 47 strains of other *Xanthomonas* species, and 14 strains representing other bacterial genera with a LOD of 10^4 cfu/ml.

3.5 Conclusion and Future Perspectives

Early detection of plant pathogens is extremely important for appropriate plant disease management to improvise healthy food to the booming world population through better yield of crop production. This chapter emphasized on the recent developments of miniaturized analytical techniques with the integration of nanomaterials with the classical techniques such as PCR, real-time PCR, immunoassays, and ELISA for the simple, reliable, economical, sensitive, and on-field detection of plant pathogens such as fungi, viruses, and bacteria in various plants. The detection of plant pathogens is mostly accomplished by optical and electrochemical detection or LFIA *via* integration of gold nanoparticles, silver nanoparticles, copper nanoparticles, quantum dots, and magnetic nanoparticles. These techniques involve the reaction or binding of target species or strains with specific antibodies labeled with nanoparticles, which results in detectable signals to provide quick, consistent, precise, and early diagnosis of pathogens for better plant disease management. However, the integrity of the developed sensors should be explored under realistic conditions at the farm level to address some constraints such as interfering constituents, nonspecific binding for the execution in the agricultural sector. Furthermore, the integration of disciplines such as metabolomics, next generation sequencing (NGS), microarray technology, microfluidic devices with unique and unexplored nanomaterials will open the door for the development of disposable, portable, accurate, and miniaturized analytical devices for the simultaneous detection of multiple pathogens in a single device. Therefore, more studies can be explored for nanomaterial-based miniaturized microfluidic devices and microelectronic platforms for *in vivo* detection of multiple pathogens in the near future.

References

1 Agrios, G.N. (2005). *Plant Pathology*, 5the, 79–103. Burlington, MA: Elsevier Academic Press.

2 Van Lenteren, J.C. (1995). Integrated pest management in protected crops. In: *Integrated Pest Management: Principles and Systems Development*. (ed. D. Dent), pp. 311–343. Chapman & Hall.

3 Giraldo, J.P., Wu, H., Newkirk, G.M., and Kruss, S. (2019). Nanobiotechnology approaches for engineering smart plant sensors. *Nat. Nanotechnol.* 14 (6): 541–553.

4 Anderson, P.K., Cunningham, A.A., Patel, N.G. et al. (2004). Emerging infectious diseases of plants: pathogen pollution, climate change and agrotechnology drivers. *Trends Ecol. Evol.* 19 (10): 535–544.

5 Khater, M., de la Escosura-Muñiz, A., and Merkoçi, A. (2017). Biosensors for plant pathogen detection. *Biosens. Bioelectron.* 93: 72–86.

6 Horsfall, J.G. (ed.) (2012). *Plant Disease: An Advanced Treatise: How Plants Suffer from Disease*. Elsevier.

7 López, M.M., Bertolini, E., Olmos, A. et al. (2003). Innovative tools for detection of plant pathogenic viruses and bacteria. *Int. Microbiol.* 6 (4): 233–243.

8 Kokoskova, B. and Janse, J.D. (2009). Enzyme-linked immunosorbent assay for the detection and identification of plant pathogenic bacteria (in particular for *Erwinia amylovora* and Clavibacter michiganensis subsp. sepedonicus). In: *Plant Pathology*, 75–87. Totowa, NJ: Humana Press.

9 Vincelli, P. and Tisserat, N. (2008). Nucleic acid–based pathogen detection in applied plant pathology. *Plant Dis.* 92 (5): 660–669.

10 Yvon, M., Thébaud, G., Alary, R., and Labonne, G. (2009). Specific detection and quantification of the phytopathogenic agent 'Candidatus Phytoplasma prunorum'. *Mol. Cell. Probes* 23 (5): 227–234.

11 Tomlinson, J.A., Dickinson, M.J., and Boonham, N. (2010). Rapid detection of *Phytophthora ramorum* and *P. kernoviae* by two-minute DNA extraction followed by isothermal amplification and amplicon detection by generic lateral flow device. *Phytopathology* 100 (2): 143–149.
12 Malin, E.M., Roth, D.A., and Belden, E.L. (1983). Indirect immunofluorescent staining for detection and identification of *Xanthomonas campestris* pv. phaseoli in naturally infected bean seed. *Plant Dis.* 67 (6): 645–647.
13 Escoffier, L., Ganau, M., and Wong, J. (2015). *Commercializing Nanomedicine: Industrial Applications, Patents, and Ethics*. CRC Press.
14 Martinelli, F., Scalenghe, R., Davino, S. et al. (2015). Advanced methods of plant disease detection. A review. *Agron. Sustain. Dev.* 35 (1): 1–25.
15 Knell, M. (2010). Nanotechnology and the sixth technological revolution. In: *Nanotechnology and the Challenges of Equity, Equality and Development*, 127–143. Dordrecht: Springer.
16 Nair, R., Varghese, S.H., Nair, B.G. et al. (2010). Nanoparticulate material delivery to plants. *Plant Sci.* 179 (3): 154–163.
17 Ghormade, V., Deshpande, M.V., and Paknikar, K.M. (2011). Perspectives for nano-biotechnology enabled protection and nutrition of plants. *Biotechnol. Adv.* 29 (6): 792–803.
18 Rajwade, J.M., Chikte, R.G., and Paknikar, K.M. (2020). Nanomaterials: new weapons in a crusade against phytopathogens. *Appl. Microbiol. Biotechnol.* 104 (4): 1437–1461.
19 Ray, M., Ray, A., Dash, S. et al. (2017). Fungal disease detection in plants: traditional assays, novel diagnostic techniques and biosensors. *Biosens. Bioelectron.* 87: 708–723.
20 Persley, D. (1993). *Diseases of Fruit Crops (No. 634.0493 D611d)*. Queensland, AU: Department of Primary Industries.
21 Knogge, W. (1996). Fungal infection of plants. *Plant Cell* 8 (10): 1711.
22 Zhan, F., Wang, T., Iradukunda, L., and Zhan, J. (2018). A gold nanoparticle-based lateral flow biosensor for sensitive visual detection of the potato late blight pathogen, *Phytophthora infestans*. *Anal. Chim. Acta* 1036: 153–161.
23 Yüksel, S., Schwenkbier, L., Pollok, S. et al. (2015). Label-free detection of *Phytophthora ramorum* using surface-enhanced Raman spectroscopy. *Analyst* 140 (21): 7254–7262.
24 Sun, L.J., Pan, Z.Q., Xie, J. et al. (2013). Electrocatalytic activity of salicylic acid on Au@ Fe_3O_4 nanocomposites modified electrode and its detection in tomato leaves infected with *Botrytis cinerea*. *J. Electroanal. Chem.* 706: 127–132.
25 Wang, Z., Wei, F., Liu, S.Y. et al. (2010). Electrocatalytic oxidation of phytohormone salicylic acid at copper nanoparticles-modified gold electrode and its detection in oilseed rape infected with fungal pathogen *Sclerotinia sclerotiorum*. *Talanta* 80 (3): 1277–1281.
26 Fang, Y., Umasankar, Y., and Ramasamy, R.P. (2014). Electrochemical detection of *p*-ethylguaiacol, a fungi infected fruit volatile using metal oxide nanoparticles. *Analyst* 139 (15): 3804–3810.

27 Luna-Moreno, D., Sánchez-Álvarez, A., Islas-Flores, I. et al. (2019). Early detection of the fungal banana black sigatoka pathogen *Pseudocercospora fijiensis* by an SPR immunosensor method. *Sensors* 19 (3): 465.
28 Van der Want, J.P. and Dijkstra, J. (2006). A history of plant virology. *Arch. virol.* 151 (8): 1467–1498.
29 Hong, S. and Lee, C. (2018). The current status and future outlook of quantum dot-based biosensors for plant virus detection. *Plant Pathol. J.* 34 (2): 85.
30 Liu, Z., Xia, X., Yang, C., and Wang, L. (2015). Visual detection of Maize chlorotic mottle virus using unmodified gold nanoparticles. *RSC Adv.* 5 (122): 100891–100897.
31 Wang, L., Liu, Z., Xia, X., and Huang, J. (2016). Visual detection of Maize chlorotic mottle virus by asymmetric polymerase chain reaction with unmodified gold nanoparticles as the colorimetric probe. *Anal. Meth.* 8 (38): 6959–6964.
32 Wang, L., Liu, Z., Xia, X. et al. (2017). Colorimetric detection of cucumber green mottle mosaic virus using unmodified gold nanoparticles as colorimetric probes. *J. Virol. Meth.* 243: 113–119.
33 Chartuprayoon, N., Rheem, Y., Ng, J.C. et al. (2013). Polypyrrole nanoribbon based chemiresistive immunosensors for viral plant pathogen detection. *Anal. Meth.* 5 (14): 3497–3502.
34 Shojaei, T.R., Salleh, M.A., Sijam, K. et al. (2016). Detection of *Citrus tristeza* virus by using fluorescence resonance energy transfer-based biosensor. *Spectrochim. Acta A: Mol. Biomol. Spectrosc.* 169: 216–222.
35 Shojaei, T.R., Salleh, M.A., Sijam, K. et al. (2016). Fluorometric immunoassay for detecting the plant virus *Citrus tristeza* using carbon nanoparticles acting as quenchers and antibodies labeled with CdTe quantum dots. *Microchim. Acta* 183 (7): 2277–2287.
36 Haji-Hashemi, H., Norouzi, P., Safarnejad, M.R., and Ganjali, M.R. (2017). Label-free electrochemical immunosensor for direct detection of *Citrus tristeza* virus using modified gold electrode. *Sensors Actuat. B: Chem.* 244: 211–216.
37 Freitas, T.A., Proença, C.A., Baldo, T.A. et al. (2019). Ultrasensitive immunoassay for detection of *Citrus tristeza* virus in citrus sample using disposable microfluidic electrochemical device. *Talanta* 205: 120110.
38 Khater, M., de la Escosura-Muñiz, A., Quesada-González, D., and Merkoçi, A. (2019). Electrochemical detection of plant virus using gold nanoparticle-modified electrodes. *Anal. Chim. Acta* 1046: 123–131.
39 Khater, M., Escosura-Muñiz, A.D., Altet, L., and Merkoçi, A. (2019). In situ plant virus nucleic acid isothermal amplification detection on gold nanoparticle-modified electrodes. *Anal. Chem.* 91 (7): 4790–4796.
40 Shokri, E., Hosseini, M., Faridbod, F., and Rahaie, M. (2016). Rapid pre-symptomatic recognition of tristeza viral RNA by a novel fluorescent self-dimerized DNA–silver nanocluster probe. *RSC Adv.* 6 (101): 99437–99443.
41 Majumder, S. and Johari, S. (2018). Development of a gold-nano particle based novel dot immunobinding assay for rapid and sensitive detection of Banana bunchy top virus. *J. Virol. Meth.* 255: 23–28.

42 Kumar, P., Arun, V., and Lokeswari, T.S. (2017). Cloning of BBTV (Banana Bunchy Top Virus) components and screening of BBTV using functionalized gold nanoparticles. *3 Biotech* 7 (3): 225.
43 Wei, J., Liu, H., Liu, F. et al. (2014). Miniaturized paper-based gene sensor for rapid and sensitive identification of contagious plant virus. *ACS Appl. Mater. Interf.* 6 (24): 22577–22584.
44 Razmi, A., Golestanipour, A., Nikkhah, M. et al. (2019). Localized surface plasmon resonance biosensing of tomato yellow leaf curl virus. *J. Virol. Meth.* 267: 1–7.
45 Wang, T.M. and Yang, J.T. (2019). Visual DNA diagnosis of tomato yellow leaf curl virus with integrated recombinase polymerase amplification and a gold-nanoparticle probe. *Sci. Rep.* 9 (1): 1–8.
46 Dharanivasan, G., Riyaz, S.M., Jesse, D.M. et al. (2016). DNA templated self-assembly of gold nanoparticle clusters in the colorimetric detection of plant viral DNA using a gold nanoparticle conjugated bifunctional oligonucleotide probe. *RSC Adv.* 6 (14): 11773–11785.
47 Dharanivasan, G., Jesse, D.M., Rajamuthuramalingam, T. et al. (2019). Scanometric detection of tomato leaf curl New Delhi viral DNA using mono- and bifunctional AuNP-conjugated oligonucleotide probes. *ACS Omega* 4 (6): 10094–10107.
48 Panferov, V.G., Safenkova, I.V., Zherdev, A.V., and Dzantiev, B.B. (2018). Post-assay growth of gold nanoparticles as a tool for highly sensitive lateral flow immunoassay. Application to the detection of potato virus X. *Microchim. Acta* 185 (11): 506.
49 Razo, S.C., Panferov, V.G., Safenkova, I.V. et al. (2018). Double-enhanced lateral flow immunoassay for potato virus X based on a combination of magnetic and gold nanoparticles. *Anal. Chim. Acta* 1007: 50–60.
50 Ivanov, A.V., Safenkova, I.V., Zherdev, A.V., and Dzantiev, B.B. (2020). Nucleic acid lateral flow assay with recombinase polymerase amplification: solutions for highly sensitive detection of RNA virus. *Talanta* 210: 120616.
51 Masoudi, N., Assareh, M.H., Rouhibakhsh, A. et al. (2019). Rapid detection of Potato virus S using antibody-coated gold nanoparticles. *Iranian J. Plant Pathol.* (2): 55. https://doi.org/10.22034/IJPP.2019.37318.
52 Tahir, M.A., Bajwa, S.Z., Mansoor, S. et al. (2018). Evaluation of carbon nanotube based copper nanoparticle composite for the efficient detection of agroviruses. *J. Hazard Mater.* 346: 27–35.
53 Byzova, N.A., Vinogradova, S.V., Porotikova, E.V. et al. (2018). Lateral flow immunoassay for rapid detection of grapevine leafroll-associated virus. *Biosensors* 8 (4): 111.
54 Razo, S.C., Panferov, V.G., Safenkova, I.V. et al. (2018). How to improve sensitivity of sandwich lateral flow immunoassay for corpuscular antigens on the example of potato virus Y? *Sensors* 18 (11): 3975.
55 Yao, K.S., Li, S.J., Tzeng, K.C. et al. (2009). Fluorescence silica nanoprobe as a biomarker for rapid detection of plant pathogens. In: *Advanced Materials Research*, vol. 79, 513–516. Trans Tech Publications Ltd.

56 Strayer, A., Ocsoy, I., Tan, W. et al. (2016). Low concentrations of a silver-based nanocomposite to manage bacterial spot of tomato in the greenhouse. *Plant Dis.* 100 (7): 1460–1465.

57 Vaseghi, A., Safaie, N., Bakhshinejad, B. et al. (2013). Detection of Pseudomonas syringae pathovars by thiol-linked DNA–gold nanoparticle probes. *Sensors Actuat. B: Chem.* 181: 644–651.

58 Panferov, V.G., Safenkova, I.V., Varitsev, Y.A. et al. (2016). Development of the sensitive lateral flow immunoassay with silver enhancement for the detection of Ralstonia solanacearum in potato tubers. *Talanta* 152: 521–530.

59 Razo, S.C., Panferova, N.A., Panferov, V.G. et al. (2019). Enlargement of gold nanoparticles for sensitive immunochromatographic diagnostics of potato brown rot. *Sensors* 19 (1): 153.

60 Huang, X., Xu, Z., Mao, Y. et al. (2015). Gold nanoparticle-based dynamic light scattering immunoassay for ultrasensitive detection of Listeria monocytogenes in lettuces. *Biosens. Bioelectron.* 66: 184–190.

61 Zhao, Y., Liu, L., Kong, D. et al. (2014). Dual amplified electrochemical immunosensor for highly sensitive detection of Pantoea stewartii sbusp. stewartii. *ACS Appl. Mater. Interf.* 6 (23): 21178–21183.

62 Zhao, W., Lu, J., Ma, W. et al. (2011). Rapid on-site detection of Acidovorax avenae subsp. citrulli by gold-labeled DNA strip sensor. *Biosens. Bioelectron.* 26 (10): 4241–4244.

63 Safenkova, I.V., Zaitsev, I.A., Varitsev, Y.A. et al. (2017). Development of a lateral flow immunoassay for rapid diagnosis of potato blackleg caused by Dickeya species. *Anal. Bioanal. Chem.* 409 (7): 1915–1927.

64 Jans, H. and Huo, Q. (2012). Gold nanoparticle-enabled biological and chemical detection and analysis. *Chem. Soc. Rev.* 41 (7): 2849–2866.

65 Lopez-Soriano, P., Noguera, P., Gorris, M.T. et al. (2017). Lateral flow immunoassay for on-site detection of Xanthomonas arboricola pv. pruni in symptomatic field samples. *PLOS ONE* 12 (4): e0176201.

4

Solvent Extraction Coupled with Gas Chromatography for the Analysis of Polycyclic Aromatic Hydrocarbons in Riverine Sediment and Surface Water of Subarnarekha River and Its Tributary, India

Balram Ambade, Shrikanta S. Sethi, Amit Kumar, and Tapan K. Sankar

Department of Chemistry, National Institute of Technology Jamshedpur, Jharkhand, 831014, India

4.1 Introduction

Polycyclic aromatic hydrocarbons (PAHs) are priority pollutants that are found in the atmosphere, soil, water bodies, organisms, and sediments [1–4]. They are organic compounds containing more than one fused aromatic ring of carbon and hydrogen atoms and are mostly formed by natural and anthropogenic processes and can enter the environment through various ways, such as anthropogenic input from oil spills, incomplete combustion, urban runoff, domestic and industrial wastewater discharges, vehicle exhaust, and industrial emissions; traffic is a very important emission source of these pollutants [5–7]. Due to their hydrophobic properties and low biodegradation rate, PAH compounds reside in sediments of the aquatic environment for a long span of time [8–10]. High-molecular-weight PAHs (4 or higher ring PAHs) are more lipophilic, less volatile, and less water soluble compared to low-molecular-weight (LMW) PAHs (3 or lesser ring PAHs) and have probable carcinogenic effects on humans ([11–13]). Investigation into the concentration of PAHs in the aquatic environment is needed, which can provide a pointer to the sources of contaminant loading. PAHs are distributed globally from dense industrialized areas to the aquatic environment with a wide range of concentration in living organisms and sediments [14–16]. PAH concentrations are generally high in more industrialized and urbanized areas, such as Canada [6], North America [16], and China [17]. The PAH concentration in sediments has been estimated to be higher than 10 000 μg/g dry weight [6]. Due to their toxic, mutagenic, and carcinogenic properties, PAHs are of serious environmental concern. Due to rapid growth in industrialization and increasing daily emission, PAH contamination has gained large attention in countries such as India and China, which are on the path of development [18, 19]. On the basis of toxicity, 16 PAHs have been listed as priority pollutants by the United States Environmental Protection Agency (USEPA). The species, exposure route, and molecular structure are responsible for the toxicity of PAHs [20].

The maximum global atmospheric emission in 2004 of 16 USEPA-listed PAHs was from China (114 Gg/y), India (90 Gg/y), and the Unites States (32 Gg/y) [19].

Miniaturized Analytical Devices: Materials and Technology, First Edition.
Edited by Suresh Kumar Kailasa and Chaudhery Mustansar Hussain.

This can be attributed to a large number of industries, rapid urbanization, and vehicular emission. Jamshedpur is a major industrial center of East India. It houses companies such as Tata Steel, Tata Motors, Tata Power. It is home to one of the largest industrial zones of India. Known as Adityapur, it houses >1200 small- and medium-scale industries and has an SEZ named AIDA in Adityapur [21]. The first planned industrial city of India, Jamshedpur, is situated on the banks of Subarnarekha River and its main tributary Kharkai River [21]. These two rivers mainly flow through the Adityapur region of Jamshedpur where >1200 industries are operating, including iron, steel, chemicals, transport equipment, and cement, which can be the cause of water pollution in both rivers. The main concern of this study was about the PAH levels in Subarnarekha and Kharkai Rivers. Therefore, the study investigated the occurrence and levels of PAHs in the surface water and sediments of both rivers and the distribution of PAHs in their rural, town, and semi-town regions. Diagnostic ratios and principal component analysis (PCA) were used to estimate the distribution and identification of sources of PAH contamination.

4.2 Materials and Methods

4.2.1 Description of Study Area

The Subarnarekha River flows across three states – Jharkhand, Odisha, and West Bengal of India. Its source is located at 23° 18′ N, 85° 11′ E and the mouth is 21° 33′ N, 87° 23′ E. Subarnarekha passes through areas with extensive mining of copper and uranium ores, and as a result it is highly polluted. The tribal communities inhabiting the Chota Nagpur region and their livelihood are threatened by water pollution of the Subarnarekha River. The Kharkai River is the main tributary of the Subarnarekha River, the length of which is 136 km and the catchment area is 6.611 km^2. It flows through Adityapur, the highly industrialized area of Jamshedpur. It arises in Mayurbhanj district, Odisha, and after flowing past Rairangpur it heads north up to Saraikela and then toward the east, and finally it enters Subarnarekha in northwestern Jamshedpur. Iron ore is mined in the mountains of the headwaters of the Kharkai River and supplied to the steel plant in Jamshedpur. These two rivers mainly flow through the Adityapur region where many industries such as sponge iron, steel, chemicals, transport equipment, and cement are fully fledged operating. The study includes the characterization of the rivers in village, town, and semi-town areas. The comprehensive sampling sites are shown in Figure 4.1.

4.2.2 Sampling and Pretreatment

As shown in Figure 4.1, 56 sampling sites along the Subarnarekha and Kharkai Rivers were selected. To effectively monitor the distribution of pollution in both rivers, sampling sites were selected so that they ranged from low contaminated (low industrialization and urbanization) areas to heavily contaminated (highly

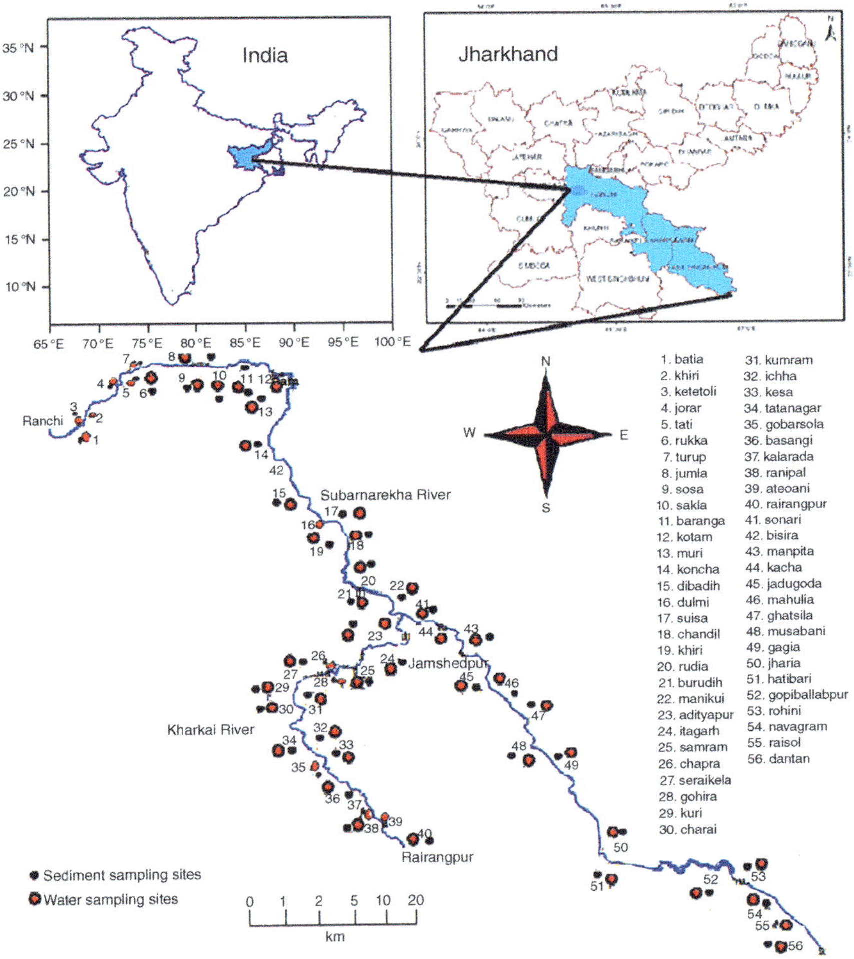

Figure 4.1 Sampling sites of the Kharkai and Subarnarekha Rivers. PAH concentration in Subarnarekha sediments PAHs concentration in Kharkai sediments. PAH concentration in water samples

industrialized and urbanized) areas. A total of 112 samples, including 56 sediment samples and 56 water samples, were collected in the post-monsoon season of August 2018. Sediment samples were collected in polyethylene bags using a grab sampler, and water samples were collected directly from the river into 2 l glass jars. The samples were kept in the ice box and sent to the laboratory, and the sediment samples were kept in the refrigerator at −18° C before analysis. The water samples were stored in the refrigerator at -4 °C before analysis. Furthermore, the water samples were carried for further extractions. All containers in contact with the sample were previously washed with diluted nitric acid and deionized water to eliminate the effect of other contaminants.

4.2.3 Extraction and Cleanup of PAHs from Samples

PAHs in sediment samples were extracted by solvent extraction procedure. The dry samples were collected, homogenized, properly sieved, and kept for the sample preparation. This process involved mixing of 2 g of dry sediment samples with 0.5 g anhydrous Na_2SO_4 and using 10 ml of dichloromethane (DCM) for 1 h of ultra-sonication followed by centrifugation. Then, 3 ml of supernatant was filtered through 2 g of silica gel column with 11 ml 1 : 1(v/v) elution of hexane and DCM [22]. Solvent fractions were evaporated by exchanging acetonitrile with a final volume of 2 ml. PAH compounds were extracted from the water samples by the procedures mentioned in the study by Chakraborty et al. [23]. DCM was used for liquid–liquid extraction for three times (50, 25, and 25 ml) with 1 l of water samples. Then, the organic phase was separated and collected by mixing 10 g sodium chloride and shaking for 4–6 minutes for three times, and finally the combined organic phase was collected and the drying process started by passing it through sodium sulfate placed on glass wool of 3 cm and was concentrated using rotary evaporator. The PAHs labeled with Deuterium (Naphthalene-D8, Pyrene-D10, Phenanthrene-D10, Anthracene-D10, and Chrysene-D12) were used as surrogate standards. For sample extraction, a silica gel column of 30 cm long and 3 cm diameter was used; top of the column was filled with 10 g activated silica gel and 5 g anhydrous Na_2SO_4. The PAHs were eluted with 100 ml of DCM and *n*-hexane mixture (1 : 1, v/v). Finally, under a gentle stream of pure nitrogen, the eluent solvent was concentrated to 20 μl. To quantify all analytes, an internal standard hexamethyl benzene was added before analysis [12].

4.2.4 Analysis

The sample extracts of water and sediments of Subarnarekha and Kharkai Rivers were analyzed in an Agilent 7890B Gas Chromatograph (GC-FID) coupled with 5977A Mass Spectrometry using a HP-5 MS capillary column (30 m × 0.32 mm × 0.25 μm) with a 7 inch cage. In splitless mode, 1 μl of each sample was injected. High-purity nitrogen gas was used as carrier gas. The carrier gas was flowing at a flow rate of 1.83 ml/min. The initial oven temperature was set to 65 °C for 3 minutes and then increased to 290 °C at a rate of 5 °C/min by holding for 20 minutes. The temperatures of injector and transfer line were set to 290 and 320 °C, respectively. Using this technique, the 16 USEPA PAHs, Naphthalene (NA), Acenaphthene (AC), Acenaphthylene (ACY), Fluorene (Fluo), Phenanthrene (Phen), Anthracene (AN), Fluoranthene (Flur), Pyrene (Pye), Benzo(a)Anthracene (BaA), Chrysene (CHRY), Benzo(b)Fluoranthene (BbF), Benzo(k)Fluoranthene (BkF), Benzo(a)Pyrene(BaP), Dibenzo [a,h]anthracene (DBA), Benzo[g,h,i]perylene (BgP), and Indeno[1,2,3-cd]pyrene (IN) compounds were identified according to their retention time and mass spectra.

4.2.5 Quality Assurance

We applied the standard solution of 16 PAHs in acetonitrile (ID-3697900), deuterium-leveled PAHs, and hexamethyl benzene (CAS Number-87-85-4; Sigma

Aldrich, USA). The chemicals DCM (CAS Number-75-09-2), anhydrous Na_2SO_4 (CAS Number-7757-82-6), sodium chloride (CAS Number-7647-14-5), and silica gel powder (CAS Number-112926-00-8) with AR grade were used in this study. The procedural blank samples, sample duplicates, samples spiked with surrogate standards Pyrene-D10 (CAS Number-1718-52-1), Anthracene-D10 (CAS Number-1719-06-8), Phenanthrene-D10 (CAS Number-1517-22-2), Naphthalene-D8 (CAS Number-1146-65-2), and Chrysene-D12 (CAS Number-1719-03-5) were used as internal standards for the calibration of all species of PAHs, and the quality control testing for PAHs analysis relied on these methods. The recovery percentages of 16 PAHs in the water and sediment samples were approximately 90–105% and 80–95%, respectively. The standard deviation was <12% for the method. The targeted PAHs comprises method detection limits (MDLs) were calculated as three times the standard deviation (SD) plus the mean concentrations of target compounds in blank samples. MDLs of the target PAHs for the sediment and surface water samples were 1 ng/g and 0.01 μg/l, respectively. The concentration values less than the MDLs were considered <DL (<detection limit).

4.3 Results and Discussion

4.3.1 PAH Concentration in Water

The concentration, mean value, and standard deviation of each PAH congener in surface water from the Subarnarekha River and its tributary are reported in Table 4.1. The total PAH concentration in the water ranged from (<DL) to 26 μg/l for the Kharkai River and (<DL) to 33 μg/l for the Subarnarekha River, which was slightly higher than (<DL), to 30 μg/l (Table 4.2) for the Hooghly and Brahmaputra Rivers. For two-ring PAHs, the concentration ranged from (<DL) to 1.01 μg/l for the Kharkai River and (<DL) to 1.23 μg/l for the Subarnarekha River. In both rivers, the concentration of three-ring PAHs was dominant, as in the Hooghly and Brahmaputra Rivers [12]. From Figure 4.2, the three-membered ring PAHs was about 92% of the total PAH concentration in the Kharkai River and about 74% in the Subarnarekha River. Among the three rings, Acenaphthene was abundant in both rivers, similarly in the Gomti River [31]. The samples collected from the Adityapur region of the Kharkai River and the Jamshedpur region of the Subarnarekha River were found to have maximum concentrations of Acenaphthene, which may be due to the vast industrial activities. The four-ring PAHs were found to be in less concentration in the Kharkai River as well as in the Subarnarekha River, but surprisingly one of the four-ring PAHs benzo(a) anthracene was found to be in more concentration in the samples collected from the town areas of the Subarnarekha River, which was 18.1 μg/l. Two five-ring PAHs BkF and BbF were detected in both rivers, whereas other two BaP and DbA were not detected in either river. Among six members, BgP was detected in a concentration of (<DL) to 0.29 μg/l in the Kharkai River and (<DL) to 0.36 μg/l in the Subarnarekha River, but another six-membered ring IN was found (<DL) in both rivers. The low concentrations of both five- and

Table 4.1 Range, mean concentration (μg/l), and standard deviation of surface water from the Kharkai and Subarnarekha Rivers.

				Kharkai River		Subarnarekha River	
S.N	Concentration in (μg/l)	Formula	Number of rings	Range	Mean ± SD	Range	Mean ± SD
1	Naphthalene	$C_{12}H_{12}$	2	<DL −1.01	0.25 ± 0.21	<DL −1.23	0.41 ± 0.29
2	Acenaphthene	$C_{12}H_{10}$	3	<DL −25	3.45 ± 6.2	<DL −32	2.57 ± 5.68
3	Acenaphthylene	$C_{12}H_8$	3	<DL −6	0.72 ± 2.06	<DL −3	0.65 ± 1.02
4	Fluorine	$C_{13}H_{10}$	3	<DL −0.57	0.12 ± 0.15	<DL −0.86	0.19 ± 0.19
5	Phenanthrene	$C_{14}H_{10}$	3	<DL −0.24	0.08 ± 0.04	<DL −0.28	0.11 ± 0.08
6	Anthracene	$C_{14}H_{10}$	3	<DL −0.61	0.86 ± 0.09	<DL −0.56	0.79 ± 0.04
7	Fluoranthene	$C_{16}H_{10}$	4	<DL −0.02	0.01 ± 0.01	<DL −0.16	0.02 ± 0.01
8	Pyrene	$C_{16}H_{10}$	4	<DL −0.04	0.01 ± 0.01	<DL −0.09	0.01 ± 0.01
9	Benzo(a) anthracene	$C_{18}H_{12}$	4	<DL −0.59	0.10 ± 0.19	<DL −18.1	0.95 ± 2.6
10	Chrysene	$C_{18}H_{12}$	4	<DL −0.11	0.05 ± 0.04	<DL −0.65	0.06 ± 0.03
11	Benzo(b) fluoranthene	$C_{20}H_{12}$	5	<DL −0.41	0.04 ± 0.07	<DL −0.41	0.04 ± 0.07
12	Benzo(k) fluoranthene	$C_{20}H_{12}$	5	<DL −0.12	0.02 ± 0.01	<DL −0.19	0.02 ± 0.03
13	Benzo(a)pyrene	$C_{20}H_{12}$	5	<DL	<DL	<DL	<DL
14	Dibenzo(a,h) anthracene	$C_{22}H_{14}$	5	<DL	<DL	<DL	<DL
15	Indeno(1,2,3-cd) pyrene	$C_{22}H_{12}$	6	<DL	<DL	<DL	<DL
16	Benzo(g,h,i) perylene	$C_{22}H_{12}$	6	<DL −0.29	0.02 ± 0.01	<DL −0.36	0.03 ± 0.02
	Σ_{16}PAHs			<DL −26	5.7 ± 8	<DL −33	5.8 ± 6

six-ring PAHs suggest that they are usually absorbed in the sediments [33]. The concentrations of PAHs are in the order town > semi-town > rural for both rivers.

4.3.1.1 PAHs Concentration in Subarnarekha Riverine Sediment

Table 4.3 illustrates the concentration of PAHs in sediments of the Subarnarekha River. The concentration of Σ_{16}PAHs ranged from (<DL) to 657 ng/g dry weight with a mean value of 359 ng/g in the town area. The concentration of Σ_{16}PAHs ranged from (<DL) to 122 ng/g with a mean value of 135 ng/g in the semi-town area and from (<DL) to 68 ng/g with a mean value of 66 ng/g in the rural area. Acenaphthylene and Pyrene were abundant in the town, semi-town, and rural areas, with a mean value of 121 ng/g (ACY) and 133 ng/g (Pyr) in the town area, 17 ng/g (ACY) and 68 ng/g (Pyr) in the semi-town area, and 12 ng/g (ACY) and 17 ng/g (Pyr) in the rural area. The concentration of 2-ring PAHs was highest (29%) in the rural area, 3-ring

Table 4.2 Comparison of PAH concentration ranges of this study with other studies.

Study area	Number of PAH studies	PAH range in water	PAH range in sediment	Reference
Aojiang River, China	15	546–1444 ng/l	404–2606 ng/g	[24]
Dali River watershed, China	18	946–13 449 ng/l	62–841 ng/g	[25]
Gao-ping River, Taiwan	16	10–9400 ng/l	8–356 ng/g	[14]
Qiantang River, China	15	70–1844 ng/l	91–614 ng/g	[26]
Hooghly and Brahmaputra Rivers, India	16	0–30 μg/l	2–1329 ng/g	[12]
Middle and lower reaches of Yellow River, China	15, 13	179–369 ng/l	31–133 ng/g	[27]
Tonghui River, China	16	192.9–2651 ng/l	127–928 ng/g	[28]
Liaohe River basin, China	16	0–615 ng/l	0–60057 ng/g	[29]
Kor River, Iran	16	45–375 ng/l	167–530 ng/g	[30]
Gomti River, India	16	0–84 μg/l	5–3723 ng/g	[31]
Songhua River, China	16	164–2746 ng/l	68–654 ng/g	[32]
Subarnarekha and Kharkai Rivers, India	16	0–33 μg/l	0–657 ng/g	This study

PAHs in the town area (43%), 4-ring PAHs in the semi-town area (58%), 5-ring PAHs in the rural area (6%), and the concentration of 6-ring PAHs (3%) was the same in both town and semi-town areas. Figure 4.2 shows that the overall PAH concentration in the Subarnarekha River was in following the order: town > semi-town > rural.

4.3.1.2 PAH Concentration in Kharkai Riverine Sediment

Table 4.4 illustrates the PAH concentration in sediments of the Kharkai River. The concentration of 2-ring PAHs was the highest in the samples collected from the town area with a mean concentration of 31 ng/g, but in percentage of the total PAH concentration, it was higher in the rural area with 7.5%. The 3-ring PAHs were highest in the rural area with 32.5% of the total PAH concentration. Among the 3-ring PAHs, Acenaphthylene was highest with a mean concentration of 24 ng/g and ranged from (<DL) to 120 ng/g in the town area samples collected from Adityapur and Jamshedpur. In this river, mostly 4-ring PAHs were found in more concentration in sediments with 50% of the total PAH contamination in the town area, 39% in the semi-town area, and 40% in the rural area. Among the 4-ring PAHs, pyrene was the highest with a mean concentration of 84 ng/g and ranged from (<DL) to 185 ng/g in the town area, 28 ng/g and ranged from (<DL) to 100 ng/g in the semi-town area, and 7 ng/g and ranged from (<DL) to 38 ng/g in the rural area. The 5- and 6-ring PAHs were comparatively less in concentration; 5-ring PAHs contribution to the total PAH

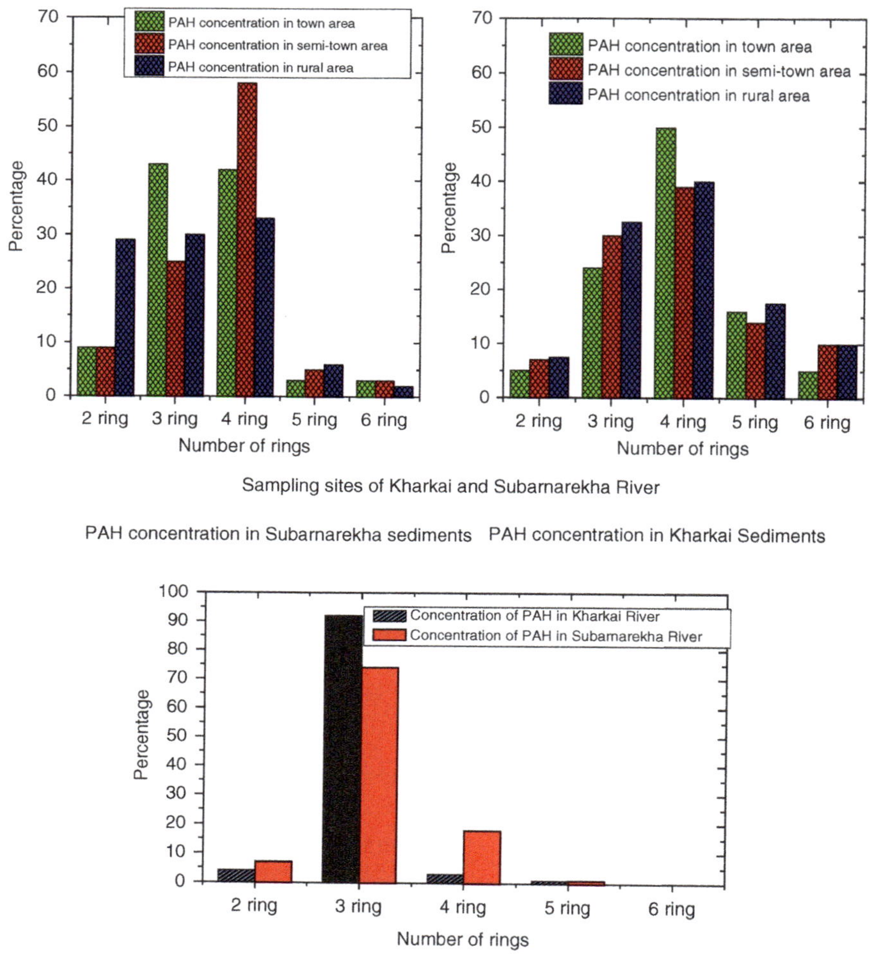

Figure 4.2 Concentration of PAHs in the sediments and surface water in town, semi-town, and rural areas of the Subarnarekha and Kharkai Rivers. Riverine sediments of the Kharkai River Riverine sediments of the Subarnarekha River. Surface water of the Kharkai

concentration was 16% in the town area, 14% in the semi-town area, and 17.5% in the rural area. Figure 4.2 suggests that the overall PAH concentration in the Kharkai River is in the order town area > semi-town area > rural area.

4.3.2 PAH Composition

The composition pattern of 16 PAHs by molecular weight and ring size in surface water and riverine sediments of both rivers is shown in Figure 4.2. The composition of PAHs in water was mainly 3-ring PAHs in the Kharkai River and 3- and 4-ring PAHs in the Subarnarekha River and in the sediment samples collected from different regions mainly 3-, 4-, and 5-ring PAHs in the Kharkai River and 2-, 3-, and

Table 4.3 Mean concentration (ng/g), range, and standard deviation of PAHs in the sediments of the Subarnarekha River.

		Town	Semi-town	Rural
S.N.	**Concentration in (ng/g) dry weight**	**Mean ± SD (range)**	**Mean ± SD (range)**	**Mean ± SD (range)**
1	Naphthalene	31 ± 27 (<DL −135)	12 ± 16 (<DL −39)	19 ± 29 (<DL −75)
2	Acenaphthene	15 ± 13 (<DL −79)	5 ± 6 (<DL −32)	4 ± 5 (<DL −29)
3	Acenaphthylene	121 ± 186 (<DL −657)	17 ± 13 (<DL −79)	12 ± 12 (<DL −68)
4	Fluorine	10 ± 7 (<DL −49)	7 ± 6 (<DL −36)	2 ± 2 (<DL −22)
5	Phenanthrene	7 ± 7 (<DL −25)	3 ± 2 (<DL −15)	1 ± 1 (<DL −8)
6	Anthracene	3 ± 3 (<DL −9)	2 ± 3 (<DL −7)	1 ± 1 (<DL −5)
7	Fluoranthene	9 ± 11 (<DL −49)	6 ± 7 (<DL −27)	2 ± 2 (<DL −9)
8	Pyrene	133 ± 120 (<DL −539)	68 ± 76 (<DL −122)	17 ± 6 (<DL −49)
9	Benzo(a)anthracene	5 ± 5 (<DL −35)	2 ± 3 (<DL −21)	2 ± 3 (<DL −19)
10	Chrysene	4 ± 5 (<DL −24)	2 ± 2 (<DL −17)	1 ± 1 (<DL −9)
11	Benzo(b)fluoranthene	3 ± 4 (<DL −18)	1 ± 1 (<DL −8)	1 ± 1 (<DL −5)
12	Benzo(k)fluoranthene	4 ± 4 (<DL −24)	3 ± 2 (<DL −18)	2 ± 1 (<DL −14)
13	Benzo(a)pyrene	3 ± 3 (<DL −27)	2 ± 1 (<DL −15)	1 ± 1 (<DL −6)
14	Dibenzo(a,h)anthracene	2 ± 1 (<DL −16)	1 ± 1 (<DL −6)	<DL
15	Indeno(1,2,3-cd)pyrene	4 ± 3 (<DL −31)	1 ± 2 (<DL −9)	<DL
16	Benzo(g,h,i)perylene	5 ± 7 (<DL −45)	3 ± 2 (<DL −26)	1 ± 1 (<DL −4)
	Σ_{16}PAHs (range)	359 ± 388 (<DL −657)	135 ± 143 (<DL −122)	66 ± 78 (<DL −68)

4-ring PAHs in the Subarnarekha River. For individual PAHs in sediment samples, almost all PAHs were found in both rivers, but for water samples large-ring PAHs were found in less concentration. Six-ring PAHs were in negligible concentration in the surface water samples. In the water samples of the Kharkai and Subarnarekha Rivers, the percentage of PAH concentration was 4% and 7% for 2 rings, 92% and 74% for 3 rings, 3% and 18% for 4 rings, 1% and 1% for 5 rings, and 0% for 6 rings for both rivers. In the sediment samples, the percentage of PAH concentration is given in Figures 4.3 and 4.4. The 2-ring PAHs were 5% in the town area, 7 % in the semi-town area, and 7.5 % in the rural area of the Kharkai River and 9 % in the town area, 9% in the semi-town area, and 29 % in the rural area of the Subarnarekha River. The high-molecular-weight PAHs (5 and 6 rings) were less in concentration in both rivers. Mostly, 3- and 4-ring PAHs were abundant in both riverine sediments.

4.3.3 Analysis for Sources of PAHs

4.3.3.1 Diagnostic Ratio

PAHs enter a river system mostly through urban runoff, municipal and industrial effluents, atmospheric fallout, and oil leakage [34, 35]. PAH sources can be

Table 4.4 Concentration (ng/g), standard deviation, and range of PAHs in the sediments of the Kharkai River.

		Town area	Semi-town	Rural
S.N.	Concentration in (ng/g) dry weight	Mean ± SD (range)	Mean ± SD (range)	Mean ± SD (range)
1	Naphthalene	11 ± 14 (<DL −64)	7 ± 8 (<DL −29)	3 ± 2 (<DL −21)
2	Acenaphthene	8 ± 9 (<DL −64)	4 ± 6 (<DL −29)	2 ± 2 (<DL −16)
3	Acenaphthylene	24 ± 23 (<DL −120)	10 ± 12 (<DL −45)	4 ± 6 (<DL −24)
4	Fluorine	9 ± 10 (<DL −64)	8 ± 9 (<DL −61)	3 ± 1 (<DL −13)
5	Phenanthrene	10 ± 9 (<DL −56)	5 ± 6 (<DL −35)	2 ± 3 (<DL −18)
6	Anthracene	6 ± 5 (<DL −28)	3 ± 4 (<DL −21)	2 ± 2 (<DL −16)
7	Fluoranthene	12 ± 12 (<DL −68)	4 ± 6 (<DL −34)	3 ± 4 (<DL −18)
8	Pyrene	84 ± 89 (<DL −185)	28 ± 31 (<DL −100)	7 ± 8 (<DL −38)
9	Benzo(a)anthracene	10 ± 12 (<DL −75)	3 ± 2 (<DL −37)	4 ± 6 (<DL −29)
10	Chrysene	14 ± 16 (<DL −67)	4 ± 6 (<DL −34)	2 ± 1 (<DL −23)
11	Benzo(b)fluoranthene	8 ± 7 (<DL −46)	3 ± 2 (<DL −22)	2 ± 1 (<DL −7)
12	Benzo(k)fluoranthene	15 ± 5 (<DL −44)	6 ± 6 (<DL −38)	3 ± 5 (<DL −18)
13	Benzo(a)pyrene	12 ± 10 (<DL −76)	3 ± 3 (<DL −26)	1 ± 1 (<DL −7)
14	Dibenzo(a,h)anthracene	4 ± 2 (<DL −34)	2 ± 3 (<DL −18)	1 ± 2 (<DL −6)
15	Indeno(1,2,3-cd)pyrene	6 ± 8 (<DL −31)	5 ± 5 (<DL −24)	2 ± 2 (<DL −7)
16	Benzo(g,h,i)perylene	7 ± 8 (<DL −41)	5 ± 5 (<DL −22)	2 ± 3 (<DL −10)
	Σ_{16}PAHs (range)	240 ± 225 (<DL −185)	100 ± 109 (<DL −100)	40 ± 49 (<DL −38)

identified by using several molecular indices such as BbF/BkF [36], Pyr/Bap [37], BaA/(BaA+Chyr) [38], Bap/B(g,h,i)P [39], AN/(AN+Phen), and Flur/(Flur+Pye) [40]. In this study, AN/(AN+Phen) and Flur/(Flur+Pye) were used. Diagnostic ratio and PCA were employed to identify the sources of PAHs in riverine sediments. The mean concentrations were used for each of the individual PAHs to calculate the diagnostic ratio for different sampling sites (town, semi-town, and rural) for both rivers. AN/(AN+Phen) < 0.1 shows the petroleum origin, while >0.1 indicates the combustion source; Flur/(Flur+Pyr) $= 0.5$, >0.5, and <0.5 indicate petroleum/combustion source, pyrolytic origin, and petrogenic origin, respectively [41]. The concentration ratios shown in Figure 4.3 were used to identify the origin of PAHs in the sediments and surface water of both rivers. From Figure 4.3, it is clear that in the sediments of both rivers the ratios of Flur/(Flur+Pye) ranged from 0.03 to 0.46, meaning 100% of the ratios of Flur/(Flur+Pye) were less than 0.5, which suggested a petrogenic origin. The ratios of AN/(AN+Phen) were all greater than 0.1, which indicates the main source to be combustion in the Kharkai River. In Subarnarekha riverine sediments, the ratios of AN/(AN+Phen) ranged from 0.16 to 0.66, which indicates that 100% of ratios were greater than 0.1, indicating

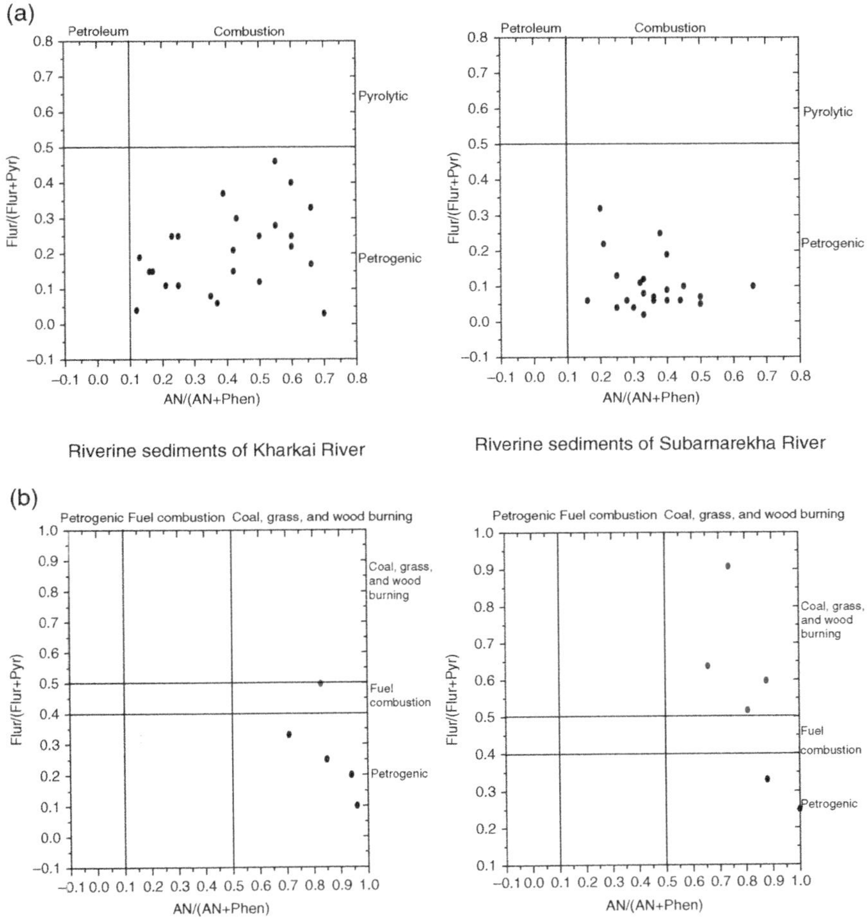

Figure 4.3 Scatter plot of the ratios AN/(AN+Phen) vs. Fluor/(Flur+Pyr) in the sediments and surface water from the Kharkai and Subarnarekha Rivers.

the main source of combustion, and the ratios of Flur/(Flur+Pye) ranged from 0.02 to 0.32 and 100% of ratios were below 0.5, suggesting a petrogenic origin. Figure 4.3 shows that sources of PAH contamination in the sediments of both rivers were of petrogenic and combustion origin. For surface water samples, the ratios AN/(AN+Phen) and Flur/(Flur+Pye) were used for source allocation with ratios <0.1, >0.1, and >0.5, indicating petrogenic, fuel combustion, and coal, grass, and wood burning, and <0.4, 0.4–0.5, and >0.5, indicating petrogenic, fuel combustion, and coal, grass, and wood burning [41]. In the surface water samples of the Kharkai River, the ratios of Flur/(Flur+Pye) ranged from 0.1 to 0.5 and 80% ratios were below 0.4, indicating a petrogenic origin, 20% ratios were between 0.4 and 0.5, indicating fuel combustion, and the ratios of AN/(AN+Phen) ranged from 0.71 to 0.96, and 100% ratios were above 0.5, indicating coal, grass and wood burning. In the surface water samples of the Subarnarekha River, the ratios of Flur/(Flur+Pye)

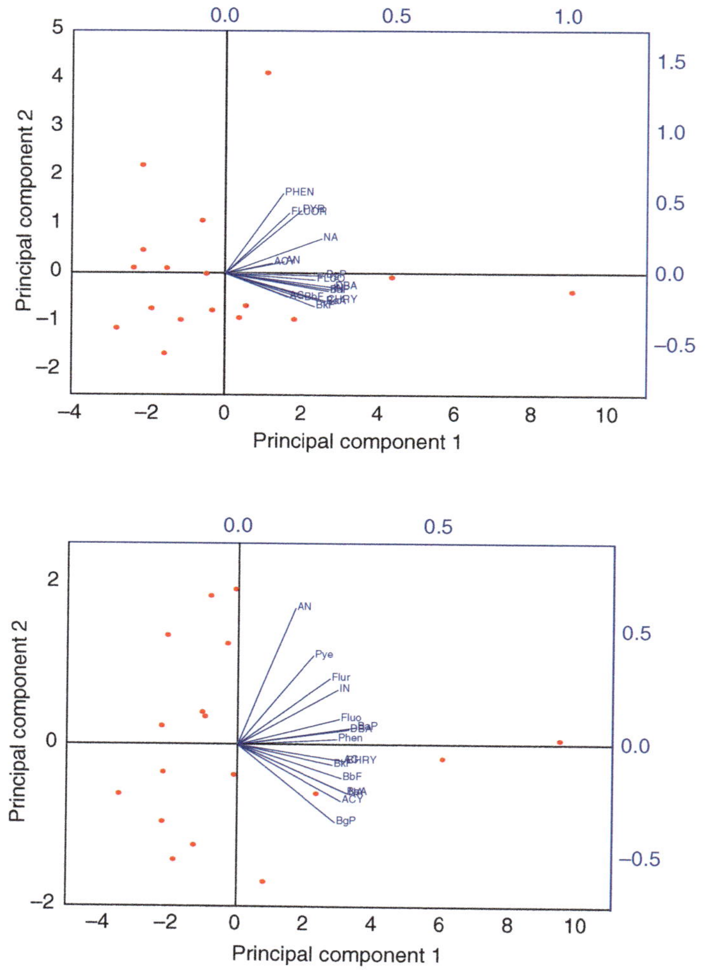

Figure 4.4 PCA biplots of the Subarnarekha and Kharkai Rivers.

ranged from 0.25 to 0.91 and 33% were below 0.4, indicating a petrogenic origin, and 67% were above 0.5, indicating coal, grass, and wood burning and the ratios of AN/(AN+Phen) ranged from 0.66 to 1 and 100% ratios were above 0.5, indicating coal, grass, and wood burning as the prime source of PAH contamination in surface water.

4.3.3.2 Principal Component Analysis

PCA is a statistical method used to simplify large data sets [14]. All data sets except the PAHs from the surface water samples were subjected to PCA analysis; due to low detection limits, surface water data sets were excluded. In Figure 4.4, the PCA biplots are shown for the Kharkai and Subarnarekha Rivers, where the bottom line shows PC-1, the top line shows the loading on PC-1, the left-most line shows PC-2, and the right line shows the loading on PC-2. From Figure 4.4 and Table 4.5, the PCA of both rivers can be explained. The Eigen values of 10.64, 1.20, and 1.01 were used for the Kharkai riverine sediments, and similarly, 8.70, 1.98, and 1.15 were used for the Subarnarekha riverine sediments to describe the source allocation, and the rotation study was avoided. The first three principal components explained about 74% of the total variability in the sediments of the Subarnarekha River and about 80% of the total variability in the sediments of the Kharkai River. In the Subarnarekha River, PC-1 explained a total variance of 54.42% (Table 4.5) and was loaded with higher 4-, 5-, and 6-ring PAHs besides the three rings (showing low loads), which are clearly shown in the PCA biplot of the Subarnarekha River (Figure 4.4). This is composed of both carcinogenic and noncarcinogenic PAHs. The sampling sites in the town areas of the Subarnarekha River are comprised of three-, four-, five-, and six-ring PAHs. In these areas, the diagnostic ratios of Flur/(Flur+Pye) and AN/(AN+Phen) suggested that the PAH contamination in the sediments was due to combustion and petrogenic sources. PC-2 and PC-3 showed a total variance of 12.39% and 7.25%, respectively. PC-2 was loaded with four-ring (Pye and Flur), three-ring (Phen), and two-ring (NA) PAHs. PC-3 was highly loaded with three-ring PAHs AC, ACY, and AN.

PCA biplot (Figure 4.4) and PC-1, PC-2, and PC-3 (Table 4.5) are shown for the Kharkai riverine sediments. PC-1 showed a total variance of 66.51% and was loaded with all PAHs. PC-2 showed a total variance of 7.52% and was loaded with high-molecular-weight PAHs (IN), four-member-ring PAHs (Pye and Flur), and three-ring PAHs (AN). High-molecular-weight PAHs were found in more concentration in some parts of the town area and semi-town area but were detected in low concentration in the rural area. PC-3 showed a total variance of 6.37% and was loaded with four-member-ring (CHRY), five-member-ring (BkF), and three-member-ring PAHs (Phen). These three PAHs were found in the town, semi-town, and rural area sampling sites of this river. From the diagnostic ratios of Flur/(Flur+Pye) and AN/(AN+Phen), it is clear that the PAH contamination in the riverine sediments was due to petrogenic and combustion sources. The loading of PAHs in PC-1 for both the riverine sediments was almost alike (Table 4.5) and was loaded with higher-ring PAHs. The higher-molecular-weight (HMW) PAHs correspond to pyrogenic sources and the PAHs such as DBA, BkF, and BgP are the typical pointers of automobile emission sources from road traffic and dust particles [42–44]. The loadings of PC-2 of the both riverine sediments were mostly by the three- and four-ring PAHs. The three-ring PAHs AN and Phen show a direct petrogenic origin [45]. The diagnostic ratios also show the petrogenic sources of PAHs for both rivers. PC-2 was also loaded with the four–ring PAHs such as Flur and Pye, and Fluoranthene is also an indicator of heavy-duty diesel combustion; also the PAHs such as four rings can be significantly emitted from light vehicles [46]. From the PCA and diagnostic ratios, combustion and petrogenic origins are the sources of PAH contamination.

4.3.3.3 Potential Ecosystem Risk Assessment

Generally, the concentration of individual PAHs in costal and surface water is 50 ng/l [47], other than highly industrially polluted rivers. The concentration of PAHs higher than 10 μg/l in water suggests the heavy contamination of water by

Table 4.5 Extracted eigenvectors and percentage of variance for PCA of Subarnarekha.

PAHs	PC-1	PC-2	PC-3
(a) Subarnarekha sediments			
NA	0.280	0.247	0.147
AC	0.182	−0.166	0.487
ACY	0.135	0.072	0.508
Fluo	0.263	−0.044	−0.190
Phen	0.167	0.563	−0.035
AN	0.170	0.082	0.464
Flur	0.185	0.426	−0.194
Pye	0.220	0.444	−0.105
BaA	0.290	−0.202	−0.254
CHRY	0.292	−0.192	−0.247
BbF	0.226	−0.174	0.136
BkF	0.260	−0.231	−0.055
BaP	0.300	−0.123	−0.095
DBA	0.316	−0.096	−0.028
IN	0.310	−0.114	−0.091
BgP	0.289	−0.015	0.115
Variance (%)	54.42	12.39	7.25
(b) Kharkai sediments			
NA	0.274	−0.215	−0.172
AC	0.261	−0.071	−0.124
ACY	0.258	−0.248	0.150
Fluo	0.253	0.113	0.089
Phen	0.247	0.024	0.338
AN	0.141	0.606	−0.019
Flur	0.229	0.290	−0.202
Pye	0.188	0.392	0.163
BaA	0.270	−0.213	0.007
CHRY	0.269	−0.073	0.403
BbF	0.259	−0.147	−0.247
BkF	0.236	−0.088	0.538
BaP	0.295	0.076	−0.039
DBA	0.277	0.066	−0.238
IN	0.249	0.240	−0.270
BgP	0.244	−0.343	−0.309
Variance (%)	66.51	7.52	6.37

PAHs [48]. The concentration of individual PAHs in surface water samples was less than 10 μg/l, except the three-ring PAH AC in both rivers. In this study, the total PAHs detected in sediments were in less concentration than in many other studies; similar levels also were detected in some studies (Table 4.2). Although PAHs do not display high toxicity to aquatic organisms, LMW PAHs exhibit higher lethal toxicity than HMW PAHs [49]. The concentration of PAHs in surface water in this study was lower than that in many other studies (Table 4.2). Since PAH concentration is less in water samples than the guideline values, it may be less harmful and toxic to aquatic organisms.

The effect-based guideline values were used to estimate the amount of contamination in sediments [50], such as effects range-low (ER-L) and effects range-median (ER-M). The probability of negative toxic effect is lower than 10% and higher than 50% suggests that the concentration of PAHs in sediments is less than ER-L and higher than ER-M values. The concentrations of individual PAHs recorded in this study ranged from (<DL) to 185 ng/g in the Kharkai River and (<DL) to 657 ng/g in the Subarnarekha River and were less than ER-L and ER-M values except for AC, ACY, and FLUO in both Kharkai and Subarnarekha Rivers, which were lower than ER-L values. From these values, it can be concluded that PAHs cause no biological effects due to contamination in the sediments of both rivers, but due to higher concentrations of AC (Acenaphthene), which was higher than 10 μg/l in surface water of both rivers, they may cause heavy contamination of water and harmful effects to aquatic life [48].

4.4 Conclusions

Our study was comprehensively carried out for the first time in the surface water and the riverine sediments of two rivers. The study found a considerable lower concentration of PAHs in both surface water and sediments than did other studies. In surface water, the concentration of PAHs ranged from (<DL) to 33 μg/l and in sediments from (<DL) to 657 ng/g. In the surface water samples of the Subarnarekha and Kharkai Rivers, the HMW PAHs were not detected or detected in a very low concentration, but the three-ring PAHs AC and ACY were dominant in both rivers. The concentration of 3-ring PAHs AC (Acenaphthene) was higher than the guideline value in the surface water samples of both rivers. Three-ring PAHs (and four-ring PAHs (3% and 18%) were abundant in the surface water samples of the Kharkai and Subarnarekha Rivers, respectively. In the sediments of both rivers, the HMW PAHs were found at all sampling sites, except some sites in rural areas of the Subarnarekha River having lower concentrations. The LMW PAHs were dominant nearly in all sediment samples, but the four-membered-ring PAHs Pye and three-membered-ring PAHs ACY were found in more concentrations in most of the sediment samples. The overall PAH concentration in the surface water and sediments of both rivers were in the order of town > semi-town > rural areas, and the samples collected in the industrial zones of both rivers were found to have PAHs in greater concentrations than the samples in the nonindustrial zones. The diagnostic ratios have shown that petrogenic and combustion origins were the sources of PAH contamination in the sediments of both rivers. PCA showed 74% of the total variance in Subarnarekha River and 80% of the total variance in Kharkai River and also confirmed that the source of contamination was petrogenic and combustion origins.

Acknowledgments

This work was financially supported (EEQ/2016/000504) by The SERB (Science and Engineering Research Board), India to provide the GC MS facilities to analyze the PAHs samples. Mr. Shrikanta Shankar Sethi thanks BRNS (Board of Research in Nuclear Science) for providing him financial support during the study.

References

1 An, N., Liu, S., Yin, Y. et al. (2016). Spatial distribution and sources of polycyclic aromatic hydrocarbons (PAHs) in the reservoir sediments after impoundment of Manwan dam in the middle of Lancang River, China. *Ecotoxicology.* 25 (6): 1072–1081.

2 Cheollee, B., Shimizu, Y., Matsuda, T., and Matsui, S. (2005). Characterization of polycyclic aromatic hydrocarbons (PAHs) in different size fractions in deposited road particles (DRPs) from Lake Biwa area, Japan. *Environ. Sci. Technol.* 39: 7402–7409.

3 Wang, L., Li, C., Jiao, B. et al. (2018). Halogenated and parent polycyclic aromatic hydrocarbons in vegetables: levels, dietary intakes, and health risk assessments. *Sci. Total Environ.* 616–617: 288–295.

4 Yao, Y., Huang, C.L., Wang, J.Z. et al. (2017). Significance of anthropogenic factors to freely dissolved polycyclic aromatic hydrocarbons in freshwater of China. *Environ. Sci. Technol.* 51 (15): 8304–8312.

5 Nadal, M., Schuhmacher, M., and Domingo, J.L. (2011). Long-term environmental monitoring of persistent organic pollutants and metals in a chemical/petrochemical area: human health risks. *Environ. Pollut.* 159: 1769–1777.

6 Simpson, C.D., Mosi, A.A., Cullen, W.R., and Reimer, K. (1996). Composition and distribution of polycyclic aromatic hydrocarbons in surficial marine sediments from Kitimat Harbour, Canada. *Sci. Total Environ.* 181: 265–278.

7 Zakaria, M.P., Takada, H., Tsutsumi, S. et al. (2002). Distribution of polycyclic aromatic hydrocarbons (PAHs) in rivers and estuaries in Malaysia: a widespread input of petrogenic PAHs. *Environ. Sci. Technol.* 36: 1907–1918.

8 Giesy, J.P., Tang, Z., and Zhao, X. (2016). Historical record of effects of human activities on absolute and relative concentrations of polycyclic aromatic hydrocarbons (PAHs) in Lake Chao, China. *Journal of Environmental Sciences.* 46: 1–4.

9 Wang, Z., Liu, Z., Xu, K. et al. (2014). Concentrations and sources of polycyclic aromatic hydrocarbons in surface coastal sediments of the northern Gulf of Mexico. *Geochemical Transactions* 15 (1): 2.

10 Zhang, J.D., Wang, Y.S., Cheng, H. et al. (2015). Distribution and sources of the polycyclic aromatic hydrocarbons in the sediments of the Pearl River estuary, China. *Ecotoxicology*. 24 (7–8): 1643–1649.
11 Hale, S.E., Lehmann, J., Rutherford, D. et al. (2012). Quantifying the total and bioavailable polycyclic aromatic hydrocarbons and dioxins in biochars. *Environ. Sci. Technol.* 46: 2830–2838.
12 Khuman, S.N., Chakraborty, P., Cincinelli, A. et al. (2018). Polycyclic aromatic hydrocarbons in surface waters and riverine sediments of the Hooghly and Brahmaputra Rivers in the Eastern and Northeastern India. *Sci. Total Environ.* 636: 751–160.
13 Tavakoly Sany, S., Hashim, R., Salleh, A. et al. (2014). Polycyclic aromatic hydrocarbons in coastal sediment of Klang Strait, Malaysia: distribution pattern, risk assessment and sources. *PLoS One* 9: e94907.
14 Doong, R. and Lin, Y. (2004). Characterization and distribution of polycyclic aromatic hydrocarbon contamination in surface sediment and water from Gao-ping River, Taiwan. *Water Research*. 38: 1773–1744.
15 Ghosh, U., Gillette, S., Luthy, R.G., and Zare, R.N. (2000). Microscale location, characterization, and association of polycyclic aromatic hydrocarbons on harbor sediment particles. *Environ. Sci. Technol.* 34: 1729–1736.
16 Soclo, H.H., Garrigues, P.H., and Ewald, M. (2000). Origin of polycyclic aromatic hydrocarbons (PAHs) in coastal marine sediments: case studies in Cotonou (Benin) and Aquitaine (France) areas. *Mar. Pollut. Bull.* 40: 387–396.
17 Shen, H., Huang, Y., Wang, R. et al. (2013). Global atmospheric emissions of polycyclic aromatic hydrocarbons from 1960 to 2008 and future predictions. *Environ. Sci. Technol.* 47 (12): 6415–6424.
18 Lang, C., Tao, S., Liu, W. et al. (2008). Atmospheric transport and outflow of polycyclic aromatic hydrocarbons from China. *Environ. Sci. Technol.* 42: 5196–5201.
19 Zhang, Y. and Tao, S. (2009). Global atmospheric emission inventory of polycyclic aromatic hydrocarbons (PAHs) for 2004. *Atmos. Environ.* 43: 812–819.
20 Douben, P.E. (2003). *PAHs: An Ecotoxicological Perspective*. Wiley.
21 Kumar, A., Ambade, B., Sankar, T.K. et al. (2020). Source identification and health risk assessment of atmospheric $PM_{2.5}$-bound polycyclic aromatic hydrocarbons in Jamshedpur, India. Sustain. *Cities Soc.* 52: 101–801.
22 Chen, Y., Zhu, L., and Zhou, R. (2007). Characterization and distribution of polycyclic aromatic hydrocarbon in surface water and sediment from Qiantang River, China. *J. Hazard. Mater.* 141: 148–155.
23 Chakraborty, P., Khuman, S.N., Selvaraj, S. et al. (2016). Polychlorinated biphenyls and organochlorine pesticides in River Brahmaputra From the outer Himalayan Range and River Hooghly emptying into the Bay of Bengal: occurrence, sources and ecotoxicological risk assessment. *Environ. Pollut.* 219: 998–1006.
24 Jianwang, L. and Xu, S. (2010). Polycyclic aromatic hydrocarbons in water, sediment, soil, and plants of the Aojiang river waterway in Wenzhou, China. *J. Hazard. Mater.* 173: 75–81.

25 Wei, G. and Mengchang, H. (2007). Distribution of poly cyclic aromatic hydrocarbons in water, suspended particulate matter and sediment from Daliao river watershed, China. *Chemosphere.* 68: 93–104.

26 Chen, Y., Zhu*, L., and Zhou, R. (2006). Characterization and distribution of polycyclic aromatic hydrocarbon in surface water and sediment from Qiantang River. *China. J. Hazard. Mater.* 141: 148–155.

27 Li, G., Xia, X., Yang, Z. et al. (2006). Distribution and sources of polycyclic aromatic hydrocarbons in the middle and lower reaches of Yellow River, China. *Environ. Pollut.* 144 (3): 985–993.

28 Huang, Z. (2004). Occurrence of PAHs, PCBs and organochlorine pesticides in the Tonghui River of Beijing, China. *Environ. Pollut.* 130: 249–261.

29 Yang, W., Wei, M., and Jian, X. (2014). Occurrence, distribution, environment risk assessment and source apportionment of PAHs in water and sediment of the Liaohe River Basin. *China, Bull. Environ. Contam. Toxicol.* 93: 744–751.

30 Farshid, K. and Amir, H.S. (2011). Determination of PAHs in water and sediment of the Kor River, Iran. *Middle-East Journal of scientific Research* 10 (1): 01–07.

31 Malik, A. and Verma, P. (2011). Distribution of PAHs in water and bed sediments of the Gomti River, India. *Environ. Monit. Assess.* 172: 529–545.

32 Xuesong, Z. and Jing, D. (2014). Spatial distribution and temporal trends of PAHs in water and sediment from Songhua River. *China. Environ. Geochem. Health.* 36: 131–143.

33 Jiries, A., Hussain, H., and Lintelmann, J. (2000). Determination of polycyclic aromatic hydrocarbons in wastewater, sediments, sludge and plants in Karak Province, Jordan. *Water, Air, Soil Pollut.* 121: 217–228.

34 Shi, Z., Tao, S., Pan, B. et al. (2005). Contamination of rivers in Tianjin, China by polycyclic aromatic hydrocarbons. *Environ. Pollut.* 134: 97–111.

35 Wetzel, D.L. and Van Vleet, E.S. (2004). Accumulation and distribution of petroleum hydrocarbons found in mussels (*Mytilus galloprovincialis*) in the canals of Venice, Italy. *Mar. Pollut. Bull.* 48: 927–936.

36 Agarwal, T., Khillare, P., Shridhar, V., and Ray, S. (2009). Pattern, sources and toxic potential of PAHs in the agricultural soils of Delhi, India. *J. Hazard. Mater.* 163: 1033–1039.

37 Cai, Q.Y., Mo, C.H., Li, Y.H. et al. (2007). Occurrence and assessment of polycyclic aromatic hydrocarbons in soils from vegetable fields of the Pearl River Delta, South China. *Chemosphere.* 68: 159–168.

38 Wang, Z., Chen, J., Qiao, X. et al. (2007). Distribution and sources of polycyclic aromatic hydrocarbons from urban to rural soils: a case study in Dalian, China. *Chemosphere.* 68: 965–971.

39 Wang, W., Simonich, S.L., Xue, M. et al. (2010). Concentrations, sources and spatial distribution of polycyclic aromatic hydrocarbons in soils from Beijing, Tianjin and surrounding areas, North China. *Environ. Pollut.* 158: 1245–1251.

40 Chakraborty, P., Zhang, G., Li, J. et al. (2016). Soil concentrations, occurrence, sources and estimation of air–soil exchange of polychlorinated biphenyls in Indian cities. *Sci. Total Environ.* 562: 928–934.

41 Tobiszewski, M. (2014). Application of diagnostic ratios of PAHs to characterize the pollution emission sources. Fifth International Conference on Environmental Science and Technology vol. 69. https://doi.org/10.7763/IPCBEE.2014.V69.9.

42 Chen, H.Y., Teng, Y.G., and Wang, J.S. (2012). Source apportionment of polycyclic aromatic hydrocarbons (PAHs) in surface sediments of the Rizhao coastal area (China) using diagnostic ratios and factor analysis with nonnegative constraints. *Sci. Total Environ.* 414: 293–300.

43 Liu, Y., Chen, L., Huang, Q.H. et al. (2009). Source apportionment of polycyclic aromatic hydrocarbons (PAHs) in surface sediments of the Huangpu River, Shanghai, China. *Sci. Tot Environ.* 407: 2931–2938.

44 Yu, X., Gao, Y., Wu, S. et al. (2006). Distribution of polycyclic aromatic hydrocarbons in soils at Guiyu area of China, affected by recycling of electronic waste using primitive technologies. *Chemosphere.* 65: 1500–1509.

45 Ramzi, A., Rahman, K.H., Gireeshkumar, T. et al. (2017). Dynamics of polycyclic aromatic hydrocarbons (PAHs) in surface sediments of Cochin estuary, India. *Mar. Pollut. Bull.* 114: 1081–1087.

46 Marr, L.C., Kirchstetter, T.W., Harley, R.A. et al. (1999). Characterization of polycyclic aromatic hydrocarbons in motor vehicle fuels and exhaust. *Emissions. Environ. Sci. Technol.* 33: 3091–3099.

47 WHO (1998). Polynuclear aromatic hydrocarbons. In: *Guidelines for Drinking-Water Quality*, 2e, vol. 2, 123–152. Geneva: Health Criteria and Other Supporting Information World Health Organization.

48 Chen, B.L., Xuan, X.D., Zhu, L.Z. et al. (2004). Distribution of polycyclic aromatic hydrocarbon in surface waters, sediment and soils of Hangzhou City, China. *Water Research.* 38: 3558–3568.

49 Law, R.J., Dawes, V.J., Woodhead, R.J., and Matthiessen, P. (1997). Polycyclic aromatic hydrocarbons (PAH) in seawater around England and Wales. *Mar. Pollut. Bull.* 34: 306–322.

50 Long, E.R., MacDonald, D.D., Smith, S.L., and Calder, F.D. (1995). Incidence of adverse biological effects within ranges of chemical concentrations in marine and estuarine sediments. *Environmental Management.* 19: 81–97.

5

Laboratory-on-a-Chip: A Multitasking Device

Mansi Mehta, Bhikhu More, Tanvi Tamakuwala, and Gaurav Shah

Veer Narmad South Gujarat University, Department of Biotechnology, Udhana Magdalla Road, Vesu, Surat 395007, India

5.1 Introduction

A lab-on-a-chip (LOC) is a miniaturized device that performs one or several laboratory functions on a single integrated circuit or chip of only few millimeters to centimeters to achieve automation and high-throughput screening such as DNA sequencing, biochemical detection· [1]. LOC devices are a subset of microelectromechanical system (MEMS) devices and sometimes called micro total analysis system (μTAS). LOC relies on two main technologies: microfluidics and molecular biology. Microfluidic technology allows manufacturing of millions of microchannels, each measuring mere micrometers, on a single chip that fits in a hand. Microchannels can handle fluids in quantities as low as a few picoliters. LOC also requires integrated pumps, electrodes, valves, electrical fields, and electronics to become complete LOC diagnostic systems. LOC indicates the scaling of single or multiple lab processes down to chip format, while μTAS uses integration of total sequence of lab processes to perform chemical analysis. The term LOC was introduced when μTAS technologies were applicable for more than only analysis purposes. Research on LOC focuses on several applications, including DNA analysis, human diagnostics, and also, to a lesser extent, synthesis of chemicals [1].

5.1.1 LOC in Multiplexing Microfabricated Devices

The term multiplexing in a microfabricated device means multiple applications or technologies are applied on a sample or multiple samples are applied on analytical techniques for rapid detection and sequencing process. Multiplex systems with around 12 and up to 96 samples were developed for rapid analysis of polymerase chain reaction (PCR) products, which share common separation channels as well as buffer reservoirs [2]. A well-developed laser-based fluorescence detector scans multiple channels and separates DNA fragments.

Miniaturized Analytical Devices: Materials and Technology, First Edition.
Edited by Suresh Kumar Kailasa and Chaudhery Mustansar Hussain.

5.1.2 LOC in Integration

Miniaturization of LOC offers a considerable reduction in cost and eliminates manual manipulations along with unintentional human errors and contamination from external sources. Integration systems consist of a controlled pressure system, temperature sensor, and fluorescence detector. At large-scale production, integration system may decrease cost with a new version of diagnostic purpose [2].

5.2 History

Microtechnologies and microfabrication were born in the early 1950s when researchers used photographic technologies to create photolithography to fabricate microsized transistors, which led to the advent of modern information technologies and telecommunications. In the 1960s, many researchers used these technologies to fabricate micromechanical structures called MEMS, which enable the production of miniaturized accelerometers for use in daily life objects such as airbags and smartphones. Using these fabrication techniques, first real LOC was made in 1979 at Stanford University by S. C. Terry for gas chromatography [3, 4]. In the 1980s, major research related to LOC began, which led to development of microfluidic and microfabrication processes, also known as soft lithography for the production of polymer chips. In the 1990s, researchers began to use microfluidics and attempted to miniaturize biochemical operations such as PCR, electrophoresis, DNA microarray [5].

Integrated research requires various steps from sample collection to final analysis, on a single chip called the μTAS. These μTAS concepts demonstrated that integration of pretreatment steps, usually done at lab scale, could extend a simple sensor functionality toward a complete laboratory analysis including additional cleaning and separation steps. A big boost in research came from the military especially from DARPA (Defense Advanced Research Projects Agency) of U.S. Department of Defense for their interest in biological and chemical warfare agent detectors. For some applications, LOC not only has the capacity of integration and parallelization but also demonstrates superior performance over conventional technologies. Integration of PCR onto LOC, which is known as micro-PCR, allows amplification of DNA ten times faster than conventional systems. Although the application of LOC is still novel and modest, there is a growing interest of companies and applied research groups in different fields such as chemical analysis, medical diagnostics, environmental monitoring, and cellomics as well as synthetic chemistry (rapid screening and microreactors for pharmaceuticals).

5.3 LOC Manufacturing Technologies

Most commonly LOC uses microfluidic manufacturing technologies, which enable the integration of microchannels in micrometer-scale sizes. Depending on the type of application, various polymers are used to construct devices [6].

5.3.1 PDMS (Polydimethylsiloxane)

It is a flexible, cheap, and transparent elastomer largely used for fabrication of polydimethylsiloxane (PDMS) LOC by casting. It is easy to integrate quake microvalves for fast switch and air permeability for cell culture studies when the LOC is made up of PDMS. As PDMS absorbs hydrophobic molecules and the material is subject to aging, it is hard to integrate electrodes into a PDMS chip, which limits its uses in industrial production. PDMS is not compatible with high-throughput chip fabrication processes such as injection molding or hot embossing.

5.3.2 Thermopolymers

Thermoplastics are a good choice for fabrication of LOC because they are transparent and compatible with micrometer-sized devices. Thermoplastics are chemically more inert than PDMS and expensive to implement. Results obtained using thermoplastic LOC are very much useful for researchers because the integration of microelectrodes in thermoplastic LOC is possible. Thermoplastic materials can be good candidates for industrialization of some LOC.

5.3.3 Glass

Characteristics such as transparency, compatibility with micrometer-sized machines, chemically inert, availability of wide chemical surface treatment, and reproducible electrode integration make glass a suitable LOC candidate. Strong knowledge of microfabrication and clean rooms are prerequisites for glass to effectively function as an LOC, which limit its use and make it unavailable to all research laboratories.

5.3.4 Silicon

As microtechnologies are based on microfabrication of silicon chips, it is the first and common choice of research scientists for the construction of LOC. First LOC was also made of silicon. However, nowadays researchers do not prefer silicon mainly because it is expensive, is optically nontransparent, and requires clean room as well as sound knowledge of microfabrication. Also, because of its semiconductivity, silicon requires a high voltage, which creates problems in the construction of LOC. Although silicon is considered an obsolete candidate for the industrialization, it may still remain a relevant choice for industrialization of some demanding LOC applications. However, construction of LOC requires high precision of silicon microfabrication, maturity of the process, investments, and the ability to integrate any kind of microelectrode and even electronics on a single chip.

5.3.5 Paper

Paper-based LOC devices are relatively cheaper and have strong outcomes for especially ultra-low-cost demanding applications. According to G. Whiteside, the most famous microfluidic researcher, paper-based LOC devices may find their market in the future and could open up the field of diagnostics especially to make it accessible to lower-income and limited-resource populations.

5.4 Advantages of LOC Compared to Conventional Technologies

There are many advantages associated with LOC compared to conventional technologies, which are listed as follows:

5.4.1 Low Cost

As microtechnologies are used to fabricate LOC, they will decrease the cost of analysis much like they decreased the cost of computed calculation. Moreover, same chips can be used to perform numerous tests and thus reducing the cost of each individual analysis [7].

5.4.2 Easy Use

LOC allows the integration of a large number of operations within a small volume. A chip of few square millimeters to centimeters in size coupled with a machine performs same functions as in analytical laboratories.

5.4.3 Reduction of Human Error

Compared to conventional analytical processes performed in laboratories, automatic diagnosis by LOC greatly reduces the risk of human error as human handling is reduced during operation.

5.4.4 Less Sample Requirement

LOC system-based diagnoses require a very small as low as picoliter amount of sample for analysis.

5.4.5 High Parallelization

LOC technology allows multiple analyses to be performed simultaneously on the same chip. This allows doctors to diagnose specific illnesses during the time of a consultation or to develop treatment strategies in order to prescribe quickly and effectively best-suited antibiotics.

5.4.6 Fast Response

Diffusion of chemicals as well as heat at a smaller scale is faster. Modifications are possible such as changes in temperature or mixing of chemicals simply by diffusion within a shorter period of time, which completes the reaction in short time.

5.4.7 Process Control and Sensitivity

Due to fast reactivity at the microscale, it is possible to control the environment of a chemical reaction in LOC, which leads to more controlled results as well.

5.4.8 Cost Effective

As the cost of production is low, it reduces diagnostic costs, and it also requires low energy consumption, LOC can be used in outdoor environments for air and water monitoring without human intervention. As a result, LOC technology will make modern medicine more accessible in developing countries at reasonable costs [6].

5.5 Limitations of LOC Compared to Conventional Technologies

5.5.1 Industrialization

LOC technologies are not yet ready for industrialization as it is not still certain that which fabrication technologies will become standard.

5.5.2 Signal/Noise Ratio

For some applications, miniaturization increases the signal-to-noise ratio, which results in poorer results compared to that of conventional techniques.

5.5.3 Additional Requirements for Efficient Work

Despite of their small size and powerful nature, LOC devices require specific machinery such as electronics and flow control systems to work properly. It is necessary to maintain a specific precise system to inject, split, and control the positioning of samples for proper working of LOC devices. Using additional devices for efficient working of LOC can increase the final size as well as cost and sometimes limit the performance.

5.5.4 Ethics

LOC-based diagnosis for infectious diseases at home by untrained people creates fear in them. Also, it can lead to sequence DNA of other persons from a single drop of saliva, which will create ethical problems.

5.6 Applications of LOC in Different Fields

5.6.1 LOC in Genomics

Although Human Genome Project was successfully completed with the prokaryotic species genome and eukaryotic species genome, current technologies cannot accelerate the processes due to reasonable tasks and cost. The first step was miniaturization of a DNA sequencing technique, which was achieved by replacing a gel slab with gel-filled capillaries having sustainable efficiency of high voltage and high surface-to-volume ratio. Nowadays, hundreds of multicapillary instruments have been used in large-scale DNA sequencing. DNA sequencing is a technique that shows the arrangement of nucleotides in a specific manner that may help in DNA-based research.

Multiplexed CGE (capillary gel electrophoresis) with microfabrication at PCR level transforms current tedious labor-intensive technology to fully automation [2].

5.6.2 LOC in Post-Genome Era

Genomics-based diagnosis, transcriptome, metabolome, and proteome will use new LOC systems to target drug, protein identification, drug discovery, development of microfluidic systems, and various technical challenges, which will give valuable platforms for analytical chemistry [2]. Microfluidic systems offer an integrated part of sample loading to detection advantage over large-scale analysis. Steady-state signals reach biosensors faster and control the environmental background as well as the diffusion profile of gas and various metabolites [8]. Using LOC technologies including electroporation on single cell stimulates cell lysis and release of proteins, which can be further studied by confocal fluorescent microscopy. Macromolecules can be separated, isolated, and positioned with microfluidic systems without disruption.

5.6.3 LOC in Immunological Assay

Immunological techniques are easy and feasible tools in health-care area. Even novices can run diagnostic kits and get results in short time at hospital, laboratory, or home itself. Rapid detection by point-of-care immunological diagnostic kits is meaningful in the treatment of acute diseases where diagnosis is considered a primary requirement. In point-of-care immunological diagnostic techniques, an antibody or specific enzyme-based analyte is coated on the membrane paper. It is completely flexible according to the requirement of health-care diagnosis [9].

Once the analyte or specific antibody is coated on the respective membrane, the sample in micro quantity is applied. The sample moves by capillary action to the membrane and binds to a specific antibody or analyte. Generally, point-of-care diagnostic techniques are qualitative as they give a colored mark or area on the membrane, but when connected to a specific reader or detector, results can also be obtained quantitatively. Certain factors affect the whole reaction such as type of membrane, type of sample, mobility rate, specificity of enzyme, and type of detector, which are used for the diagnosis. As per the current market status, blood sugar monitoring devices and home pregnancy rapid diagnostic kits have been successfully developed and accepted by the public due to rapid detection and user-friendliness [9].

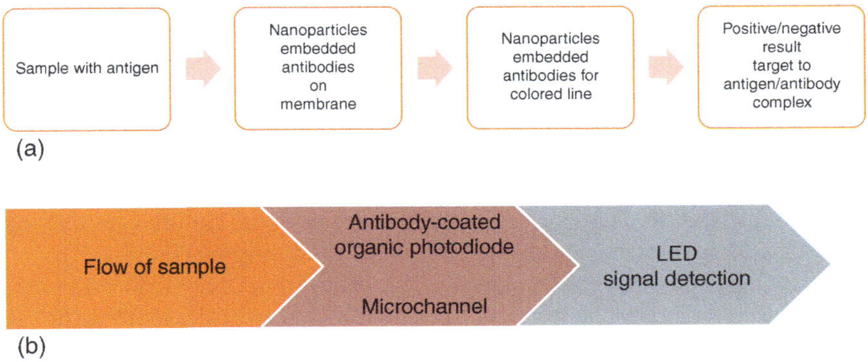

Figure 5.1 (a) Diagrammatic representation of LFA. (b) Diagrammatic representation of ELISA in LOC. Source: (b) Based on Yoon and Kim [10].

A simple format of immunological assay is lateral flow assay (LFA) or lateral flow immunochromatographic assay (LFIA) in which a membrane is embedded with specific antibodies. Through capillary action, the sample is loaded and pathogen-specific antigens are detected by giving a colored line. Gold/fluorescent nanoparticles are generally used and embedded with antibodies on the membrane and give a clear line of positive results as described in Figure 5.1a.

Example of LFA is pregnancy detection kit that can be easily used by nontechnical persons and enables the users to interpret results without any difficulty.

Few instances where sample requirement is very high will require amplification processes such as PCR, due to which , LFA is not popular in applications where nucleic acid is needed. To overcome this shortcoming, reflectometer with latest software and a cell phone camera gives sufficient sensitivity and makes it cost effective [10].

Enzyme-linked immunosorbent assay (ELISA) is an immunosensor technique that uses a microtiter plate with microwells for reading. Microwells have been replaced by microchannels on a single plate to create microchips, which is called ELISA–LOC technique. The antibodies are immobilized on the walls of microchannels and run for optical detection. From the other side of the plate, light-based sensors are used to capture the signals.

Figure 5.1b shows innovative ELISA–LOC, which consists of microchannels instead of microwells where process operates on complete automation from sample loading to immunosensor reading taken by LED [10].

Latex immunoagglutination assay (LIA) is another form of immunological assay, which is rinse-free and based on the micro-agglutination reaction between antigens and antibodies in a liquid medium and detected visually. LOC can detect micro- or nanoparticles without precipitation. In this type of immunological assay, a known concentration of antigen standards is used to compare and the analyte is detected from the sample [10].

5.6.4 Organ-on-a-Chip

In tissue culture technology, due to lack of three-dimensional surface for cell-to-cell interactions, there is a huge demand for microfluidic cell culture platforms. Microfluidic cell culture platforms provide precise microenvironment for cellular interactions and provide in vivo conditions for drug testing. For a specific organ, software is designed by including different parameters, and microfabrication is done. After cell culture is performed, tissue culture or organ culture is mimicked on a biochip for various biological, biophysiological, and drug testing [9].

Skin-on-chip (Combination of Cutaneous Vasculature and Dermal Blood Vessels) is developed as it is highly important for cosmetic, pharmaceutical, and toxicological fields. Microfluidic systems are used to provide different media and reagents to target a specific site with a fixed amount in particular time. Biosensors are incorporated to check the responses in real time.

A three-dimensional microfluidic model has been also developed for gut-on-a-chip. Two lumens separated by a porous membrane have been developed to check mechanical movement of fluids, epithelial behavior of gut cells, and the role of various elements in cell morphology. Gut-on-a-chip model is an efficient platform to mimic in vivo models for drug testing and analysis.

The pluripotent cells developed on a microfluidic chip provide a platform to check and cure neurodegenerative diseases and drug testing for brain-on-a-chip models. Proximal tubular epithelial cells are developed on a microfluidic perfusion chip to check selective permeability of glomerulus cells and study hydrodynamic pressure with shear stress to emulate brain and kidney diseases, respectively. Because all organs in the body are very crucial to routine and basic metabolic activities, co-culture models have been developed recently rather than a single organ chip. Body-on-a-chip (whole miniature body on a chip) models will help recent innovative technologies to develop more applications and to yield better outcome [9].

5.6.5 LOC in Food Safety

Due to the lack of real sensors for foodborne pathogens, it takes long time between the first outbreak and identification of pathogens in. Lab-on-a-chip technology is a perfect medium for portable and real-time sensitivity data that integrates a number of platforms in centimeters or millimeters in size. Microfluidic systems involve a small area that enables mixing, dilution, electrophoresis, staining, and detection on a single chip material. Sometimes, in food samples, pathogens found on food surface are also washed with washing buffer in pretreatment steps. Hence, it is advisable to eliminate complicated steps and use small and simple equipment such as mini centrifuges to incorporate LOC materials, which is called lab-on-CD for single-step centrifugation process shown in Figure 5.2 [10].

In a lab-on-CD technique (Figure 5.2), microchannels are loaded on a compact disc (CD) from its outer surface to the inner wall of the disc. Reagents, chemicals, and buffers are channelized through the microchannels by applying a centrifugal field to the disc. Lab-on-CD microcentrifuges can be handled by nonexpert users without any operation difficulty [10].

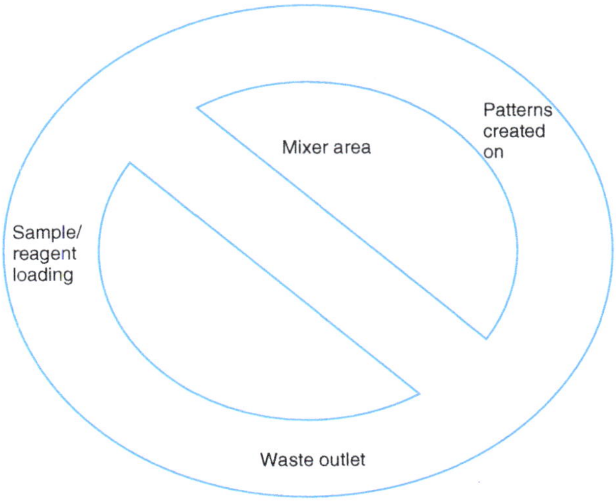

Figure 5.2 Lab-on-CD.

Y- or T-type junction channels for sample inlets are preferred for complete automation in mixing, transport, and separation steps by applying voltage or syringe pressure. Incorporation of micropump or microvalve fabrication on chip enhances the efficacy of overall procedure.

5.6.6 LOC in Environmental Monitoring

Chips integrated with electronic equipment are used to track environmental pollutants, monitor nutrients in food or water, and analysis of body fluids and identification of explosives. Hydrodynamic chromatographic chip (HDC) is one type of LOC technique similar to size exclusion chromatography in which analytes move on thin separated microchannel layers with different velocities. Flow injection analysis (FIA) system is one kind of miniaturization form of chip that measures ammonia in water sample. FIA is a very simple and versatile method based on complete chemical automation analysis from injection of sample flow in a carrier channel to detect downstream assay. Instead of different fluidic systems, single-flow channel is connected where the sample and multiple reagents are added step by step in the mixer area for ammonia detection with reference to a specific wavelength [11].

5.6.7 LOC in Cancer Diagnosis

LOC or μTAS integrated on silicon chips help in evaluating and diagnosing various diseases with controlled drug delivery and monitoring. Microfluidic systems would be able to analyze a small portion of biopsy or a small sample of saliva, blood sample, lung aspirations, urine, breast fluid and provide readouts of the presence of specific pathogens, toxins, or cancer markers. The major benefits of using microfluidic systems in the medical field are automation, rapid processing time, reduced amounts of agents, reproducibility, minimal risk of handling, minimally trained personnel, and low cost.

A fully integrated LOC for cancer diagnosis requires cell sorting between cancer and noncancerous cells, cell lysis, nucleic acid isolation, and multiplex RT-PCR on a single credit-card-sized chip with full automation. The collector fitted in a single credit-card-sized chip made up of a microfluidic system called cassette is inert with magnetic beads conjugated with antibodies would assist in the removal of lymphocytic cells. Glycoprotein surfaces presenting cancer cells are than captured and targeted to specific antibodies for detection. Detection of cancer cells indicates the initiation of disease. Isolated and captured cells would be lysed, and RNA would be extracted and functionalized with poly T tail to determine gene-specific transcripts [12].

The transcription profiles of cancer cells can be compared with standards to determine the type of cancer cells as well as to differentiate between cancer and normal cells. A whole cancer diagnostic format on a chip is shown in Figure 5.3 for better understanding. Microfluidic devices need more advanced developments in cancer diagnosis and monitoring techniques.

5.6.8 LOC in COVID-19 Detection

In the current pandemic era of COVID-19, diagnosis at the early stage of infection is the primary goal for health agencies across the world. Clinical and biological manifestations of SARS-CoV-2/COVID-19 make it very difficult to design a diagnostic tool in rapid time with good sensitivity. Miniaturization of any microfluidic platform will reduce cost and provide automation for viral detection [13]. Nucleic acid or protein parts of the viral capsids are used as biomarkers for detection.

In LOC tool application, the nucleic and protein parts are used for the detection of specific viral diseases. Microfluidic LOC technology provides complete

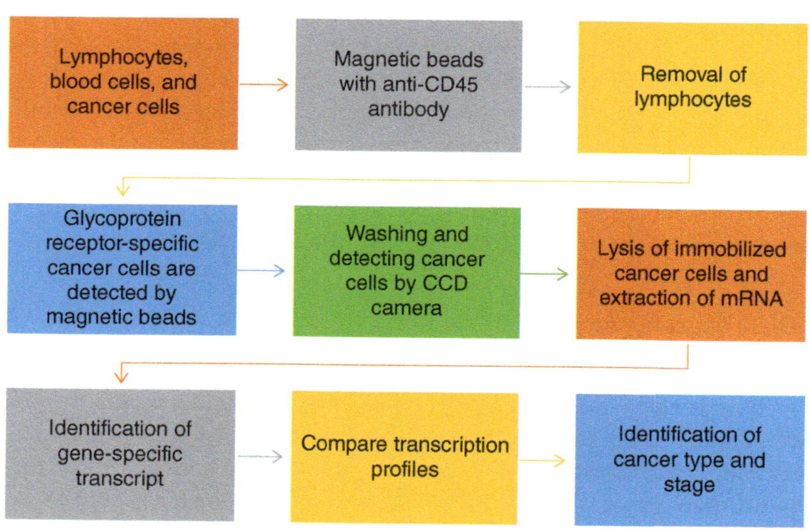

Figure 5.3 Cancer diagnostics format on a chip.

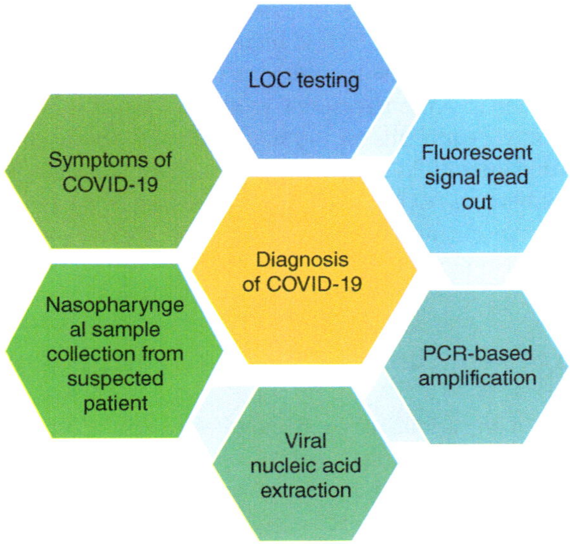

Figure 5.4 Diagnosis of COVID-19.

automation from sample collection to detection. The system is capable of performing viral DNA extraction, purification, RT-PCR-based amplification, and finally fluorescence-based signal detection on a single microfluidic chip, which is shown in Figure 5.4. On the single chip, chemicals, reagents, and washing buffers are added, due to which unskilled persons find it easy to operate. Recently, CRISPER-based fluid preparation and detection method is designed to avoid complex sample preparation and amplification process. However, sensitivity and specificity could not be achieved without amplification step in viral nucleic acid detection [14].

Magnetic nanoparticles coated with poly-(amino-ester) carboxyl group are efficient in RNA extraction process. They provide very strong bonding between RNA viral particles and magnetic nanoparticles and can be directly introduced into the RT-PCR step without any elution step. This method is more efficient than the traditional silica-based spin column method in which the sample requires lysis and release of nucleic acid via multiple centrifugation steps [14].

The applicability of microfluidic biochips is that they can be integrated into smartphones and allow simple operation, in-built optic and, more important, confidentiality of a patient.

In the future, new advanced technologies will emerge to improve the public health in a pandemic and to fulfill the increasing public demand [13].

5.7 Present Challenges

Many LOC devices are already available on the market that provide easy, time-saving, and cost-effective diagnosis such as blood glucose monitoring, pathogen detection, pregnancy test. However, for industrialization to increase market value, development of specific fabrication processes and specific surface treatment and flow control systems are needed. Research has been focused on to improve technologies for various applications such as cell separation [15], DNA sequencing, micro-PCR to increase PCR efficiency and to reduce false positive and false negative results, microreactors to perform chemical transformations. Still further advanced research is needed to develop LOC devices that can perform multiple functions on the same chip and can detect hundreds of pathogens in the same microfluidic cartridge. Research on increasing the ease of use of LOC such as enabling use of basic LOC functions using smartphones for cholesterol testing [16], anemia diagnosis [17], cardiovascular disease monitoring [16], and ELISA assays [18] is underway.

5.8 Conclusion and Future Perspectives

Based on various research and effective use of different products which are already available on the market, LOC concept will definitely bring changes to diagnostic testing in the near future. Many lab-on-chip devices are commercialized for key applications such as HIV detection, glucose monitoring, pregnancy test, detection of typhoid or syphilis. Research in this technology focuses on DNA analysis, chemical synthesis, environmental monitoring including testing of water as well as food quality, and human diagnostics. LOC devices automate laboratory processes such as sample transport, dispensing, mixing and thus significantly reducing time for laboratory tests. Challenges for industrial research include the use of the same chip for many individual operations in order to decrease cost, to increase ergonomics, to increase speed of analysis, and to develop high-throughput diagnostics.

References

1 Volpatti, L.R. and Yetisen, A.K. (2014). Commercialization of microfluidic devices. *Trends Biotechnol* 32 (7): 347–350. https://doi.org/10.1016/j.tibtech.2014.04.010.

2 Figeys, D. and Pinto, D.A. (2000). Lab-on-a-chip: a revolution in biological and medical sciences. *Analytical Chemistry* 72 (9): 330A–335A.

3 Terry, S.C., Herman, J.H., and Angell, J.B. (1979). A gas chromatographic air analyzer fabricated on a silicon wafer. *IEEE Trans Electron Devices.* 26 (12): 1880–1886.

4 Angell, J.B., Terry, S.C., and Barth, P.W. (1997). Silicon micromechanical devices. Micromechanics MEMS Class Semin Pap to 1990. *Scientific American* 248 (4): 38–49.

5 Yilmaz, B. and Yilmaz, F. (2018). Lab-on-a-Chip technology and its applications. In: *Omics Technologies and Bio-Engineering: Towards Improving Quality of Life*, vol. 1, 145–153. Elsevier Inc. http://dx.doi.org/10.1016/B978-0-12-804659-3.00008-7.

6 Casquillas, G. and Houssin, T. (2020). Introduction to lab on a chip 2020: Review, history and future, Microfluidics reviews.

7 Medical Device Developments (2014). Where the chips fall: lab-on-chip diagnostics. http://www.medicaldevice-developments.com/features/featurewhere-the-chips-fall-lab-on-chip-diagnostics-4212684/ (accessed 10 April 2014).
8 Sedgwick, H., Caron, F., Monaghan, P.B. et al. (2008). Lab-on-a-chip technologies for proteomic analysis from isolated cells. *J R Soc Interface.* 5 (SUPPL.2): S123–S130.
9 Azizipour, N., Avazpour, R., Rosenzweig, D.H., and Sawan, M. (2020). Evolution of biochip technology: a review from lab-on-a-chip to organ-on-a-chip. *Micromachines (Basel)* 11: 599.
10 Yoon, J. and Kim, B. (2012). Lab-on-a-chip pathogen sensors for food safety. *Sensors (Basel)* 12: 10713–10741.
11 Gardeniers, J.G.E. and van den Berg, A. (2004). Lab-on-a-chip systems for biomedical and environmental monitoring. *Anal Bioanal Chem* 378: 1700–1703.
12 Ziober, B.L., Mauk, M.G., Falls, E.M. et al. (2008). Lab-on-a-chip for oral cancer screening and diagnosis. *Head Neck* 30: 111–121.
13 Tymm, C., Zhou, J., Tadimety, A. et al. (2020). Scalable COVID-19 detection enabled by lab-on-chip biosensors. *Cell Mol Bioeng.* 13 (4): 313–329.
14 Qin, Z., Peng, R., Baravik, I.K., and Liu, X. (2020). Fighting COVID-19: integrated micro- and nanosystems for viral infection diagnostics. *Matter* 3 (3): 628–651. https://doi.org/10.1016/j.matt.2020.06.015.
15 Shields, C.W., Ohiri, K.A., Szott, L.M., and López, G.P. (2017). Translating microfluidics: cell separation technologies and their barriers to commercialization. *Cytom Part B – Clin Cytom.* 92 (2): 115–125.
16 Oncescu, V., Mancuso, M., and Erickson, D. (2014). Cholesterol testing on a smartphone. *Lab Chip.* 14 (4): 759–763.
17 Plevniak K, Campbell M, and He M. (2016) 3D printed microfluidic mixer for point-of-care diagnosis of anemia. Proceedings of the Annual International Conference IEEE Engineering in Medicine and Biology Society EMBS (2016 October), 267–270.
18 Chen, A., Wang, R., Bever, C.R.S. et al. (2014). Smartphone-interfaced lab-on-a-chip devices for field-deployable enzyme-linked immunosorbent assay. *Biomicrofluidics* 8 (6): 1–11. http://dx.doi.org/10.1063/1.4901348.

6

Microscopic Tools for Cell Imaging

Parveen Parasar[1] *and Vivek K. Singh*[2]

[1] *Henry Ford Hospital, Henry Ford Health Systems, Department of Neurology Research, 2799 West Grand Blvd, Detroit 48202, MI, USA*
[2] *Junagadh Agricultural University, Department of Veterinary Physiology and Biochemistry, Junagadh, Gujarat 362001, India*

6.1 Introduction

Live cell imaging has emerged as a preferred method for cell biology research and drug screening. Many cellular phenomena of interest are highly dynamic processes, and much information is lost when cells are fixed. Furthermore, fixation makes cells prone to artifacts in both localization patterns and gross morphology. With the emergence of live imaging techniques and tools, scientists now can visualize processes and molecules in vivo in cells, tissues, and whole organisms greatly enhancing our understanding of cellular processes and developmental events.

The first step in designing a live cell imaging experiment is to determine and select a probe that addresses the question of interest. Deciding which fluorescent probe should be used depends on the microscope setup and researchers' needs. In determining and selecting these probes, researchers need to consider parameters such as autofluorescence, source of illumination, lens elements, filters, and detectors.

The use of fluorescent probes in microscopy has facilitated a detailed study of cellular components and complex architecture of cells. Discovery of green fluorescent protein (GFP) in the 1990s allowed transfection of cells with fluorescent proteins, marking a new era of microscopy in the study of living cells. However, there are some technical difficulties in imaging of living cells, which can be overcome by confocal and widefield deconvolution microscopes. To obtain meaningful data, general factors should be considered such as appropriately preparing (label) the cells, maintaining the prepared sample in a condition that will support normal cell and tissue health before imaging, and analyzing the cells with sufficient temporal and spatial resolution without perturbing and compromising the cell and tissue health.

Another advancement in live-cell imaging was the discovery of confocal microscopy and its applications in the 1990s to characterize developmental changes in intact animals by imaging neuronal axons and their growth cones in developing

Miniaturized Analytical Devices: Materials and Technology, First Edition.
Edited by Suresh Kumar Kailasa and Chaudhery Mustansar Hussain.

brains of tadpoles [1]. Since then, confocal microscopy has been widely used in myriads of biological preparations.

For many researchers, phototoxicity and photobleaching are serious limitations in their observations. Therefore, all steps of imaging should be optimized to achieve maximum signal-to-noise (S/N) ratio, maximum spatial and temporal resolution, and minimal phototoxic effects. The art of successful live-cell imaging is that finding the balance between image quality and cell viability. In this chapter, we review the different types of microscopic tools for live-cell imaging, general considerations for preparing the samples for those imaging techniques, and some limitations still present and the strategies to overcome those limitations.

6.2 Microscopy – History and Development

Early microscope was invented by Zacharias Janssen in 1590 by placing multiple lenses in a tube. Galileo Galilei developed a compound microscope in 1609 by using a convex and a concave lens. The term "microscope" was first used by Giovanni Faber in 1625. Later in 1665, Robert Hooke published *Micrographia* and coined the term "cells" describing the tissues. He included images of the honeycomb structure of a cork in his book. He used a simple, single-lens microscope illuminated by a candle. First live-cell imaging was performed by Antonie van Leeuwenhoek in 1676 with his single-lens simple microscope. He first described bacteria and cells and examined blood, yeast, and insects [2].

A major breakthrough of much more enhanced and improved resolution of visualization was achieved by the invention of electron microscopy in 1931 by Ernst Ruska and Max Knoll referred to as transmission electron microscopy (TEM); however, this technique was not used with live cells. Ernst Ruska was awarded the Nobel Prize in physics for his eminent work on electron microscopy. EM relies on electrons to view an object and not on light. He developed, in 1942, the first scanning electron microscope (SEM), which transmits a beam of electrons across the surface of a specimen.

Imaging of transparent samples was made possible by development of phase-contrast illumination by Frits Zernike in 1932. This technique uses interference of light and not absorption of light, thus enabling imaging of transparent cells without their staining.

The principle of confocal imaging was developed by Marvin Minsky in 1957, which uses the scanning point of light and gives slightly higher resolution than light microscopy. Commonly known as confocal laser scanning microscope, it uses multiple mirrors (typically 2 or 3 scanning linearly along the x- and y-axes) to scan the laser across the sample and "descan" the image across a fixed pinhole and detector. Confocal microscopy overcomes some limitations of traditional widefield fluorescence microscopy.

GFP was discovered in the jelly fish *Aequorea victoria,* which exhibits green fluorescence when exposed to blue light. GFP was successfully cloned in 1992, and GFP and its derivatives were widely used for labeling the cells for fluorescence microscopy.

Super-resolution microscopy was pioneered by Stefan Hell in 1993, and it bestowed opti imaging world with techniques which improve the resolution to a much larger factor. Laser scanning fluorescence microscope integrated with super-resolution techniques led to development of 4i microscope with an improved axial resolution. In 2014, Stefan Hell, Eric Betzig, and William Moerner received the Nobel Prize in chemistry for their contribution to development of super-resolved fluorescence microscopy that can resolve a matter smaller than 0.2 micrometer (μM) [3].

6.2.1 Live-cell Imaging Microscopy

6.2.2 Fluorescent Microscopy

GFP was discovered in the course of bioluminescence studies on the jellyfish *A. victoria*. This protein is a 28kD protein and emits green light upon stimulation by blue or ultraviolet (UV) light. Since then, GFP has been used as a marker of gene activity and to label proteins and subcellular compartments within living cells. Mutagenesis studies produced yellow- and blue-shifted derivatives of GFP with variations optimized for biological experimentations allowing other studies in conjunction. This significant discovery of GFP earned the Nobel Prize in chemistry for its discoverers Osamu Shimorura, Martin Chalfie, and Roger Tsien in 2008.

6.2.2.1 Principle

Fluorescence microscopy relies on illuminating fluorescently labeled proteins or intracellular molecules with a wavelength of light close to the peak of the excitation spectrum of fluorophores and detecting light emitted at a longer wavelength. Fluorescence microscope uses fluorescence to generate an image whether in a simple widefield epifluorescence microscope or a complex design such as confocal microscope, which uses optical sectioning to create high-resolution images of specimens.

A general optical train of a fluorescence microscope includes a light source (a xenon arc lamp or mercury vapor lamp or more advanced high-power light emitting diodes (LED) or LASER), an excitation filter, a dichroic mirror, and an emission filter. The filters are selected based on the spectral emission and excitation characteristics of the fluorophore used to label proteins or cellular molecules. The objective lens should be chosen based on the magnification needed for the experimental question. For low fluorescence signals to be recorded, the brightest possible objective lens should be used. A detector or camera is used to detect and record the images. Figure 6.1 shows a general sketch of the parts of a fluorescence microscope.

6.2.2.2 Photobleaching

High-intensity light is toxic and damaging to cells especially in the near UV range, which damages DNA molecules. Each time a fluorophore is irradiated by light, a portion of the fluorophore is irreversibly destroyed. With repeated light exposure, the fluorescence signal gets reduced and free radicals and breakdown products are generated. Photobleaching and photodamage of the cells can be reduced by minimizing

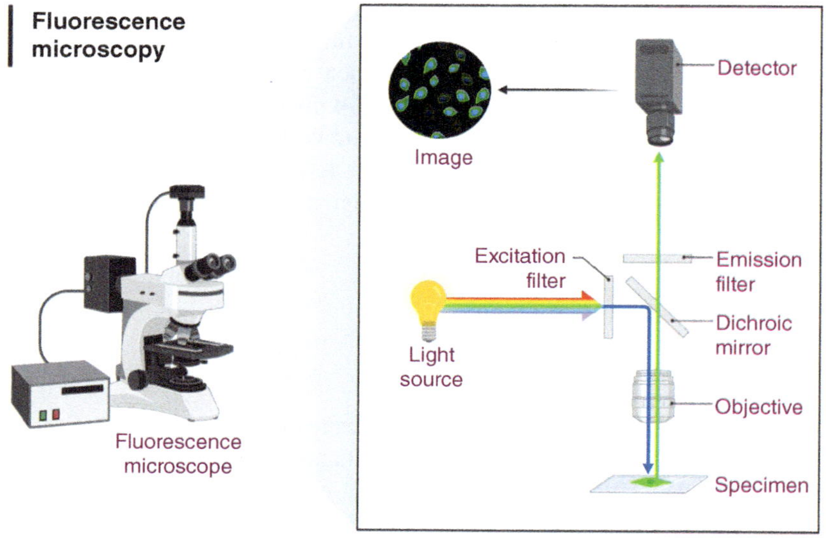

Figure 6.1 Fluorescence microscopy.

excitation light exposure, which can further be minimized by reducing time of exposure and intensity of light and by retaining an S/N ratio required for the specific experimental question. In single-wavelength experiments, long-pass emission filters can be used to collect as many photons as possible. Similarly, narrower bandpass filters are a good choice for specimens with high autofluorescence. It is vital to detect emitted light efficiently. With inefficient detection, higher intensity of excited light will be needed to obtain the same S/N ratio, which can lead to increased photobleaching and phototoxicity.

Lasers are most widely used light sources for more complex fluorescence microscopes such as confocal microscopes and total internal reflection fluorescence microscopes, while mercury, xenon lamps, and LEDs are routinely used for widefield epifluorescence microscopes.

6.2.2.3 Fluorescence Microscopy and Dynamics of Cellular Processes

Although imaging live cells using confocal microscopy is difficult, it has successfully been used to discover glutamate-stimulated, transcellular Ca^{2+} waves in astroglia [1] and to characterize developmental changes by imaging neuronal axons and their growth cones in the developing brain of a tadpole [1, 4]. Usually, repeated imaging of a labeled specimen at defined time points (time-lapse microscopy) is performed to monitor movements of cells. However, other techniques are combined with time-lapse microscopy to understand dynamic organization of cellular compartments, molecules, and cells in biological specimens. Some of the major techniques are listed as follows:

a. Time-lapse fluorescence imaging: In this technique, a single optical section or a field of view in a live specimen is repeatedly imaged over time, which allows to study dynamic distribution of fluorescently labeled components in living systems. Using this technique, dynamic changes in fluorescently labeled Golgi membranes were imaged [5, 6]. Additionally, vital fluorescent probes such as GFP and its derivatives are available, which give specific labeling patterns with cellular and subcellular molecules and are therefore suitable for repeated imaging needed for time-lapse experiments.

b. Multichannel or ratiometric fluorescence imaging: This technique allows to simultaneously or sequentially image in two or more fluorescent channels over time as shown by Keller et al. in 2001 where they performed monitored sorting of CFP- and YFP-tagged proteins through the Golgi apparatus [7].
c. Three-dimensional time-lapse (4D) imaging: This technique allows collecting z-series stacks of images repeatedly over time. This technique has been used to image neuronal dendritic spines in brain slice cultures [8], T-cell-dendritic cell interactions in lymph nodes [9], and microglial phagocytosis in the brain [10].
d. Three-dimensional multichannel (5D) time-lapse imaging: This allows collecting z-stacks in two or more fluorescent channels repeatedly over time.
e. Spectral imaging and linear unmixing: This technique enables discriminating distinct fluorophores with strongly overlapping emission spectra [11]. Linear unmixing is a computational approach where deconvolution uses spectra of each dye as a point-spread function at a fixed location in order to unmix the component signals. This tool can be used to discriminate distinct fluorophores with highly overlapping spectra [11–14].
f. Fluorescence resonance energy transfer (FRET): This is used to measure interaction between two fluorescently tagged proteins. The nonradiative energy transfer from a donor to an acceptor fluorophore with overlapping emission and excitation spectra is measured. This method has been used to study activation of small G proteins during phagocytosis [15, 16].

6.2.2.4 Confocal Microscopy of Living Cells: General Approach

Live-cell confocal imaging involves some factors to be considered for proper and appropriate labeling and visualization of cells. Temperature, pH, humidity, oxygenation, and photobleaching should be monitored, and extra care should be taken to keep these factors at a range suitable and optimal for the cells or tissues [17]. Confocal microscopy is used to evaluate the embryo quality [18] after tagging the embryos, which could be detrimental to embryo survival.

6.2.2.5 Minimizing Photodynamic Damage

When the cells are labeled and are on the microscope, they are prone to light-induced damage and this is a big challenge to researchers to preserve the natural state of the cells during imaging. Standard culture media also have been reported to contribute to light-induced adverse effects on cells. Other studies reported that riboflavin/vitamin B2 may also mediate phototoxic effects [19, 20]. Many of these adverse effects are reduced by antioxidants, so it is useful to maintain antioxidants in the specimen chamber.

6.2.2.6 Improving Photon Efficiency

Higher numerical aperture (NA) objective lenses collect more fluorescence emission; therefore, for a given lens, an optimal setting should be employed that best matches the resolution required. For the focal planes more than 5 μM, water-immersion lenses should be used to avoid signal loss caused by spherical aberration when the oil lens is used [17]. As previously mentioned, exposure time should be reduced to reduce photodamage to cells and tissues. Microscopists should act promptly and focus as keen as possible and be able to turn off the light source once the focus point is chosen.

6.2.2.7 Use of Antioxidants

Adding antioxidants to the medium reduces phototoxicity to cells. Antioxidants such as Oxyrase at a rate of 0.3 units/ml [21] and ascorbic acid of 0.1–1.0 mg/ml [22] are routinely used to reduce phototoxicity to cells. In the latter study, they reported that frequency of Ca^{2+} oscillations in chondrocytes was associated with laser intensity. Ca^{2+} events were more frequent and cell viability decreased with higher laser intensity used. Ascorbic acid treatment reduced the Ca^{2+} events and improved cell viability.

6.2.3 Fluorescence Imaging Modalities

6.2.3.1 Light Sheet Fluorescence Microscopy (LSFM)

Light sheet fluorescence microscopy illuminates and detects the specimen along two orthogonal, typically horizontal directions, allowing to illuminate only the part of the specimen that is in the focus of the objective lens and thus recording a well-focused, high-contrast optical section of the fluorescent specimen. With orthogonal excitation and imaging planes, LSFM drastically improves imaging speed, while minimizing photodamage to living samples [23]. In other fluorescence imaging techniques such as **widefield**, **spinning disk**, or **confocal microscopy**, the specimen is illuminated and detected along the same direction and through one objective lens. In LSFM, only the section of the specimen that is currently imaged is illuminated and other sections are unexposed, which drastically reduces the risk of photobleaching and phototoxicity, especially in thick specimens where many sections need to be imaged. LSFM is faster and can record a section consisting of millions of pixels in a few milliseconds and a stack of thousands of such images in a few seconds [24, 25]. LSFM is a tool of choice for large, delicate, and dynamic specimens such as cleared tissue, organs and organisms, and cultured organoids [26].

Other modalities include widefield, laser scanning, and spinning disk fluorescence microscopy techniques. Optical sectioning is the primary purpose of confocal microscopy techniques; however, due to their slow acquisition techniques and repeated exposure of the sample to the excitation light intensity, confocal modalities are limited in live-cell applications. With the introduction of spinning disk, the confocal imaging speed has considerably increased. A spinning disk is a set of two corotating disks, one of microlenses and the other of pinholes that spin at high revolutions to acquire the pixels simultaneously [27]. More recently, developments have been made to reduce acquisition times even further to milliseconds. The advent of more sensitive detectors such as electron-multiplying CCD (EM-CCD) and scientific complementary metal-oxide semiconductor (sCMOS) cameras have allowed users to improve acquisition times. Furthermore, point scanning confocal microscopes have combined more advanced detectors such as hybrid and GaAsP photomultipliers along with resonant and programmable array scanning (PAM) capabilities leading to enhanced acquisition speeds comparable to that of spinning-disk systems [28–31].

Confocal [32] and two-photon fluorescence microscopy are important tools for imaging 3-D thick specimens. Although they provide excellent sectioning through tissue, the amount of power used could be harmful to samples due to the focused and short wavelength use [33]. Label-free microscopy overcomes most of these limitations [34]. Phase-contrast and differential interference contrast (DIC) microscopy, which are label-free microscopy techniques, generate the contrast by visualizing the modifications in the light or wave front when it propagates through the sample. Both phase-contrast and DIC are mostly qualitative in nature and measure the wave front qualitatively and not quantitatively. Quantitative phase imaging (QPI) measures this phase shift quantitatively. In the following sections, we briefly cover these techniques that measure qualitative and quantitative aspects of the wave front.

6.2.4 Phase-contrast Microscopy

Introduction of phase-contrast microscope facilitated observing living cells to a much higher degree. Earlier, traditional light or optical microscopes used visible light and a system of lenses to generate magnified images of small objects. Living cells, being translucent, must be stained and fixed in order to be imaged under the simple light microscope. Staining renders the cells nonviable. With the phase-contrast microscope, high-contrast images of living cells such as cultured cells, microorganisms, thin tissue slices, and subcellular fragments can be studied in their natural state. Furthermore, dynamics of ongoing processes can be observed, recorded, and analyzed including minute details of specimens.

6.2.4.1 Principle

When light waves travel or pass through a medium, due to absorption and diffraction in the medium, they cause changes in amplitude and phase depending on the refractive index of the medium. Phase-contrast microscope translates these minute phase changes into the corresponding amplitude variations, which are visualized as differences in image contrast. These phase changes are made visible by separating and manipulating the illuminating light (background) from the specimen-scattering light.

Figure 6.2 shows the visual diagram of the optical path in a phase-contrast microscope. Partially coherent illumination produced by a light source such as halogen or tungsten–halogen lamp is directed by a collector lens and focused on an annular ring (annulus) positioned below the stage. Wave fronts passing through the annulus illuminate the specimen and either pass nondiffracted or diffracted and retarded in phase by the structures present in the specimen. These undeviated and diffracted

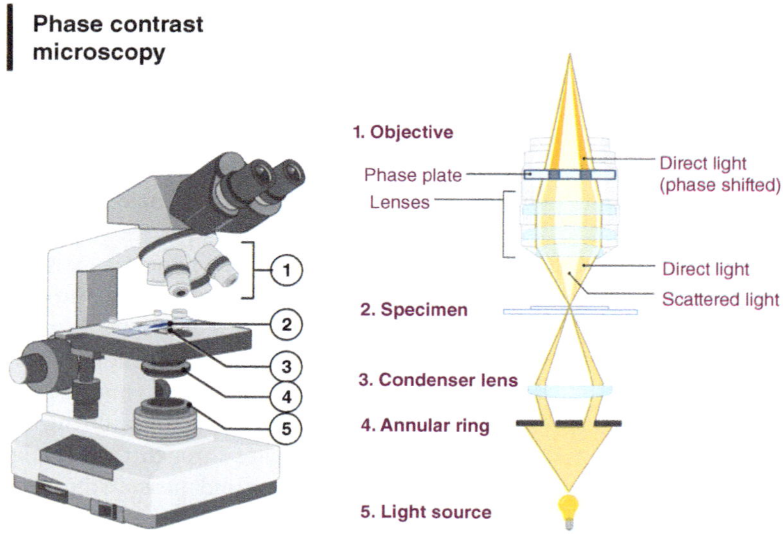

Figure 6.2 Phase-contrast microscopy.

light waves are collected by the objective lenses, and the wave front is segregated at the rear focal plane by a phase plate and focused on the intermediate image plane to form the final high-contrast image that is visualized in the eyepieces.

Modern phase-contrast microscopes are very efficient and refined well enough to detect specimens containing even just a few protein molecules especially when combined with electronic enhancement and post-acquisition image processing techniques. In this microscope, the intensity of image does not show a linear relationship to the optical path difference produced by the specimen for the entire thickness and refractive index range. Image intensity, instead, depends on a variety of factors such as absorption and phase retardation or advancement at the phase plate and relative sign of the phase shift.

6.2.5 Quantitative Phase-contrast Microscopy

Traditional phase-contrast techniques capture qualitative information of an image, which makes it difficult to automate the process. QPI is a more advanced form of the traditional phase-contrast microscopy where a digital image sensor is used to record the interference pattern or hologram. A computer algorithm is then used to create a phase-shift image and intensity from the recorded hologram therefore providing high-throughput screening.

6.2.5.1 Principle

The principle of QPI relies on two beams, a beam that travels through the sample (sample beam) and bears the information and the second does not (reference beam). Subtle variations in thickness and RI of the sample cause local phase shifts in the two beams. QPI generates a quantitative map of the local phase shifts in the sample.

This technique is mostly used to observe unstained living cells. By analyzing phase delay images of biological cells, quantitative information about the morphology can be obtained and dry mass of cells can be measured [35]. The QPI tool has been used in medical diagnosis such as hematological diagnosis and cancer pathology [36]. Unlike conventional microscopy, where cells are observed in a single plane, QPI allows for images to be created and focused at different focal planes with single exposure enabling a three-dimensional live-cell imaging by fluorescence techniques [37]. Studies have demonstrated that combining QPI with rotational scanning allows capturing of high-resolution 3-D images (W. [38, 39]).

There are additional features known as add-ons that have been added to QPI, namely, spatial interference light microscopy (SLIM), gradient light interference microscopy (GLIM), and whitelight diffraction phase microscopy (wDPM). SLIM provides 4-D quantitative, real-time, and nondestructive imaging of live cell cultures, microorganisms (mycoplasma), and reconstituted protein solutions [40]. GLIM provides quantitative, high-resolution 4-D tomogram of optical thick specimens such as cell aggregates, bacterial colonies, 3D organoids, brain slices, embryos, and modal multicellular organisms. Due to its sectioning capabilities, GLIM has been used to generate tomogram of optically thick specimens such as embryos [34]. GLIM is superior when imaging optically thick specimens such as embryos. wDPM provides 3D quantitative, nondestructive live cell imaging and can monitor material changes generated by its femtosecond laser pulses focused in the volume of transparent materials.

6.2.6 Holotomography (HT) or Optical Diffraction Tomography

This live-cell imaging technique relies on measuring 3-D refractive index (RI) tomography of cells and tissues, thus providing label-free quantitative images of transparent or phase objects. HT or optical diffraction technique [41] was proposed by Emil Wolf and later demonstrated experimentally by Fercher et al. in 1979 [42, 43].

6.2.6.1 Principle

RI is an intrinsic optical parameter through which speed of light passing through a medium can be explained. Light passing through a cell diffracts and slows down compared to that in the surrounding medium. HT employs a laser beam to measure 3-D RI distribution of cells. HT system measures 2-D holographic images of cells in various illumination angles and then uses the inverse scattering algorithm to reconstruct 3-D RI tomogram.

HT provides label-free imaging of subcellular and intracellular organelles and molecules, which poses no risk of photodamage to cells or tissues and thus eliminates artifacts caused by dyes and fluorescent proteins. Furthermore, quantitative measuring of RI and thus imaging of 3D RI of samples can provide information about dry mass, cell volume, surface area, protein concentration, morphology, and dynamics of the cell membrane. This technique uses low-intensity laser and is a precise and high-throughput imaging technique in 2-D, 3-D, and 4-D with high resolution. Recently, HT has integrated 3-D fluorescence imaging technique to acquire 3-D images of both holotomography and 3-D fluorescence with molecular specific for each sample as well as it allows long-time tracking of specific targets in live cells. Applications of HT are seen in cell biology, medicine, infectious diseases, biotechnology, etc. It has been used to quantify chromosomes [44] and used to study proteasomal catabolism and degradation and regulatory pathways in autophagy in mammalian cells ([45]).

6.3 Other Considerations

6.3.1 Oil Immersion and Water Immersion Lenses

Live cells mostly require aqueous media or suspension. Oil immersion objective lenses are not recommended for live cells as oil–water can cause spherical aberration. However, silicon immersion oil is used in some live cell imaging techniques due to its refractive index proximity to that of living cells. Silicon immersion is particularly used in the areas of developmental biology and regenerative biology, e.g. micro- and macro-observation of embryos, zebrafish, and other model organisms, differentiation of embryonic stem (ES) and induced pluripotent stem (iPS) cells. Silicon immersion has been also used to visualize endoreplication in trophoblast giant cells and megakaryocytes [46].

On the other hand, water immersion results in improved resolving power due to its RI closer to that of aqueous medium of cells [47]. Water has higher refractive index; therefore, water immersion lenses have a high NA and can produce images superior to that of oil immersion objective lenses. The use of water immersion has also demonstrated that it can overcome spherical aberrations caused by oil immersion and has been combined with a fluorescence super-resolution technique stimulated emission depletion (STED), which increases the microscope's resolution in the lateral and axial directions. In one study, water immersion led to superior imaging when used at penetration depths of 5–180 μm [48].

6.3.2 Dry Lenses

Specialized correction collar objective lenses should be used when extra space is required between the objective and the specimen for correcting spherical aberrations. 4×, 10×, and 40× are all dry lenses and cannot be used with immersion oil. For 100× dry, which can also be used with immersion oil, the dried oil must be removed with lens paper as the dried oil severely reduces the optical performance of the instrument.

6.3.3 Photodamage of Cells

In normal physiological environment, cells and tissues are exposed to no or very low levels of light. The ultraviolet (UV) light damages DNA, focused infra-red (IR) light can cause localized heating, and fluorescence light induces phototoxicity to cells and tissues [49]. During excitation of fluorescent proteins and dyes, free radicals are produced, which react with surrounding cellular components in an oxygen-dependent manner, causing photodamage to cells. Therefore, users should ensure to minimize exposure of cells to light by using highly efficient microscopes to optimize the light path and by the use of detectors that are able to detect all of fluorescence emission [50].

6.3.4 Specimen Environment

Key factors should be considered such as type of culture media (phenol-free media), temperature of the cells (37 °C for mammalian cells), pH (must be maintained at physiological level), and osmolarity of the medium (minimize the evaporation of media). All media should contain HEPES (~25 mM) to avoid harmful effects of pH change. To prevent change in osmolarity, the air above the sample should be humidified. Usually, large chambers where microscopes are enclosed are equipped with CO_2 gas through water reservoirs to humidify the environment [50].

6.3.5 Improve S/N Ratio

Use of phenol-free media reduces the nonspecific background intensity and thus improves the S/N ratio. Image deconvolution using the algorithmic software can greatly increase the S/N. A microscopic image is generated by convolution of the optical elements of the microscope with the emitted light or light transmitted through the sample. Deconvoluting software that use quantitative algorithms mathematically reverse the blurring effects of the optics and reassign the captured light back to its point of origin, thereby increasing the S/N ratio and generate higher resolution images with retained quantification of fluorescence [51–54].

6.4 Conclusions

Imaging tools discussed in this chapter can be optimized for live cell imaging; however, each tool has its use according to the sample type and cellular dynamics of interest. Live-cell imaging is advancing at a rapid pace with more understanding of the cellular processes and details. Discovery of FPs and its derivatives has enabled researchers to follow and pinpoint the dynamic proteins and molecules within cells. Furthermore, development of high-sensitivity and high-efficiency detection devices such as PMTs has provided more diverse imaging options. 3-, 4-, and 5-dimensional imaging are useful to follow moving cells and fine cellular processes within complex tissues as well as to determine dynamic relationships between different cell or tissue components. Introduction of super-resolution microscopy will make rapid cell imaging feasible and will enhance the understanding of intricate dynamics and complex protein and molecular interactions within cells. Each of the techniques has its advantages and limitations on time, specimen preparation, and maintenance of cell and tissue health, thus implying that a careful consideration should be given when choosing an approach most suitable and useful for the biological question being asked. An enduring future research will lead to development of newer techniques facilitating the advancement of microscopy.

References

1 O'Rourke, N.A. and Fraser, S.E. (1990). Dynamic changes in optic fiber terminal arbors lead to retinotopic map formation: an in vivo confocal microscopic study. *Neuron.* 5: 159–171. https://doi.org/10.1016/0896-6273(90)90306-Z.

2 Richards, O.W. and Bradbury, S. (1968). The evolution of the microscope. *Transactions of the American Microscopical Society.* 87: 113–114. https://doi.org/10.2307/3224347.

3 Presley, J.F., Cole, N.B., Schroer, T.A. et al. (1997). ER-to-Golgi transport visualized in living cells. *Nature.* 389: 81–85. https://doi.org/10.1038/38001.

4 Fraser, S.E. and O'Rourke, N.A. (1990). In situ analysis of neuronal dynamics and positional cues in the patterning of nerve connections. *Journal of Experimental Biology.* 153: 61–70.

5 Cooper, M.S., Cornell-Bell, A.H., Chernjavsky, A. et al. (1990). Tubulovesicular processes emerge from trans-Golgi cisternae, extend along microtubules, and interlink adjacent trans-Golgi elements into a reticulum. *Cell.* 61 (1): 135–145. https://doi.org/10.1016/0092-8674(90)90221-Y.

6 Cooper, M.S., D'Amico, L.A., and Henry, C.A. (1999). Confocal microscopic analysis of morphogenetic movements. *Methods in Cell Biology.* 59: 179–204. https://doi.org/10.1016/s0091-679x(08)61826-9.

7 Keller, P., Toomre, D., Díaz, E. et al. (2001). Multicolour imaging of post-Golgi sorting and trafficking in live cells. *Nature Cell Biology.* 3: 140–149. https://doi.org/10.1038/35055042.

8 Marrs, G.S., Green, S.H., and Dailey, M.E. (2001). Rapid formation and remodeling of postsynaptic densities in developing dendrites. *Nature Neuroscience.* 4: 1006–1013. https://doi.org/10.1038/nn717.

9 Stoll, S., Delon, J., Brotz, T.M., and Germain, R.N. (2002). Dynamic imaging of T cell-dendritic cell interactions in lymph nodes. *Science.* 296 (5574): 1873–1876. https://doi.org/10.1126/science.1071065.

10 Petersen, M.A. and Dailey, M.E. (2004). Diverse microglial motility behaviors during clearance of dead cells in hippocampal slices. *GLIA.* 46: 195–206. https://doi.org/10.1002/glia.10362.

11 Dickinson, M.E., Bearman, G., Tille, S. et al. (2001). Multi-spectral imaging and linear unmixing add a whole new dimension to laser scanning fluorescence microscopy. *BioTechniques.* 31 (6): 1272, 1274–1276, 1278. https://doi.org/10.2144/01316bt01.

12 Hiraoka, Y., Shimi, T., and Haraguchi, T. (2002). Multispectral imaging fluorescence microscopy for living cells. *Cell Structure and Function.* 27: 367–374. https://doi.org/10.1247/csf.27.367.

13 Tsurui, H., Nishimura, H., Hattori, S. et al. (2000). Seven-color fluorescence imaging of tissue samples based on Fourier spectroscopy and singular value decomposition. *Journal of Histochemistry and Cytochemistry.* 48: 653–662. https://doi.org/10.1177/002215540004800509.

14 Zimmermann, T., Rietdorf, J., Girod, A. et al. (2002). Spectral imaging and linear un-mixing enables improved FRET efficiency with a novel GFP2-YFP FRET pair. *FEBS Letters.* 531 (2): 245–249. https://doi.org/10.1016/S0014-5793(02)03508-1.

15 Hoppe, A.D. and Swanson, J.A. (2004). Cdc42, Rac1, and Rac2 display distinct patterns of activation during phagocytosis. *Molecular Biology of the Cell.* 15: 8. https://doi.org/10.1091/mbc.E03-11-0847.

16 Sekar, R.B. and Periasamy, A. (2003). Fluorescence resonance energy transfer (FRET) microscopy imaging of live cell protein localizations. *Journal of Cell Biology.* 160 (5): 629–633. https://doi.org/10.1083/jcb.200210140.

17 Dailey, M.E., Manders, E., Soll, D.R., and Terasaki, M. (2006). *Confocal microscopy of living cellsHandbook of Biological Confocal Microscopy*, 3e (ed. J.B. Pawley). New York: Springer Science+Business Media, LLC https://doi.org/10.1007/978-0-387-45524-2_19.

18 Zucker, R.M. (2006). Whole insect and mammalian embryo imaging with confocal microscopy: Morphology and apoptosis. *Cytometry Part A.* https://doi.org/10.1002/cyto.a.20343.

19 Lucius, R., Mentlein, R., and Sievers, J. (1998). Riboflavin-mediated axonal degeneration of postnatal retinal ganglion cells in vitro is related to the formation of free radicals. *Free Radical Biology and Medicine.* 24: 798–808. https://doi.org/10.1016/S0891-5849(97)00358-4.

20 Zigler, J.S., Lepe-Zuniga, J.L., Vistica, B., and Gery, I. (1985). Analysis of the cytotoxic effects of light-exposed hepes-containing culture medium. *In Vitro Cellular & Developmental Biology.* 21 (5): 282–287. https://doi.org/10.1007/BF02620943.

21 Waterman-Storer, C.M., Sanger, J.W., and Sanger, J.M. (1993). Dynamics of organelles in the mitotic spindles of living cells: membrane and microtubule interactions. *Cell Motility and the Cytoskeleton.* 26 (1): 19–39. https://doi.org/10.1002/cm.970260104.

22 Knight, M.M., Roberts, S.R., Lee, D.A., and Bader, D.L. (2003). Live cell imaging using confocal microscopy induces intracellular calcium transients and cell death. *American Journal of Physiology - Cell Physiology.* 284: C1083–C1089. https://doi.org/10.1152/ajpcell.00276.2002.

23 Huisken, J., Swoger, J., Del Bene, F. et al. (2004). Optical sectioning deep inside live embryos by selective plane illumination microscopy. *Science.* 305: 1007–1009. https://doi.org/10.1126/science.1100035.

24 Gustavsson, A.-K., Petrov, P.N., and Moerner, W.E. (2018). Light sheet approaches for improved precision in 3D localization-based super-resolution imaging in mammalian cells [Invited]. *Optics Express.* 26 (10): 13122–13147. https://doi.org/10.1364/oe.26.013122.

25 Santi, P.A. (2011). Light sheet fluorescence microscopy: a review. *Journal of Histochemistry and Cytochemistry.* 59 (2): 129–138. https://doi.org/10.1369/0022155410394857.

26 Rakotoson, I., Delhomme, B., Djian, P. et al. (2019). Fast 3-D imaging of brain organoids with a new single-objective planar-illumination two-photon microscope. *Frontiers in Neuroanatomy.* 13: 77. https://doi.org/10.3389/fnana.2019.00077.

27 Nakano, A. (2002). Spinning-disk confocal microscopy – a cutting-edge tool for imaging of membrane traffic. *Cell Structure and Function.* 27: 349–355. https://doi.org/10.1247/csf.27.349.

28 Becker, W., Su, B., Holub, O., and Weisshart, K. (2011). FLIM and FCS detection in laser-scanning microscopes: increased efficiency by GaAsP hybrid detectors. *Microsc Res Tech.* 74 (9): 804–811. https://doi.org/10.1002/jemt.20959.

29 Hanley, Q.S., Verveer, P.J., Gemkow, M.J. et al. (1999). An optical sectioning programmable array microscope implemented with a digital micromirror device. *Journal of Microscopy.* 196 (Pt 3): 317–331. https://doi.org/10.1046/j.1365-2818.1999.00602.x.

30 Hendriks, B.H.W., Bierhoff, W.C.J., Horikx, J.J.L. et al. (2011). High-resolution resonant and nonresonant fiber-scanning confocal microscope. *Journal of Biomedical Optics* 16: 026007. https://doi.org/10.1117/1.3534781.

31 Michalet, X., Cheng, A., Antelman, J. et al. (2008). Hybrid photodetector for single-molecule spectroscopy and microscopy. *Single Molecule Spectroscopy and Imaging.* 6862: 68620F. https://doi.org/10.1117/12.763449.

32 Panula, P.A.J. (2003). Handbook of Biological Confocal Microscopy. *Journal of Chemical Neuroanatomy.* 25: 228–229. https://doi.org/10.1016/s0891-0618(03)00007-3.

33 Masters, B.R., So, P.T.C., and Mantulin, W.W. (2009). Handbook of biomedical nonlinear optical microscopy. *Journal of Biomedical Optics.* 14: 019901. https://doi.org/10.1117/1.3077566.

34 Nguyen, T.H., Kandel, M.E., Rubessa, M. et al. (2017). Gradient light interference microscopy for 3D imaging of unlabeled specimens. *Nature Communications.* 8: 210. https://doi.org/10.1038/s41467-017-00190-7.

35 Zangle, T.A. and Teitell, M.A. (2014). Live-cell mass profiling: an emerging approach in quantitative biophysics. *Nature Methods.* 11 (12): 1221–1228. https://doi.org/10.1038/nmeth.3175.

36 Majeed, H., Sridharan, S., Mir, M. et al. (2017). Quantitative phase imaging for medical diagnosis. *Journal of Biophotonics.* 24: 798–808. https://doi.org/10.1002/jbio.201600113.

37 Rosen, J. and Brooker, G. (2008). Non-scanning motionless fluorescence three-dimensional holographic microscopy. *Nature Photonics.* 2: 190–195. https://doi.org/10.1038/nphoton.2007.300.

38 Choi, W., Fang-Yen, C., Badizadegan, K. et al. (2007). Tomographic phase microscopy. *Nature Methods.* 4 (9): 717–719. https://doi.org/10.1038/nmeth1078.

39 Pollaro, L., Equis, S., Dalla Piazza, B., and Cotte, Y. (2016). Stain-free 3D nanoscopy of living cells. *Optik & Photonik.* 11: 38–42. https://doi.org/10.1002/opph.201600008.

40 Wang, Z., Ding, H., and Popescu, G. (2010). Spatial light interference microscopy (SLIM). *Optics InfoBase Conference Papers.* 19: 1016. https://doi.org/10.1364/oe.19.001016.
41 Lauer, V. (2002). New approach to optical diffraction tomography yielding a vector equation of diffraction tomography and a novel tomographic microscope. *Journal of Microscopy.* 205: 165–176. https://doi.org/10.1046/j.0022-2720.2001.00980.x.
42 Fercher, A.F., Bartelt, H., Becker, H., and Wiltschko, E. (1979). Image formation by inversion of scattered field data: experiments and computational simulation. *Applied Optics.* 18 (14): 2427–2439. https://doi.org/10.1364/ao.18.002427.
43 Wolf, E. (1969). Three-dimensional structure determination of semi-transparent objects from holographic data. *Optics Communications.* 1: 153. https://doi.org/10.1016/0030-4018(69)90052-2.
44 Kim, S., Kim, N.H., Park, J.E. et al. (2020). PRMT6-mediated H3R2me2a guides Aurora B to chromosome arms for proper chromosome segregation. *Nature Communications.* 11: 612. https://doi.org/10.1038/s41467-020-14511-w.
45 Choi, W.H., Yun, Y., Park, S. et al. (2020). Aggresomal sequestration and STUB1-mediated ubiquitylation during mammalian proteaphagy of inhibited proteasomes. *Proceedings of the National Academy of Sciences of the United States of America.* 117 (32): 19190–19200. https://doi.org/10.1073/pnas.1920327117.
46 Sakaue-Sawano et al. (2013). Visualizing developmentally programmed endoreplication in mammals using ubiquitin oscillators. *Development* 140: 4624–4632.
47 Hibbs, A.R. (2004). *Confocal Microscopy for Biologists.* Kluwer Academic/Plenum Publishers.
48 Heine, J., Wurm, C.A., Keller-Findeisen, J. et al. (2018). Three dimensional live-cell STED microscopy at increased depth using a water immersion objective. *The Review of Scientific Instruments* 89 (5): 53701. https://doi.org/10.1063/1.5020249.
49 Pattison, D.I. and Davies, M.J. (2006). Actions of ultraviolet light on cellular structures. *EXS.* 96: 131–157. https://doi.org/10.1007/3-7643-7378-4_6.
50 Frigault, M.M., Lacoste, J., Swift, J.L., and Brown, C.M. (2009). Live-cell microscopy – tips and tools. *Journal of Cell Science.* 122 (Pt 6): 753–767. https://doi.org/10.1242/jcs.033837.
51 Holmes, T.J., Biggs, D., and Abu-Tarif, A. (2006). Blind deconvolution. In: *Handbook of Biological Confocal Microscopy*, 3e. New York: Springer Science+Business Media, LLC https://doi.org/10.1007/978-0-387-45524-2_24.
52 McNally, J.G., Karpova, T., Cooper, J., and Conchello, J.A. (1999). Three-dimensional imaging by deconvolution microscopy. *Methods* 19: 373–385. https://doi.org/10.1006/meth.1999.0873.
53 Swedlow, J.R. (2013). Quantitative fluorescence microscopy and image deconvolution. *Methods in Cell Biology.* 114: 407–426. https://doi.org/10.1016/B978-0-12-407761-4.00017-8.
54 Wallace, W., Schaefer, L.H., and Swedlow, J.R. (2001). A workingperson's guide to deconvolution in light microscopy. *BioTechniques.* 296 (5574): 1873–1876. https://doi.org/10.2144/01315bi01.

Section 2

Functionalized Nanomaterial for Miniaturized Devices

7

Ionic Liquid–Assisted Single-Drop Microextraction: A Miniaturized Sample Preparation Tool for Various Analytes

Janardhan R. Koduru and Lakshmi P. Lingamdinne

Department of Environmental Engineering, Kwangwoon University, 20-Kwangwoon-ro, Nowon-gu, Seoul 01897, South Korea

7.1 Introduction

Chemical processes have impacts on the environment. To reduce their impacts on the environment and human beings, development of safer and cleaner alternatives is encouraged. In this connection, Anastas introduced the concept of green chemistry with 12 principles that provides a roadmap for development and application of safer, more benign chemical processes [1, 2]. Thus, the introduction of single-drop microextraction (SDME) in the mid-1990s was an important development to dramatically reduce solvent consumption in analytical techniques involving conventional solvent extraction methods [3–5]. SDME is a miniaturized solvent extraction technique that has been widely used for isolation and preconcentration of several analytes prior to analytical instrumental analysis [6]. However, it is necessary to search for alternative solvents that could replace the conventional toxic solvents such as benzene used in SDME according to the fifth principle of green chemistry.

In this sense, ionic liquids (ILs) are considered green solvents due to their unique properties such as very low vapor pressure under ambient conditions and are therefore known as nonvolatile, liquid state over a wide range of temperatures, excellent hydraulic and lubrication properties, highly polar nature, low viscosity, higher solubility, and biodurability [7, 8]. The introduction of eco-friendly ILs in SDME offers significant advantages including reduced toxicity, higher selectivity, and improved extraction efficiency with sustainable environment [9].

The term ILs was familiar to only a few number of specialized research groups and they were not aware of the fact that ILs could be the best solvent for reactions including polymerization, alkylation, Beckmann rearrangement, carbonization, esterification, depolymerization, pre-extraction, etc. [8, 10–18]. Moreover, the introduction of ILs in SDME not only replaces toxic organic solvents but also adheres to the green analytical chemistry principles such as reduction of waste generation, use of miniaturized equipment, and increased safety measurements [1].

Miniaturized Analytical Devices: Materials and Technology, First Edition.
Edited by Suresh Kumar Kailasa and Chaudhery Mustansar Hussain.

This chapter briefly covers the basic fundamentals of ILs including their chemistry and functionality, their classification, and their applications in SDME for analysis of various analytes prior to instrumental analysis.

7.2 Ionic Liquids

7.2.1 Background

ILs are a new type of ionic salts that are liquid at low temperatures. Usually, these ionic compounds are liquid below 100 °C (the boiling point of common solvent, water). Most of these compounds have melting points below room temperature; some have melting points <0 °C. In addition, these compounds are liquid over a wide range of temperatures. That is, these ILs have a high decomposition temperature, which leads to ILs being very stable compounds thermally, mechanically, and electrochemically, demonstrating IL as a typical material for different applications [1]. They are classified as process chemicals (e.g., processing solvents and separation media) and performance chemicals (e.g. electrolytes, and lubricants) based on their use. ILs have appealing solvent properties and are miscible with water and organic solvents. The corresponding applications of these ionic fluids can be divided based on their use as process chemicals and performance chemicals [1, 7]. In 1948, this class of electrically conducting materials (1-butylpyridinium chloride/$AlCl_3$) first drew attention for their use as electrolytes for electrodeposition of aluminum [19–21]. This field is still the subject of current research efforts. Other important application in this context is their use as electrolytes in batteries and fuel cells.

In 1992, a major breakthrough occurred with the development of less corrosive, air-stable materials. Wilkes and Zaworotko introduced 1-ethyl-3-methylimidazolium tetrafluoroborate and focused on solvent applications [22]. Although 1-butyl-3-methylimidazolium hexafluorophosphate and -tetrafluoroborate still dominate the current literature, there are better choices with respect to performance and handling. In aqueous media, hexafluorophosphate and tetrafluoroborate anions will degrade, forming HF, toxic and strong acid. ILs have very low viscosity, low vapor pressure or non-volatility under ambient conditions, tunable solubility, acidity, basicity, long-range thermal stability, and very low corrosivity relative to mineral acids and bases [23]. Due to their negligible vapor pressure, ILs do not cause explosion unlike volatile organic solvents, and they also do not affect atmospheric photochemistry [24]. Moreover, the nonvolatile nature of most ILs make then nonflammable under ambient conditions, which could be seen with some low-boiling solvents such as pet ether, dichloromethane, acetone [25]. Hence, ILs are considered good molecular and/or green solvents replacing commonly used volatile organic solvents [26].

7.2.2 Chemistry and Functionality of Ionic Liquids

ILs are liquids over a wide temperature range (300–400 °C), which have high thermal stability at low melting points. Comparing the melting point of a typical IL (1-ethyl-3-methylimidazolium ethyl sulfate) (<−20 °C) with a typical inorganic salt (NaCl) (801 °C) shows an obvious difference. ILs have a significant lower

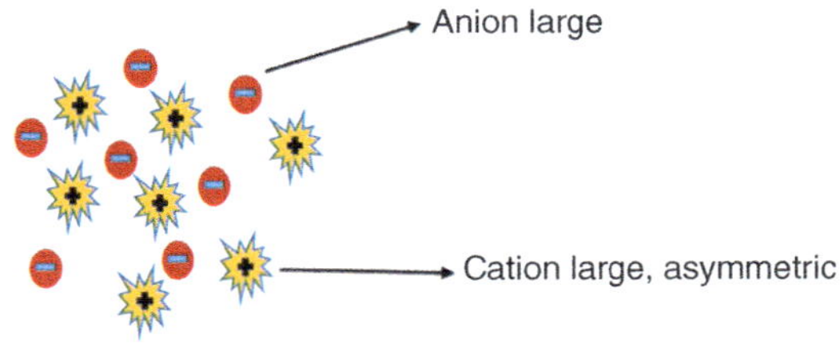

Figure 7.1 Charge distribution in ILs.

symmetry, with the distribution of anion and cation charges over a larger volume of the molecule by resonance as shown in Figure 7.1. As a result, the solidification of the ILs will take place at lower temperatures. Due to the strong columbic interaction (ionic interaction) within ILs, they have negligible vapor pressure, nonflammability, and high stability. In addition to these very interesting combination properties, ILs exhibit other properties such as very appealing solvent properties and immiscibility with water and organic solvents that result in biphasic systems.

The choice of cations and anions of ILs has a strong impact on their properties. Usually, the cation defines the stability of ILs, while the anion controls the chemistry and functionality of ILs. Typical structures of ILs combine organic cations with inorganic or organic anions as shown in Figure 7.2.

A comprehensive report on the physicochemical properties of ILs is available in the literature [27, 28]. The ILs displays a variety of physical and chemical properties such as density, acidity, polarity, viscosity, vapor pressure, melting point, crystallization temperature, surface tension, conductivity, refractive index, isentropic compressibility, expansibility, thermodynamic functions, phase equilibria, and many more remarkable properties. A small amount of an impurity (water or any organic solvent) will largely affect the properties of ILs [7]. For example, addition of a small fraction of water to ILs could bring tremendous changes in their properties such as biocatalytic activity, acidity, density, viscosity, electrical conductivity, enthalpy, surface and interfacial tensions, and molecular behavior [29, 30]. In addition, water present in ILs will lead to strong intermolecular interactions by van der Waals forces, hydrogen bonds, or electrostatic forces. However, at high concentrations, ILs will decompose into ion pairs or individual ions [29]. Rivera-Rubero and Baldelli have characterized the surface orientation properties of ILs and water using vibrational spectroscopy [31]. Widegren et al. have reported the effect of binary mixtures, such as solubilized water and hydrophobic room temperature ILs, on viscosity [32]. Compared to other materials, ILs show intermolecular and intramolecular interactions that can be significantly influenced by the structural size of cations and anions [33]. Further, the unique properties of ILs and how they influence the use of ILs are discussed in Section 4.1.

7.2.3 Classification of ILs

Based on the nature of cation and anion combinations and their physical and chemical behaviors, ILs are classified into various types including task-specific ILs (TS-ILs), chiral ILs (C-ILs), switchable polarity ILs (SP-ILs), Bio ILs (B-ILs), poly ILs (P-ILs), energetic ILs (E-ILs), neutral ILs (N-ILs), protic ILs (Pr-ILs), metal salt ILs (M-ILs), basic ILs, supported ILs (S-ILs), etc. Functionalized ILs are called task-specific ILs [7].

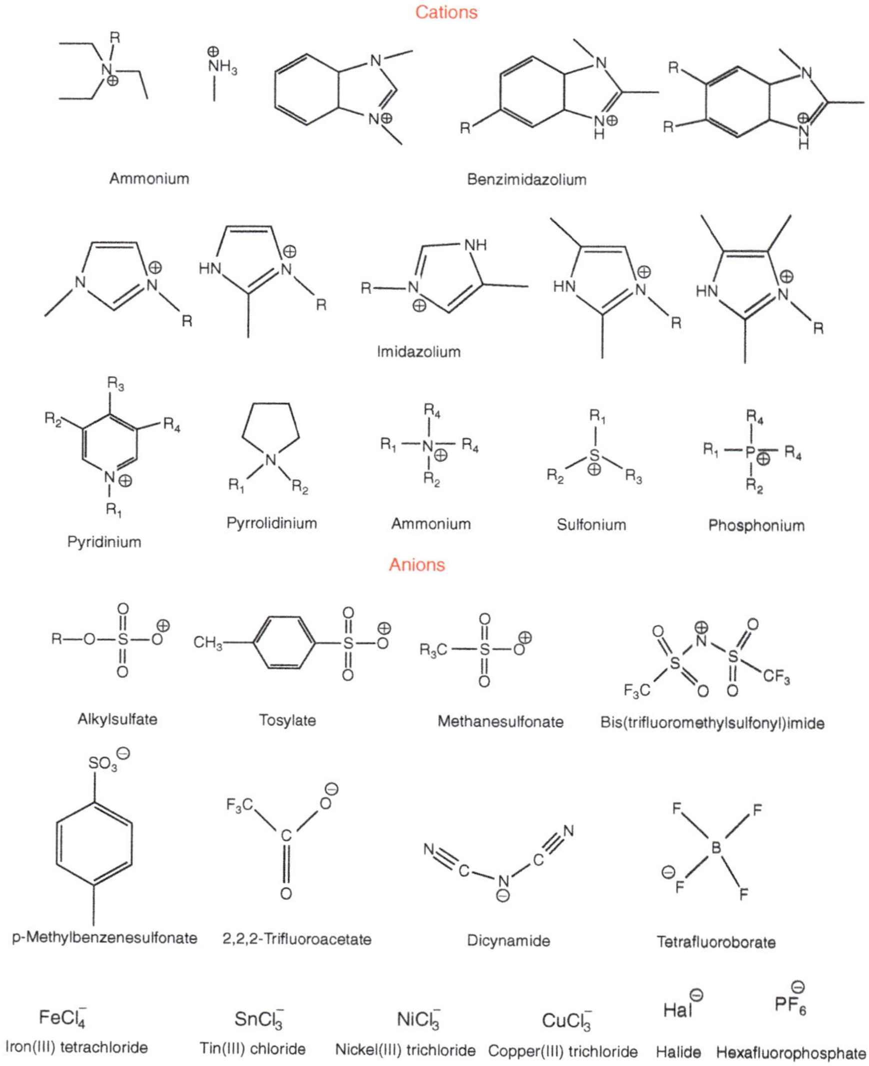

Figure 7.2 Typical structures combining organic cations with inorganic or organic anions.

Properties of task-specific ILs are tuned by altering the anion and cation combinations according to their use as solvent in applications. TS-ILs are often used in organic synthesis, catalytic reactions, and chiral compound synthesis [34–38]. 3-Sulfopropyl tri-phenyl phosphonium *p*-toluene sulfonate is the best example for TS-ILs, which is used in organic catalytic reactions [13].

C-ILs are one of the TS-ILs classes that have been specifically used in chiral liquid chromatography, stereoselective polymerization, chiral compound synthesis, liquid crystals, discrimination of chiral compounds using NMR, and other functional activities [39–41]. C-ILs are particularly used as the solvent for synthesis of asymmetric compounds [42]. Due to their chiral nature, the synthesis of C-ILs is very difficult [41, 43]. The compound [*N*-(3′-oxobutyl)-*N*-methylimidazolium][(+)-camphorsulfonate] is one example of C-IL.

SP-ILs are generally prepared using an activator that equilibrates into a wide range of lower and higher polarities for both anions and cations, respectively. Typically, secondary amines are activated using CO_2 to get SP-ILs in the carbamate salt reaction in Equation (7.1) [44]:

$$NHR_2 + CO_2 \Leftrightarrow R_2NCOOH + NHR_2 \Leftrightarrow [R_2NH_2][R_2NCOO_2] \tag{7.1}$$

For example, 1,8-diazabicyclo[5.4.0]undec-7-ene (DBU) and alcohol equilibrated with CO_2 get a high range of polarity ILs that switched from lower to higher polarities [45]. The long-chain alkyl imidazolium- and bezimidazolium-based ILs are highly toxic and less biodegradable. To overcome these issues, B-ILs have been developed using bio-precursors, which are biodegradable and less toxic. According to the European Standards, (2-hydroxyethyl)-ammonium lactate is found to have highest biodegradable levels (95%) [46].

P-ILs are nothing but polymers having repeating units of respective monomers of ILs that may exist as dimers, trimers, oligomers, higher molecular weight polymers, or copolymers [47–51]. A typical P-IL was prepared using *N*-(2-(dimethylamino) ethyl) methacrylate (DMAEMA) combined with a series of natural carboxylic acids (RCOOH). Due to their tunable properties, P-ILs are widely used in solid-state Li-ion batteries [52].

Due to the presence of Brønsted acidic proton(s) in Pr-ILs, they are used as either solvent or catalysts in reactions including acidic hydrolysis, dehydration, fuel cell chemistry, etc. [16]. The design and facile synthesis of protic or Brønsted acidic ILs (BAILs) is well illustrated in two steps: first step is zwitterion preparation and second or final step is neutralization or IL synthesis [7, 17].

Most of the metal salt ILs (M-ILs) are prepared using imidazolium- or pyridium-based cations or using direct metal halides. To enhance their Brønsted and Lewis acidic properties, bromometalate or chlorometalate salts (e.g. $[Al_2Br_7]^-$, $[CuCl_3]^-$, $[FeCl_4]^-$, $[AlCl_3]^-$, $[NiCl_4]^-$, $[SnCl_3]^-$) are applied as counteranions of ILs. Metal halides or Lewis acidic.

ILs are highly viscous than other ILs. Moreover, the stability of the low nucleophilic $[Al_2Br_7]$ anion was capitalized by the $[(CH_3)_3C]$ cation at ambient temperature.

Thus, it could be applied as a highly stabilized $[(CH_3)_3C]$ cation [53].

Room-temperature ILs are primarily classified into neutral, acidic, and basic ILs. The acidic and neutral ILs are broadly recognized and are potentially used in many organic transformation applications [54–64]. Basic ILs have replaced commonly used inorganic bases and are also applied in organic transformation reactions such as Michael addition, Markovnikov addition, aldol condensation, aza-Michael reactions [65–68]. Moreover, Basic ILs potentially offer eco-friendly basic catalysts and have better advantages compared to inorganic bases. Thus, the use of Basic ILs provide more potential benefits for the replacement of typical inorganic bases since they exhibit high flexibility, non-corrosiveness, non-volatility, solubility compared to many organic solvents [69, 70].

7.2.4 Various Applications of Ionic Liquids (ILs)

The property behavior of ILs is a combination of the cationic and anionic nature that emerged with good chemical, physical, biological, and thermal stability, minimal environmental release, noticeable task specificity, with a diverse range of applications in electrochemistry, separation and extraction, solvent and catalysts, analytical, physical, and synthetic chemistry, biological engineering, and so on [71]. Scientists and engineers have been encouraged to develop less-toxic materials, sustainable and eco-friendly sources for chemicals, energy, fuels, etc. ILs have been used in a wide range of applications as they are regarded as green solvents or catalysts [72, 73]. Some examples of potential applications are provided below in Figure 7.3 [16, 72, 74–77].

The use of ILs in academic and industry has been growing, which is evident from the increasing number of publications annually and this increase can be attributed to the unique properties of this new class of materials. Ultimately, the possible combinations of organic cations and anions places chemists in the position to design and fine-tune physical and chemical properties by introducing or combining structural motifs and, thereby, making tailor-made materials and solutions possible. The chart shown in Figure 7.3 summarizes the important properties and current applications of ILs in various fields such as lubricants [78], MALDI–TOF–matrices [79], GC–headspace solvents [80, 81], protein crystallization [82], biocatalysis, organic synthesis and catalysis [83–85], deep desulfurization [86], nanoparticle synthesis [87, 88], liquid crystals [89, 90], thermal fluids [91], fuel cells [92], sensors

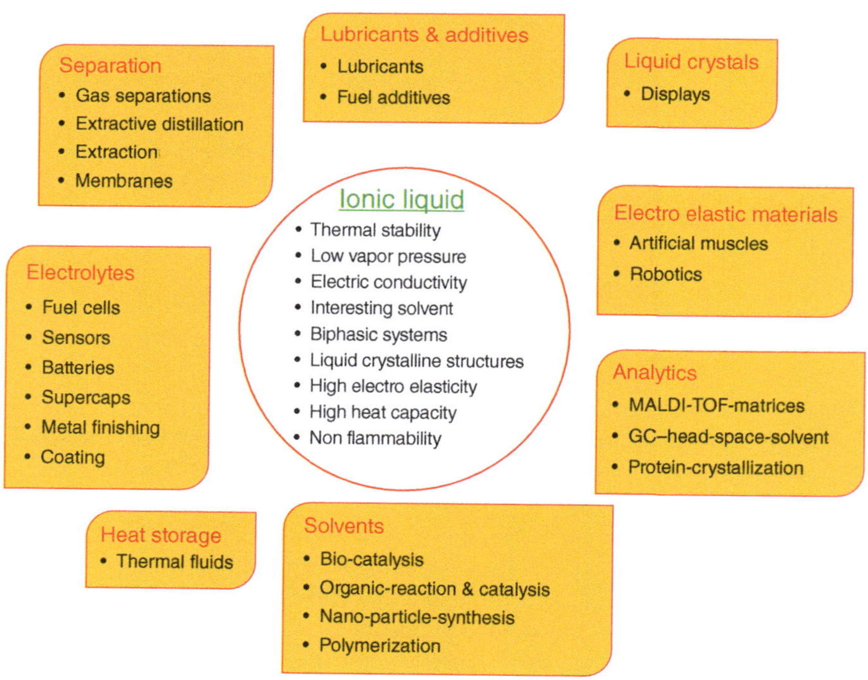

Figure 7.3 Various applications of ILs.

[73], supercapacitors [93], metal finishing [94–96], gas separation [97], extractive distillation [98], extraction [99], and membranes [100, 101].

In addition, due to their unique characteristics and eco-friendly nature, ILs are used in various fields of analytical chemical processes, including chromatography, spectrometry, sensing [7, 102, 103], isolation, extraction, electroanalysis and are used as a subclass in SDME [5, 104, 105], bioanalytical applications [106, 107]. Moreover, the use of ILs in SDME is of more interest. Thus, the reported results of IL-assisted SDME coupled with various analytical instruments are discussed in the following sections.

7.3 Ionic Liquid-Assisted SDME for Analytes

Many miniaturized strategies have been developed for green sample preparation. In this regard, the introduction of SDME is an important boost toward a dramatic reduction of solvent consumption in analytical methodologies involving conventional liquid–liquid extraction [3–5]. SDME has evolved from LPME in which the extraction phase is a drop of solvent, which is suspended from a syringe needle, directly immersed in a stirred aqueous solution (DI-SDME) or in close contact with its headspace (HS-SDME) [108–111] as shown in Figure 7.4. However, conventional organic solvents may generate waste toxic to the environment and humans. To substantially minimize the use of toxic organic solvents such as benzene in SDME, it is necessary to search for alternative green solvents. In this sense, ILs are considered green solvents and have unique properties that make them suitable to be used in SDME [23, 112]. ILs in SDME not only serve as green solvents but also develop green analytical chemistry with more safety.

SDME has been combined with a variety of analytical techniques for the determination of organic, organometallic, and metallic analytes using ILs as extractant phase. More than 75% of the publications are devoted to SDME that makes use of [CnC1IM][PF_6] ($n = 4, 6, 8$) as the IL of choice. Advantageous physicochemical properties, commercial availability, and relatively low price have made this type of IL a preferred choice [1].

7.3.1 Factors Influencing Ionic-Liquid-Assisted SDME

The special properties of ILs make them valuable extraction solvents in SDME. Here are described the properties of ILs that influence their use in SDME.

7.3.1.1 Vapor Pressure and Thermal Stability of ILs

ILs possess low vapor pressure. Because of their low vapor pressure, ILs have high thermal stability that prevents reduction in drop volumes through evaporation in SDME. This property is more important in headspace-SDME where the drop volume should be stable and not decrease. Moreover, the low vapor pressure of ILs ensures that they may be employed at relatively high temperatures for

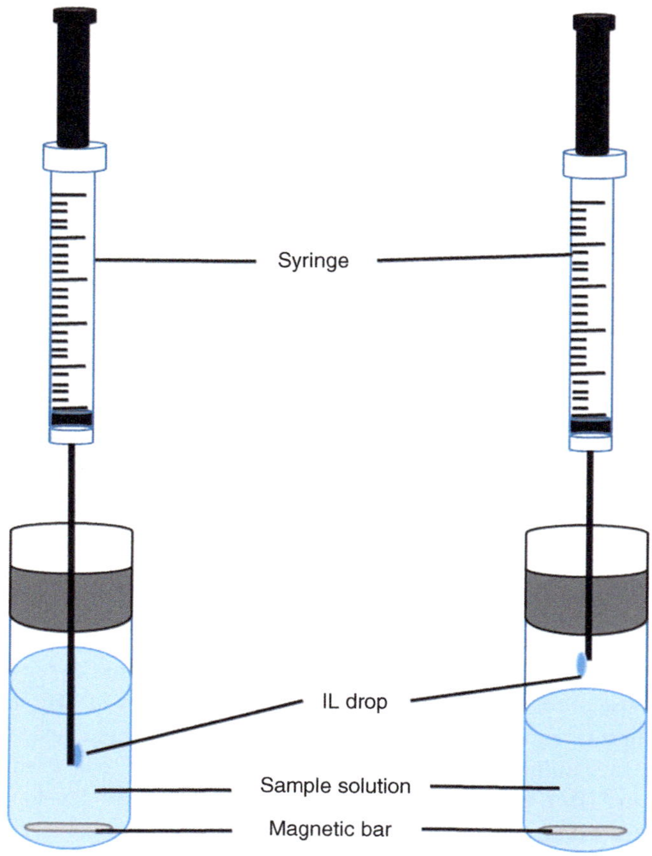

Figure 7.4 The modes of SDME employed with ionic liquids (ILs): (a) direct SDME (D-SDME); (b) headspace SDME (HS-SDME).

long extinction times, which has a beneficial effect on the target extractability as compared to organic solvents [1].

Based on the studies on ILs in the literature, thermal decomposition occurs at above 200 °C, and in some cases it does not occur even at above 450 °C [110]. The literature results also show that anions have a decisive influence on thermal stability of ILs [89]. For example, ILs containing trifluoromethanesulfonate [CF_3SO_3] and bis(trifluoromethylsulfonyl)imide [NTf_2] anions that have coordinating or less nucleophilic anions show highest thermal stability, while ILs containing highly nucleophilic nature halide anions show lowest thermal stability especially in S_N1 or S_N2 nucleophilic reactions [113]. Thus, for the ILs having same cation, 1-benzyl-3-methylimidazolium, combined with various anions [X], thermal stability increased in the following sequence (based on nucleophilic nature): thiocyanate [SCN] < *trifluoroacetate* [CF_3COO] < dicyanamide [$N(CN)_2$] < [NTf_2] ≈ [CF_3SO_3]. The influence of cations on thermal stability of ILs has been studied [114]. Hence, for a series of the ILs that contain the same anion,

[NTf_2], combined with various cations, *1-butyl-3-methylpyridinium* [C4C1Py], 1-butyl-1-methylpyrrolidinium [C4C1Pyrr], [N4,1,1,1], *butyltrimethylammonium* [CnC1IM], and *1-alkyl-3-methylimidazolium,* thermal stability increased in the following sequence: [C4C1Pyrr] > [C4C1IM] > [N4,1,1,1] > [C4C1Py]. It has also been reported that isothermal heating of ILs for long time could decrease their thermal stability. For example, the alkyl-substituted imidazolium ILs combined with the chloride anion show lower decomposition temperatures (200 °C) than those specified in the literature (350 °C).

7.3.1.2 The ILs are Liquids in a Broad Range

ILs are in the liquid state in a broad range of temperature even below room temperature (<100 °C), an advantageous property to be used in SDME where the extractant drop must be in the liquid state. But this property mainly depends on the degree of nonsymmetry of the IL cations. Thus, low symmetrical 1-ethyl-3-methylimidazolium or 1-butyl-3-methylimidazolium salts of ILs exhibit lower melting points than high symmetrical 1,3-dimethylimidazolium and 1,3-diethylimidazolium salts [115]. In the case of anions, it is not easy to understand the relationship between the structure and melting point [1].

7.3.1.3 Viscosity and Surface Tension of ILs

Viscosity is an important property that ensures the use of ILs in SDME process. High-viscosity ILs can reduce the rate of mass diffusion that extends the extraction time [1]. Usually, ILs exhibit higher viscosity (10 to <2000 mPa s) at room temperature than commonly used solvents–water, hexane, and ethylene glycol exhibit a viscosity of 0.9, 0.3, and 16.1 mPa S, respectively, at 25 °C. From the literature results, the viscosity of ILs is mainly influenced by their structure and cations. However, the viscosity of a given cation of ILs is correlated with combined anion nature and size. Thus, the viscosity of nonsymmetrical and large size (e.g. [NTf_2]) anions is relatively lower than that of nonpolar and symmetrical anions. Surface tension of solvents can favor the stability of phase separation especially multiphase systems. Higher surface tension solvents give more stability and larger drops in SDME. Usually, surface tension of ILs ranges from 18 to 73 mN/m [116]. The ILs having halide ions possess higher surface tension [117].

7.3.2 ILs in SDME Coupled with Various Analytical Detectors for Analysis of Various Analytes

To overcome the limitations of organic solvents in SDME, ILs are a promising and alternative solvent system in SDME because of their unique physicochemical properties (viscosity, thermal stability, solubility, and chemical functionality). Moreover, with respect to green technology and environmental sustainability, IL-SDME is considered an eco-friendly sample extraction technique for various analytes [118]. ILs, which comprise cations (organic) and anions (inorganic or organic), exhibit outstanding physicochemical properties including high thermal stability, negligible vapor pressure, good stability, and hydrophilicity/hydrophobicity [119], allowing them to act as ideal solvents in SDME for the preconcentration of numerous chemical species [1]. Thus, ILs have proven to be efficient solvents in SDME for the isolation and concentration of trace-level target analytes.

Various ILs have been used as solvents in SDME for the separation and pre-concentration of a wide variety of chemical species prior to their identification by analytical tools. These include 1-octyl-3-methylimidazolium hexafluorophosphate ([OMIM][PF_6]) [120–124], 1-butyl-3-methylimidazolium hexafluorophosphate ([BMIM][PF_6]) [125], tetradecyl(trihexyl)phosphonium chloride (CYPHOS-IL 101) [126, 127], [BMIM][PF_6] with 20 mM borate [128], 1-hexyl-3-methylimidazolium bis(trifluoromethylsulfonyl)-imide ([HMIM][NTf_2]) [129], [BMIM][PF_6] [130], CdSe/ZnS quantum dots dispersed in 1-hexyl-3-methylimidazolium hexafluorophosphate ([HMIM][PF_6]) [131], 1-octyl-3-methylimidazolium hexafluorophosphate [OMIM][PF_6] [132], [BMIM][PF_6] with ammonium pyrrolidinedithiocarbamate (APDC) [133], 1-hexyl-3-methylimidazolium tetrafluoroborate [HMIM][BF_4] [134], [HMIM][BF_4]–nanocellulose–MWCNTs [135], trihexyl(tetradecyl) phosphonium tetrachloromanganate(II) ([$P_{6,6,6,14}{}^+$]$_2$[$MnCl_4{}^{2-}$]) [136], aliquat tetrachloromanganate(II) ([Aliquat$^+$]$_2$[$MnCl_4{}^{2-}$]) [137], trihexyl (tetradecyl)phosphonium tris(hexafluoroacetylaceto)manganate(II) ([$P_{6,6,6,14}{}^+$] [Mn(hfacac)$_3{}^-$]) [138], [EMIM]$_2$[Co(NCS)$_4$] with a small amount of neodymium magnet [139], 1-butyl-3-methylimidazolium bis(trifluoromethylsulfonyl)imide ([BMIM][TF_2N]) [140], [$P_{6,6,6,14}{}^+$]$_2$[$MnCl_4{}^{2-}$] [141], [HMIM][PF_6] [142], three ILs ([BMIM][PF_6], [HMIM][PF_6], and [OMIM][PF_6]) [143], Fe_3O_4 sphere–[HMIM][BF_4] [144], [HMIM][NTf_2] [145], [BMIM][PF_6] [146], ZnO nanoparticles–[HMIM][PF_6] [147], and Pt nanoparticles–[BMIM][PF_6] [148]. This section summarizes the use of ILs in SDME coupled with various analytical techniques for the analysis of organic/bioorganic and inorganic molecules since last decade.

7.3.2.1 Analysis of Organic/Bioorganic Molecules

The developed IL-SDME ([OMIM][PF_6]-SDME) coupled with GC was used to successfully detect trihalomethanes in various water samples [121, 122]. IL-based SDME studies have revealed that the greater length of the alkyl-methyl-imidazolium chain in ILs increased the extraction efficiency, enhancing the selectivity and sensitivity of SDME–GC toward halocarbons [121]. Moreover, the target analytes were well extracted from the sample matrices by the addition of NaCl as ionic strength in ILs [121]. In another study, the new IL [C_6MIM][PF_6] (10 μl IL with 20 ml EtOH 1% (v/v), pH 2.0, 37 minutes) was used as a solvent in the SDME of six organic compounds (BP-3, MBC, IMC, EDB, OCR, and EMC) employed as UV filters from surface water, which were then identified by LC with UV detection [124]. The developed method showed good recovery in the range 92.00–115.00% with low LOD (μg/l).

Three aromatic amines (3-CA, 2-NA, and 4-BA) were successfully extracted and determined by SDME–LC with UV detection [141]. In this method, [HMIM][PF_6] was used as the solvent in SDME, exhibiting high EFs (132–186) for the three aromatic amines and offering LOD in the range 1.00– 2.50 μg/l. The new IL [BMIM][PF_6] with borate (40 nl) was used as the solvent for the SDME of three phenolic compounds (4-CP, 3-NP, and 2-NP), where the drop was directly connected to a CE instrument for their separation and identification [128]. To find the best IL for this SDME, three ILs, i.e. [HMIM][PF_6], [BMIM][PF_6], and [OMIM][PF_6], were investigated. Among them, [BMIM][PF_6] was found to be the best solvent for the SDME of the three phenolic compounds, owing to its lowest viscosity that enhances the phenol diffusion coefficient. Volatile and semivolatile phytochemicals (β-pinene, γ-terpinene, cuminal, *p*-cymene, and cumin alcohol) were successfully extracted from *Cuminum cyminum L.* using [OMIM][PF_6] as the solvent (1.5 μl) in SDME [143]. Similarly, ultrasonic nebulization–HS-SDME–GC–MS was developed for the analysis of limonene, hexyl acetate, and geranyl acetate from fruit juices using [HMIM][BF_4] as the solvent (12.5 μl and 5 minutes) [134]. The developed method exhibited good recovery in the range of 80.40–115.00% with RSD of <9.1%. The LOD were 0.00437, 0.0971, and 0.0989 μg/ml for limonene, hexyl acetate, and geranyl acetate, respectively. To enhance the extraction ability of ILs, nanocellulose–MWCNTs were suspended in [HMIM][BF_4] and subsequently used as the solvent for the selective isolation of HCA and MeIQx from fried food samples [135]. The extraction efficiency of the SDME was greatly improved using [HMIM][BF_4] as the solvent system owing to the dispersion of nanocomposites in the IL. This allowed extraction and preconcentration of the target analytes to be performed in a single step, and an LOD of 0.29 mg/l was achieved. Another study illustrated the use of ZnO NPs dispersed in [HMIM][PF_6] as a novel solvent in SDME coupled with HPLC for the analysis of three fungicides (kresoxim-methyl, chlorothalonil, and famoxadone) in water samples [147]. The practicability of the method for the detection of the three pesticides in water samples was explored, and good recoveries in the range of 74.94–96.11% with LOD of 0.13–0.19 ng/ml were achieved. The analytical performances of HS-SDME and LLME were evaluated for the analysis of 12 aromatic compounds by HPLC [129]. In this approach, $[P_{6,6,6,14}{}^{+}]_2[MnCl_4{}^{2-}]$ was used as the solvent for the SDME of the 12 organic compounds. The analytical results revealed that LLME exhibited rapid extraction with higher EFs for the low-vapor-pressure analytes, whereas HS-SDME presented higher extraction efficiencies for the compounds with higher vapor pressures. This HS-SDME–HPLC method showed recoveries ranging from 70.20% to 109.60% with LOD of 0.04–1.0 μg/l for all 12 aromatic compounds.

Several reports have described the procedure for the preparation of magnetic ILs, which are efficient solvent systems for the SDME of several trace-level analytes. For example, MIL ($[P_{6,6,6,14}{}^{+}]$ $[Mn(hfacac)_3{}^{-}]$) was used as a promising solvent for the SDME of short-chain fatty acids from milk prior to their identification by GC–MS [138]. In this method, 20 μl MIL was inserted in the sample solution and the SDME of the fatty acids was achieved within 60 minutes. The extraction efficiency was greatly improved by using MIL as the solvent system, with LOD as low as 14.5 μg/l and good accuracy and precision. Similarly, Fe_3O_4 sphere–[HMIM][BF_4] was successfully used as an efficient solvent for the SDME of essential oils from dried lavender [144]. The SDME was successfully completed within 20 min using 2.0 μl IL. Further, an IL ([Hmim][NTf_2])-based SMDE was successfully coupled with IMS for the selective and sensitive analysis of 2,4,6-TCA in wine and water [145]. This method exhibited an impressive LOD of 0.2 ng/l with high precision and accuracy.

The HS-SDME system was developed using the MIL $[Emim]_2[Co(NCS)_4]$ as the solvent for the SDME of nine chlorobenzenes (1,2-DCB, 1,3-DCB, 1,4-DCB, 1,2,3-TCB, 1,2,4-TCB, 1,3,5-TCB, 1,2,3,4-TCB, 1,2,3,5-TCB, and pentachlorobenzene) from water samples prior to GC–MS analysis [141]. The best extraction was observed under the following conditions: MIL (1.0 μl), 20 ml sample volume, 10.0 minutes extraction time, 1500 rpm stirring rate, and 15% NaCl. This method exhibited low LOD of 8 ng/l for 1,2,4,5-TCB and 4 ng/l for both 1,2,3,4-TCB and 1,4-DCB.

The use of ILs in SDME was also expanded to the analysis of the pharmaceutical ingredients in three drugs (MMS, IPMS, and EMS) by HPLC–DAD [141]. The extracted compounds were derivatized using sodium iodide to form iodoalkanes. Subsequently, these iodoalkanes were treated with *N*,*N*-diethyldithiocarbamate to form the corresponding derivatives, which were analyzed by HPLC–DAD. This method showed excellent recoveries in the range 86.20–107.50% with RSD of <3.5%, confirming that IL-SDME–HPLC can be used as an efficient method for the identification of MMS impurities in pharmaceutical samples. In a recent study, MIL $[P_{6,6,6,14}{}^+]_2[MnCl_4{}^{2-}]$ was successfully used as a solvent for the development of a novel high-throughput SDME–HPLC method for the parallel identification of various organic compounds (MP, EP, PP, BP, BPA, BzP, and TCC) by HPLC [141]. In this method, neodymium magnet rods were fixed in a 96-well plate system comprising pipette tips and epoxy glue, and a known amount of MIL was immersed in 1.5 ml of the sample (pH 6.0) at 500 rpm for 90 minutes. This novel method exhibited low LOD in the range 1.50–3.00 μg/l, with higher recoveries of 63.00–126.00%. Two ILs, namely $[BMIM][PF_6]$ and $[BMIM][BF_4]$, were successfully used as solvents for the SDME of MCMT from water and gasoline samples [146], of which $[BMIM][PF_6]$ displayed better extraction efficiency. Indeed, 2.5 μl $[BMIM][PF_6]$ was successfully used to extract MCMT and subsequently quantify it by ETAAS. ZnO NPs were dispersed in $[HMIM][PF_6]$ and used as a novel solvent for the SDME of three fungicides (kresoxim-methyl, chlorothalonil, and famoxadone) in water samples [147]. The extracted fungicides were successfully analyzed by HPLC, exhibiting impressive LOD in the range of 0.13–0.19 ng/ml. Another report described Pt NP–dispersed $[BMIM][PF_6]$ as an outstanding solvent in SDME for the preconcentration of bacterial proteins prior to their identification by MALDI–MS [148]. This method successfully detected bacterial species (*Escherichia coli* and *Serratia marcescens*) in concentrations as low as 106 cfu/ml. Since last decade, ILs used in SDME coupled with different analytical detectors for the analysis of organic/bioorganic molecules are listed in Table 7.1.

7.3.2.2 Inorganic Analysis

In another study, 4 μl of $[HMIM][PF_6]$ was used as the solvent for the SDME of various mercury species (Hg^{2+}, $PhHg^+$, $EtHg^+$, and $MeHg^+$) from water samples (river, tap, and wastewater) within 20 minutes [123]. The drop was easily transferred to HPLC for the separation and identification of various mercury species using tetrahydrofuran/MeOH/0.1 M HAc/AcNa + 50 μM EDTA [36/32/32, % (v/v/v); pH 4.0] as the mobile phase [123]. Three ILs, $[HMIM][PF_6]$, $[BMIM][PF_6]$, and $[OMIM][PF_6]$, were tested for their extraction ability toward mercury species,

Table 7.1 Overview of ILs in SDME coupled with various analytical detectors for the analysis of organic/bioorganic molecules.

Name of ionic solvent	Volume of ILs (μl)	Enrichment factor	Name of analyte	Linear range	LOD	Analytical detector	Real samples	References
[OMIM][PF_6]	2.0	—	Trihalomethanes	0.50–200.00 μg/l	0.50–0.90 μg/l	GC–MS	Tap, river, swimming pool, and drinking water	[121]
[OMIM][PF_6]	2.0	—	Trihalomethanes	0.10–50.00 μg/l	0.10–0.90 μg/l	GC–ECD	Tap, river, swimming pool, and drinking water	[122]
[HMIM][PF_6] 20 ml of sample containing 1% (v/v) EtOH and NaCl free adjusted to pH 2	10	98.00	BP-3	1.00–150.00 μg/l	0.11 μg/l	LC–UV	Surface water (river, sea, canal water)	[124]
		32	IMC	1.00–150.00 μg/l	0.16 μg/l			
		48.00	MBC	1.00–150.00 μg/l	0.06 μg/l			
		39.00	EDB	1.00–150.00 μg/l	0.07 μg/l			
		14.00	EMC	1.00–150.00 μg/l	0.19 μg/l			
		8.00	OCR	10.00–150.00 μg/l	3.00 μg/l			
[BMIM][PF_6] with 20.0 mM of borate (pH 10.0) including 10% methanol (v/v) as the run buffer	0.0024	156.00	4-CP	0.05–50.00 μg/ml	0.05 μg/ml	CE–UV–Vis	Yellow river water and tap water	[128]
		107.00	3-NP		0.05 μg/ml			
		257.00	2-NP		0.05 μg/l			
[Hmim][NTf_2]	2.0	—	2,4,6-TCA	0.05–25.00 ng/l	0.01 ng/l	IMS	Wine	[128]
CdSe/ZnS QDs–[HMIM][PF_6]	20	—	TMA	0.05–2.50 mg/l	0.014 mg/l	FS	Fish	[131]

(Continued)

Table 7.1 (Continued)

Name of ionic solvent	Volume of ILs (μl)	Enrichment factor	Name of analyte	Linear range	LOD	Analytical detector	Real samples	References
[OMIM][PF_6]	1.0	—	DPMI	0.050–10.00 ng/ml	0.010 ng/ml	IT–MS/MS	Water samples	[132]
			ADBI					
			AHMI					
			ATII					
			HHCB	0.070–10.00 ng/ml	0.024 ng/ml			
			AHTN	0.070–10.00 ng/ml	0.021 ng/ml			
			MX	0.10–10.00 ng/ml	0.030 ng/ml			
			MM	0.10–10.00 ng/ml	0.030 ng/ml			
			MK	0.10–10.00 ng/ml	0.030 ng/ml			
			HHCB-lactone	0.10–10.00 ng/ml	0.030 ng/ml			
[HMIM][BF_4]	12.5	—	Hexyl acetate	0.35–7.05 μg/l	0.097 μg/l	GC–MS	Fruit juices	[134]
			Limonene	0.02–1.72 μg/l	0.0044 μg/l			
			Geranyl acetate	0.36–7.16 μg/l	0.099 μg/l			
[HMIM][BF_4]–nanocellulose–MWCNTs	4.0	—	HCAMeIQx	0.1–10.00 mg/l	0.29 mg/l	DAD	Food fried samples	[135]

$[P_{6,6,6,14}^+]_2[MnCl_4^{2-}]$	2500	25.30	Acetophenone	1.00–300.00 µg/l	1.00 µg/l	HPLC–UV	Lake water	[136]
		87.00	2-Chloroaniline	1.00–300.00 µg/l	0.20 µg/l			
		36.70	α,α,α-6-Tetrafluoro-*m*-toluidine	1.00–300.00 µg/l	0.50 µg/l			
		36.30	2-Bromo-4-fluorobenzaldehyde	0.50–300.00 µg/l	0.50 µg/l			
		36.80	1-Chloro-4-nitrobenzene	1.00–300.00 µg/l	0.20 µg/l			
		38.10	4-Chlorobutyrophenone	2.00–300.00 µg/l	0.20 µg/l			
		41.20	Benzophenone	2.00–300.00 µg/l	1.00 µg/l			
		36.00	2-Nitronaphthalene	2.00–300.00 µg/l	1.00 µg/l			
		46.40	Biphenyl	0.10–60.00 µg/l	0.04 µg/l			
		44.60	Fluorene	0.20–300.00 µg/l	0.20 µg/l			
		55.60	Phenanthrene	1.00–600.00 µg/l	0.20 µg/l			
		64.30	Anthracene	1.00–300.00 µg/l	0.50 µg/l			
$[Aliquat^+]_2[MnCl_4^{2-}]$	8.0 mg	110.90	Ascorbic acid	300.0–900.0 nM	0.43 µM	MIL–SDME–voltammetry	Vitamin C and orange juice	[137]
$[P_{6,6,6,14}^+][Mn(hfacac)_3^-]$	20	7.60	Propionic acid	0.80–7.60 mg/l	216.00 µg/l	Vacuum–MIL–HS-SDME–GC–MS	Milk samples	[138]
		13.00	Isobutyric acid	0.20–8.70 mg/l	70.10 µg/l			
		11.00	*n*-Butyric acid	0.80–8.70 mg/l	47.30 µg/l			
		20.00	iso-Valeric acid	0.30–10.00 mg/l	14.50 µg/l			
		18.00	*n*-Valeric acid	0.30–5.20 mg/l	55.30 µg/l			
		33.00	*iso*-Hexanoic acid	0.30–11.00 mg/l	70.300 µg/l			
		34.00	*n*-Hexanoic acid	0.10–11.00 mg/l	21.30 µg/l			
		37.00	*n*-Heptanoic acid	0.10–13.00 mg/l	17.50 µg/l			

Table 7.1 (Continued)

Name of ionic solvent	Volume of ILs (μl)	Enrichment factor	Name of analyte	Linear range	LOD	Analytical detector	Real samples	References
$[EMIM]_2[Co(NCS)_4]$ is located on one end of a small neodymium magnet	1.0	—	1,2-DCB, 1,3-DCB, 1,4-DCB, 1,2,3-TCB, 1,2,4-TCB, 1,3,5-TCB, 1,2,3,4-TCB, 1,2,3,5-TCB and pentachlorobenzene	0.05–5.00 μg/l	4.00–8.00 ng/l	GC–MS	Tap water, pond water, and wastewater	[139]
$[BMIM][TF_2N]$	7.0	—	MMS, EMS and IPMS	0.04–1.00 μg/ml	15.00 ng/ml	HPLC–UV	Drug matrix after conversion to iodoalkanes	[140]
$[P_{6,6,6,14}{}^+]_2[MnCl_4{}^{2-}]$	10	10.00–14.00	MP, EP, PP, BP, BPA, BzP and TCC	5.00–500.00 μg/l	8.00–20.00 μg/l	HPLC–DAD	Lake, stream, and swamp water samples	[141]
$[HMIM][PF_6]$; extraction at alkaline medium (0.01 M of NaOH)	15	132.40	2-NA	10.00–1000.00 μg/l	2.50 μg/l	LC–UV	Rain and tap water	[142]
		175.80	3-CA	10.00–1000.00 μg/l	2.50 μg/l			
		186.50	4-BA	5.00–1000.00 μg/l	1.00 μg/l			
$[Hmim][NTf_2]$	2.0	—	2,4,6-TCA	1.00–10.00 ng/l	0.20 ng/l	IMS	Water	[145]
				5.00–50.00 ng/l	0.10 ng/l		Wine	
$[BMIM][PF_6]$ and $[BMIM][BF_4]$	2.5	124.00	MCMT	0.30–9.00 μg/l	10.00 ng/l	EAAS	Gasoline and water	[146]
ZnO NPs–$[HMIM][PF_6]$	10	764.00–984.00	Chlorothalonil, kresoxim-methyl, and famoxadone	—	0.13–0.19 ng/ml	HPLC–UV	Lake water, river water, and wastewater	[147]

Table 7.2 Overview of ILs in SDME coupled with various analytical detectors for the analysis of inorganic molecules.

Name of ionic solvent	Volume of ILs (μl)	Enrichment factor	Name of analyte	Linear range (μg/l)	LOD (μg/l)	Analytical detector	Real samples	References
$[HMIM][PF_6]$	4.0	—	Methyl-Hg^+	5.00–120 .00	1.00	HPLC–PAD	Tap, river, and wastewater	[123]
			Ethyl-Hg^+	10.00–200.00	1.60			
			Phenyl-Hg^+	40.00–240.00	7.10			
			Hg^{2+}	80.00–480.00	22.80			
$[BMIM][PF_6]$	7.0	76.00	Pb^{2+}	0.025–0.80	0.015	ETAAS	NIST SRM1643e Trace elements in water and NIST SRM 1549 nonfat milk powder samples	[125]
CYPHOS® IL 101 from aqueous solution at pH 9.0	4.0	—	Pb^{2+}	0.00–4.50	3.20	AAS	River, well, tap, and mineral water samples	[126]
CYPHOS- IL 101	6.0	75.00	Inorganic and organic mercury	—	10.00	ETAAS	Seawater, hair, fish tissues, and wine	[127]
$[BMIM][PF_6]$	20.0	33.00	Cu^{2+}	0.00–50.00	0.15	UV–Vis spectrometry	Food and water	[130]
$[BMIM][PF_6]$	20.0	—	Cd^{2+}	2.00–80.00	0.015	ETAAS	Water and standard reference materials	[133]
$[BMIM] [PF_6]$ with APDC	50.0	—	Cd^{2+}, As^{3+}, Pb^{2+} and Ni^{2+}	1.00–20.00	0.086, 0.049, 0.126, and 0.262	ETAAS	Eye makeup products	[149]
$[C_4C_1IM][PF_6]$	7.0	76	Pb^{2+}	0.025–0.8.00	0.015	ETAAS	Fresh water and nonfat milk powder	[150]
$[P_{14,6,6,6}][Cl]$	4.0	32	Pb^{2+}	-	0.0032	ETAAS	Tap, river, well, mineral, and fresh water	[151]

(Continued)

Table 7.2 (Continued)

Name of ionic solvent	Volume of ILs (µl)	Enrichment factor	Name of analyte	Linear range (µg/l)	LOD (µg/l)	Analytical detector	Real samples	References
$[C_4C_1IM][PF_6]$	4.0	30.3	Mn^{2+}	0.10–3.00	0.024	ETAAS	Underground water, river, tap, spring, and subterranean canal water	[152]
$[C_6C_1IM][PF_6]$	6.0	120	Co^{2+}	0.10-2.00	0.004	ETASS	Underground water, river, tap, and subterranean canal water	[153]
$[P_{14,6,6,6}][Cl]$	6.0	75	Hg^{2+} and total Hg	—	0.010	ETASS	Seawater, fish tissues, hair, and wine	[154]
$[C_4C_1IM][PF_6]$	20.0	42	Cd^{2+}	0.20–2.00	0.015	W-coil ETASS	Rice, lake and river water	[155]
$[C_4C_1IM][PF_6]$	20.0	33	Cu^{2+}	—	0.15	UV–Vis	Tea, nonfat milk powder, tap and lake water	[156]
$[C_4C_1IM][PF_6]$	2.5	50–350	Co^{2+}, Hg^{2+}, Pb^{2+}	—	—	ETV–ICP–MS	Human serum, human hair, river and lake water	[157]

with [HMIM][PF_6] exhibiting the highest extraction efficiency [123]. Similarly, 7.0 μl [BMIM][PF_6] was used as the solvent in SDME for the selective isolation of Pb^{2+} ions from water and milk samples [125]. This extraction was based on the formation of a Pb^{2+}–APDC complex and subsequent extraction using an IL as the SDME solvent. This IL-based SDME–EAAS technique exhibited a good linear range (0.025–0.80 μg/l) with an LOD of 0.015 μg/l. A trace level of Pb^{2+} ions was successfully extracted from water samples using CYPHOS® IL 101 as the solvent in SDME [126]. The Pb^{2+} ion concentration was estimated by injecting the solvent drop directly into the ETAAS system. The extraction mechanism was based on the formation of a Pb^{2+} complex with 5-Br-PADAP, offering an impressive LOD of 3.2 ng/l. Ultra-trace Cu^{2+} was successfully detected in food and water samples by [BMIM][PF_6]-assisted SDME coupled with UV–vis spectrometry [130]. This method displayed an EF of 33 with an LOD of 0.15 μg/l. Recently, [BMIM] [PF_6] with APDC was used as the solvent in the SDILNDμE of As^{3+}, Cd^{2+}, Ni^{2+}, and Pb^{2+} ions in eye makeup products [149]. The extracted elements were successfully analyzed by ETAAS. The recoveries were greatly improved, owing to the addition of APDC as a complexing agent in ILs, which allows the formation of hydrophobic metal complexes. This method effectively detected the target ions in eye makeup products, revealing its potential for the analysis of toxic elements in cosmetics. Applications of ILs in SDME coupled with different analytical detectors for the analysis of inorganic molecules are listed in Table 7.2.

The overall literature results demonstrate that methods comprising IL-based SDME coupled with various analytical detectors exhibit outstanding analytical advantages, including a very low sample volume, parallel extraction of several compounds, drop stability, and rapid extraction. Thus, such methods show great potential as an analytical tool for the analysis of various chemical compounds.

7.4 Conclusion and Future Prospects

The use of SDME coupled with green solvents ILs can reduce waste generation and enable miniaturized sample preparation and safer and rapid operation. Thus, IL-SDME is a powerful approach to develop green methodologies. However, IL-SDME has not gained popularity and is in its infancy. It may be associated with various reasons, such as difficulty in finding appropriate combinations of ILs and SDME with analytical instruments and the discouraging results derived from an inadequate selection of experimental conditions. The selection of unique properties of ILs is very important when combined with SDME that could make invaluable analytical tools, because properties of ILs largely influence their use in SDME in the most of applications studied here.

The IL-SDME coupled with various analytical instruments is a very convenient tool for preconcentration and quantification of various analytes. Also, the combination of ILs and SDME not only minimize waste production but also improve accuracy of the techniques. Moreover, with IL-SDME, the analytical performances (sensitivity, recovery, precision, and accuracy) of various analytical techniques have been greatly improved. Notably, the extraction efficiency of SDME is often superior to those of conventional extraction methods. Thus, IL-SDME is a novel, potential approach for development of green analytical methodologies for various analytes. We anticipate that further design and developments will continue to increase the use of IL-SDME for on-site extraction and preconcentration of trace-level target analytes directly from sampling fields.

References

1 Marcinkowski, Ł., Pena-Pereira, F., Kloskowski, A., and Namieśnik, J. (2015). Opportunities and shortcomings of ionic liquids in single-drop microextraction. *TrAC - Trends Anal. Chem.* 72: 153–168.
2 Anastas, P.T. and Warner, J.C. (1998). *Green Chemistry: Theory and Practice*. New York: Oxford University Press.
3 Liu, S. and Dasgupta, P.K. (1995). Liquid droplet. A renewable gas sampling interface. *Anal. Chem.* 67: 2042–2049.
4 Liu, H. and Dasgupta, P.K. (1996). Analytical chemistry in a drop. Solvent extraction in a microdrop. *Anal. Chem.* 68: 1817–1821.
5 Jeannot, M.A. and Cantwell, F.F. (1996). Solvent microextraction into a single drop. *Anal. Chem.* 68: 2236–2240.
6 Tang, S., Qi, T., Ansah, P.D. et al. (2018). Single-drop microextraction. *TrAC - Trends Anal. Chem.* 108: 306.
7 Singh, S.K. and Savoy Singh, A.W. (2020). Ionic liquids synthesis and applications: an overview. *J. Mol. Liq.* 297: 112038.
8 Kore, R. and Srivastava, R. (2011). Synthesis and applications of highly efficient, reusable, sulfonic acid group functionalized Brönsted acidic ionic liquid catalysts. *Catal. Commun.* 12: 1420–1424.
9 Hansen, F.A. and Pedersen-Bjergaard, S. (2020). Emerging extraction strategies in analytical chemistry. *Anal. Chem.* 92: 2–15.
10 Peter, W. and Wilhelm, K. (2000). Ionic liquids—new "solutions" for transition metal catalysis. *Angew. Chem. Int. Ed.* 39: 3772–3789.
11 Huang, J.-F., Baker, G.A., Luo, H. et al. (2006). Bronsted acidic room temperature ionic liquids derived from *N,N*-dimethylformamide and similar protophilic amides. *Green Chem.* 8: 599–602.
12 Fang, D., Yang, J., and Jiao, C. (2011). Dicationic ionic liquids as environmentally benign catalysts for biodiesel synthesis. *ACS Catal.* 1: 42–47.
13 Cole, A.C., Jensen, J.L., Ntai, I. et al. (2002). Novel Brønsted acidic ionic liquids and their use as dual solvent–catalysts. *J. Am. Chem. Soc.* 124: 5962–5963.
14 Matsagar, B.M. and Dhepe, P.L. (2015). Bronsted acidic ionic liquid-catalyzed conversion of hemicellulose into sugars. *Catal. Sci. Technol.* 5: 531–539.
15 Zhang, Z., Song, J., and Han, B. (2017). Catalytic transformation of lignocellulose into chemicals and fuel products in ionic liquids. *Chem. Rev.* 117: 6834–6880.

16 Singh, S.K. and Dhepe, P.L. (2016). Ionic liquids catalyzed lignin liquefaction: mechanistic studies using TPO-MS, FT-IR, RAMAN and 1D, 2D-HSQC/NOESY NMR. *Green Chem.* 18: 4098–4108.

17 Singh, S.K. and Dhepe, P.L. (2018). Effect of structural properties of organosolv lignins isolated from different rice husks on their liquefaction using acidic ionic liquids. *Clean Techn. Environ. Policy* 20: 739–750.

18 Singh, S.K., Banerjee, S., Vanka, K., and Dhepe, P.L. (2018). Understanding interactions between lignin and ionic liquids with experimental and theoretical studies during catalytic depolymerisation. *Catal. Today* 309: 98–108.

19 Hurley, F.H. (1948). US Patent 2,446,331.

20 Wier, T.P. and Hurley, F.H. (1948). US Patent 2,446,349.

21 Wier, T.P. (1948). US Patent 2,446,350.

22 Wilkes, J.S. and Zaworotko, M.J. (1992). Air and water stable 1-ethyl-3-methylimidazolium based ionic liquids. *Chem. Commun.* 965: 965–967.

23 Welton, T. (1999). Room-temperature ionic liquids. Solvents for synthesis and catalysis. *Chem. Rev.* 99: 2071–2084.

24 Hallett, J.P. and Welton, T. (2011). Room-temperature ionic liquids: solvents for synthesis and catalysis. *Chem. Rev.* 111: 3508–3576.

25 Fox, D.M., Awad, W.H., Gilman, J.W. et al. (2003). Flammability, thermal stability, and phase change characteristics of several trialkylimidazolium salts. *Green Chem.* 5: 724–727.

26 Singh, S.K. (2019). Solubility of lignin and chitin in ionic liquids and their biomedical applications. *Int. J. Biol. Macromol.* 132: 265–277.

27 Greaves, T.L. and Drummond, C.J. (2008). Protic ionic liquids: properties and applications. *Chem. Rev.* 108: 206–237.

28 Marcus, Y. (2016). *Ionic Liquid Properties: From Molten Salts to RTILs*. Springer.

29 Ma, C., Laaksonen, A., Liu, C. et al. (2018). The peculiar effect of water on ionic liquids and deep eutectic solvents. *Chem. Soc. Rev.* 47: 8685–8720.

30 Yang, Z. and Pan, W. (2005). Ionic liquids: green solvents for nonaqueous biocatalysis. *Enzym. Microb. Technol.* 37: 19–28.

31 Rivera-Rubero, S. and Baldelli, S. (2004). Influence of water on the surface of hydrophilic and hydrophobic room-temperature ionic liquids. *J. Am. Chem. Soc.* 126: 11788–11789.

32 Widegren, J.A., Laesecke, A., and Magee, J.W. (2005). The effect of dissolved water on the viscosities of hydrophobic room-temperature ionic liquids. *Chem. Commun.*: 1610–1612. https://doi.org/10.1039/B417348A.

33 Huddleston, J.G., Visser, A.E., Reichert, W.M. et al. (2001). Characterization and comparison of hydrophilic and hydrophobic room temperature ionic liquids incorporating the imidazolium cation. *Green Chem.* 3: 156–164.

34 Giernoth, R. (2010). Task-specific ionic liquids. *Angew. Chem. Int. Ed.* 49: 2834–2839.

35 Sawant, A.D., Raut, D.G., Darvatkar, N.B., and Salunkhe, M.M. (2011). Recent developments of task-specific ionic liquids in organic synthesis. *Green. Chem. Lett. Rev.* 4: 41–54.

36 Bates, E.D., Mayton, R.D., Ntai, I., and Davis, J.H. (2002). CO_2 capture by a task-specific ionic liquid. *J. Am. Chem. Soc.* 124: 926–927.

37 Quijada-Maldonado, E., Sánchez, F., Pérez, B. et al. (2018). Task-specific ionic liquids as extractants for the solvent extraction of molybdenum(VI) from aqueous solution using different commercial ionic liquids as diluents. *Ind. Eng. Chem. Res.* 57: 1621–1629.

38 Nockemann, P., Thijs, B., Pittois, S. et al. (2006). Task-specific ionic liquid for solubilizing metal oxides. *J. Phys. Chem. B* 110: 20978–20992.

39 Avtar, S. and Harish Kumar, C. (2017). Chiral ionic liquids: design, synthesis and applications in asymmetric organo-catalysis. *Curr. Org. Synth.* 14: 488–510.

40 Suzuki, Y., Wakatsuki, J., Tsubaki, M., and Sato, M. (2013). Imidazolium-based chiral ionic liquids: synthesis and application. *Tetrahedron* 69: 9690–9700.

41 Rodríguez-Cárdenas, E., Cardoso-Martínez, J., Nieto-Camacho, A., and Frontana-Uribe, B.A. (2017). Physical-chemical properties of chiral ionic liquids derived from the phenylethylamine enantiomers. *J. Mol. Liq.* 236: 435–444.

42 Baudequin, C., Baudoux, J., Levillain, J. et al. (2003). Ionic liquids and chirality: opportunities and challenges. *Tetrahedron: Asymmetry* 14: 3081–3093.

43 Singh, A. and Chopra, H.K. (2018). Synthesis, characterization and applications of some novel DMAP-based chiral ionic liquids. *J. Mol. Liq.* 266: 106–111.

44 Yinghuai, Z., Yuanting, K.T., and Hosmane, N.S. (2013). Applications of ionic liquids in lignin chemistry, ionic liquids - new aspects for the future. In: *Ionic Liquids - New Aspects for the Future* (ed. D.J.-I. Kadokawa). InTech.

45 Boyd, A.R., Jessop, P.G., Dust, J.M., and Buncel, E. (2013). Switchable polarity solvent (SPS) systems: probing solvatoswitching with a spiropyran (SP)-merocyanine (MC) photoswitch. *Org. Biomol. Chem.* 11: 6047–6055.

46 Pavlovica, S., Zicmanis, A., Gzibovska, E. et al. (2011). (2-Hydroxyethyl)ammonium lactates—highly biodegradable and essentially non-toxic ionic liquids. *Green Sustain. Chem.* 01, 8 (03).

47 Qian, W., Texter, J., and Yan, F. (2017). Frontiers in poly(ionic liquid)s: syntheses and applications. *Chem. Soc. Rev.* 46: 1124–1159.

48 Kausar, A. (2017). Research progress in frontiers of poly(ionic liquid)s: a review. *Polym. Plast. Technol. Eng.* 56: 1823–1838.

49 Shaplov, A.S., Ponkratov, D.O., and Vygodskii, Y.S. (2016). Poly(ionic liquid)s: synthesis, properties, and application. *Polym. Sci., Ser. B* 58: 73–142.

50 Yuan, J. and Antonietti, M. (2011). Poly(ionic liquid)s: polymers expanding classical property profiles. *Polymer* 52: 1469–1482.

51 Mehta, R., Kumar, V., Bhunia, H., and Upadhyay, S.N. (2005). Synthesis of poly(lactic acid): a review. *J. Macromol. Sci., Part C* 45: 325–349.

52 Vasko, J., German, C., Hans, G., and David, M. (2012). Comparison between two different synthetic routes of pyrrolidinium functional polymeric ionic liquids. *Macromol. Symp.* 311: 77–82.

53 Tindale, J.J., Hartlen, K.D., Alizadeh, A. et al. (2010). Maleimide-modified phosphoniumionic liquids: a template towards (multi)task-specific ionic liquids. *Chem. Eur. J.* 16: 9068–9075.

54 Zhu, H.-P., Yang, F., Tang, J., and He, M.-Y. (2003). Bronsted acidic ionic liquid 1- methylimidazolium tetrafluoroborate: a green catalyst and recyclable medium for esterification. *Green Chem.* 5: 38–39.

55 Driver, G. and Johnson, K.E. (2003). 3-Methylimidazolium bromohydrogenates(I): a room temperature ionic liquid for ether cleavage. *Green Chem.* 5: 163–169.

56 Joseph, T., Sahoo, S., and Halligudi, S.B. (2005). Brönsted acidic ionic liquids: a green, efficient and reusable catalyst system and reaction medium for Fischer esterification. *J. Mol. Catal. A Chem.* 234: 107–110.

57 Zheng, Y., Zheng, Y., Yang, S. et al. (2017). Esterification synthesis of ethyl oleate catalyzed by Brønsted acid–surfactant-combined ionic liquid. *Green Chem. Lett. Rev.* 10: 202–209.

58 Kemperman, G.J., Roeters, T.A., and Hilberink, P.W. (2003). Cleavage of aromatic methyl ethers by chloroaluminate ionic liquid reagents. *Eur. J. Org. Chem.* 2003: 1681–1686.

59 Yadav, J.S., Reddy, B.V.S., and Baishya, G. (2003). Green protocol for conjugate addition of thiols to α,β-unsaturated ketones using a [Bmim]PF_6/H_2O system. *J. Organomet. Chem.* 68: 7098–7100.

60 Ranu, B.C. and Dey, S.S. (2004). Catalysis by ionic liquid: a simple, green and efficient procedure for the Michael addition of thiols and thiophosphate to conjugated alkenes in ionic liquid, [pmIm]Br. *Tetrahedron* 60: 4183–4188.

61 Zhang, F.-Y. and Corey, E.J. (2000). Highly enantioselectiveMichael reactions catalyzed by a chiral quaternary ammonium salt. Illustration by asymmetric syntheses of (S)-ornithine and chiral 2-cyclohexenones. *Org. Lett.* 2: 1097–1100.

62 Corey, E.J. and Zhang, F.-Y. (1999). Mechanism and conditions for highly enantioselective epoxidation of α,β-enones using charge-accelerated catalysis by a rigid quaternary ammonium salt. *Org. Lett.* 1: 1287–1290.

63 McNulty, J., Cheekoori, S., Nair, J.J. et al. (2005). A mild esterification process in phosphonium salt ionic liquid. *Tetrahedron Lett.* 46: 3641–3644.

64 Jorapur, Y.R., Lee, C.-H., and Chi, D.Y. (2005). Mono- and dialkylations of pyrrole at C2 and C5 positions by nucleophilic substitution reaction in ionic liquid. *Org. Lett.* 7: 1231–1234.

65 Yang, L., Xu, L.-W., Zhou, W. et al. (2006). Highly efficient aza-Michael reactions of aromatic amines and N-heterocycles catalyzed by a basic ionic liquid under solvent-free conditions. *Tetrahedron Lett.* 47: 7723–7726.

66 Ranu, B.C. and Banerjee, S. (2005). Ionic liquid as catalyst and reaction medium. The dramatic influence of a task-specific ionic liquid, [bmIm]OH, in Michael addition of active methylene compounds to conjugated ketones, carboxylic esters, and nitriles. *Org. Lett.* 7: 3049–3052.

67 Xu, J.-M., Qian, C., Liu, B.-K. et al. (2007). A fast and highly efficient protocol for Michael addition of N-heterocycles to α,β-unsaturated compound using basic ionic liquid [bmIm]OH as catalyst and green solvent. *Tetrahedron* 63: 986–990.

68 Mehnert, C.P., Dispenziere, N.C., and Cook, R.A. (2002). Preparation of C9-aldehyde via aldol condensation reactions in ionic liquid media. *Chem. Commun.*: 1610–1611. https://doi.org/10.1039/B203068C.

69 Hajipour, A.R. and Rafiee, F. (2015). Recent progress in ionic liquids and their applications in organic synthesis. *Org. Prep. Proced. Int.* 47: 249–308.

70 Olivier-Bourbigou, H., Magna, L., and Morvan, D. (2010). Ionic liquids and catalysis: recent progress from knowledge to applications. *Appl. Catal. A Gen.* 373: 1–56.

71 Roy, K., Kar, S., and Das, R.N. (2015). Chapter 12 - Future Avenues. In: *Understanding the Basics of QSAR for Applications in Pharmaceutical Sciences and Risk Assessment* (eds. K. Roy, S. Kar and R.N. Das), 455–462. Boston: Academic Press.

72 Zhao, D., Wu, M., Kou, Y., and Min, E. (2002). Ionic liquids: applications in catalysis. *Catal. Today* 74: 157–189.

73 Plechkova, N.V. and Seddon, K.R. (2008). Applications of ionic liquids in the chemical industry. *Chem. Soc. Rev.* 37: 123–150.

74 Matsagar, B.M., Munshi, M.K., Kelkar, A.A., and Dhepe, P.L. (2015). Conversion of concentrated sugar solutions into 5-hydroxymethyl furfural and furfural using Bronsted acidic ionic liquids. *Catal. Sci. Technol.* 5: 5086–5090.

75 MacFarlane, D.R., Tachikawa, N., Forsyth, M. et al. (2014). Energy applications of ionic liquids. *Energy Environ. Sci.* 7: 232–250.

76 Liang, L., Gan, Q., and Nancarrow, P. (2014). Composite ionic liquid and polymer membranes for gas separation at elevated temperatures. *J. Membr. Sci.* 450: 407–417.

77 Weidmann, S., Kemmerling, S., Madler, S. et al. (2012). Ionic liquids as matrices in microfluidic sample deposition for high-mass matrix-assisted laser desorption/ionization mass spectrometry. *Eur. J. Mass Spectrom. (Chichester, Eng.)* 18: 279–286.

78 Wang, H., Lu, Q., Ye, C. et al. (2004). Friction and wear behaviors of ionic liquid of alkylimidazolium hexafluorophosphates as lubricants for steel/steel contact. *Wear* 256: 44–48.

79 Armstrong, D.W., Zhang, L.K., He, L., and Gross, M.L. (2001). Ionic liquids as matrixes for matrix-assisted laser desorption/ionization mass spectrometry. *Anal. Chem.* 73: 3679–3686.

80 Andre, M., Loidl, J., Schottenberger, H. et al. (2005). Ionic liquids as advantageous solvents for headspace gas chromatography of compounds with low vapor pressure. *Anal. Chem.* 77: 702–705.

81 Weyershausen, B. and Lehmann, K. (2005). Industrial application of ionic liquids as performance additives. *Green Chem.* 7: 15–19.

82 Moriguchi, T., Yanagi, T., and Kunimori, M. (2000). Synthesis and properties of aminoacylamido-AMP: chemical optimization for the construction of an N-acyl phosphoramidate linkage. *J. Org. Chem.* 65: 8229–8238.

83 Hayakawa, Y., Kawai, R., and Hirata, A. (2001). Acid/azole complexes as highly effective promoters in the synthesis of DNA and RNA oligomers via the phosphoramidite method. *J. Am. Chem. Soc.* 123: 8165–8176.

84 Liu, Y., Wang, M., Li, J. et al. (2005). Highly active horseradish peroxidase immobilized in 1-butyl-3-methylimidazolium tetrafluoroborate room-temperature ionic liquid based sol–gel host materials. *Chem. Commun.* 13: 1778–1780.

85 Picquet, M., Tkatchenko, I., Tommasi, I. et al. (2003). Ionic liquids, 3. Synthesis and utilisation of protic imidazolium salts in homogeneous catalysis. *Adv. Synth. Catal.* 345: 959–962.

86 Bösmann, A., Datsevich, L., Jess, A. et al. (2001). Deep desulfurization of diesel fuel by extraction with ionic liquids. *Chem. Commun.* 23: 2494–2495.

87 Stathatos, E., Lianos, P., Zakeeruddin, S.M. et al. (2003). A quasi-solid-state dye-sensitized solar cell based on a sol–gel nanocomposite electrolyte containing ionic liquid. *Chem. Mater.* 15: 1825–1829.

88 Zhou, Y. and Antonietti, M. (2003). Synthesis of very small TiO_2 nanocrystals in a room-temperature ionic liquid and their self-assembly toward mesoporous spherical aggregates. *J. Am. Chem. Soc.* 125: 14960–14961.

89 Holbrey, J.D. and Seddon, K.R. (1999). The phase behaviour of 1-alkyl-3-methylimidazolium tetrafluoroborates; ionic liquids and ionic liquid crystals. *Dalton Trans.* 13: 2133–2139.

90 Haristoy, D. and Tsiourvas, D. (2003). Novel ionic liquid-crystalline compounds bearing oxadiazole and pyridinium moieties as prospective materials for optoelectronic applications. *Chem. Mater.* 15: 2079–2083.

91 Reddy, R.G., Zhang, Z., Arenas, M.F., and Blake, D.M. (2003). Thermal stability and corrosivity evaluations of ionic liquids as thermal energy storage media. *High Temp. Mater. Process* 22: 87–94.

92 Yanes, E.G., Gratz, S.R., and Baldwin, M.J. (2001). Capillary electrophoretic application of 1-alkyl-3-methylimidazolium-based ionic liquids. *Anal. Chem.* 73: 3838–3844.

93 Sato, T., Masuda, G., and Takagi, K. (2004). Electrochemical properties of novel ionic liquids for electric double layer capacitor applications. *Electrochim. Acta* 49: 3603–3611.

94 Zell, C.A. and Freyland, W. (2003). In situ STM and STS study of Co and Co–Al alloy electrodeposition from an ionic liquid. *Langmuir* 19: 7445–7450.

95 Scheeren, C.W., Machado, G., Dupont, J. et al. (2003). Nanoscale Pt (0) particles prepared in imidazolium room temperature ionic liquids: synthesis from an organometallic precursor, characterization, and catalytic properties in hydrogenation reactions. *Inorg. Chem.* 42: 4738–4742.

96 He, L., Zhang, W., Zhao, L. et al. (2003). Effect of 1-alkyl-3-methylimidazolium-based ionic liquids as the eluent on the separation of ephedrines by liquid chromatography. *J. Chromatogr. A* 1007: 39–45.

97 Scovazzo, P., Kieft, J., Finan, D.A. et al. (2004). Gas separations using non-hexafluorophosphate $[PF6]^-$ anion supported ionic liquid membranes. *J. Membr. Sci.* 238: 57–63.

98 Jork, C., Seiler, M., Beste, Y.-A., and Arlt, W. (2004). Influence of ionic liquids on the phase behavior of aqueous azeotropic systems. *J. Chem. Eng. Data* 49: 852–857.

99 Uerdingen, M. (2004). Entschwefelung von Dieselkraftstoff. *Chem. Unserer Zeit* 38: 212–213.

100 Wang, J., Luo, J., Feng, S. et al. (2016). Recent development of ionic liquid membranes. *Green Energy Environ.* 1: 43–61.

101 Fortunato, R., Afonso, C.A.M., Reis, M.A.M., and Crespo, J.G. (2004). Supported liquid membranes using ionic liquids: study of stability and transport mechanisms. *J. Membr. Sci.* 242: 197–209.

102 Opallo, M. and Lesniewski, A. (2011). A review on electrodes modified with ionic liquids. *J. Electroanal. Chem.* 656: 2–16.

103 Shiddiky, M.J.A. and Torriero, A.A.J. (2011). Application of ionic liquids in electrochemical sensing systems. *Biosens. Bioelectron.* 26: 1775–1787.

104 Wen, X., Deng, Q., and Guo, J. (2011). Ionic liquid-based single drop microextraction of ultratrace copper in food and water samples before spectrophotometric determination. *Spectrochim. Acta A: Mol. Biomol. Spectrosc.* 79: 1941–1945.

105 Márquez-Sillero, I., Cárdenas, S., and Valcárcel, M. (2011). Direct determination of 2,4,6-tricholoroanisole in wines by single-drop ionic liquid microextraction coupled with multicapillary column separation and ion mobility spectrometry detection. *J. Chromatogr. A* 1218: 7574–7580.

106 Ahmad, F. and Wu, H.-F. (2011). Characterization of pathogenic bacteria using ionic liquid via single drop microextraction combined with MALDI-TOF MS. *Analyst* 136: 4020–4027.

107 Carrillo-Carrion, C., Simonet, B.M., and Valcarcel, M. (2012). (CdSe/ZnS QDs)-ionic liquid-based headspace single drop microextraction for the fluorimetric determination of trimethylamine in fish. *Analyst* 137: 1152–1159.

108 Psillakis, E. and Kalogerakis, N. (2002). Developments in single-drop microextraction. *TRAC - Trends Anal. Chem.* 21: 53–63.

109 Han, X. and Armstrong, D.W. (2007). Ionic liquids in separations. *Acc. Chem. Res.* 40: 1079–1086.

110 Maton, C., De Vos, N., and Stevens, C.V. (2013). Ionic liquid thermal stabilities: decomposition mechanisms and analysis tools. *Chem. Soc. Rev.* 42: 5963–5977.

111 Xia, L., Li, X., Yunli, W. et al. (2008). Ionic liquids based single drop microextraction combined with electrothermal vaporization inductively coupled plasma mass spectrometry for determination of Co, Hg and Pb in biological and environmental samples. *Spectrochim. Acta Part B: Atom. Spectrosc.* 63 (11): 1290–1296.

112 Sun, X., Luo, H., and Dai, S. (2012). Ionic liquids-based extraction: a promising strategy for the advanced nuclear fuel cycle. *Chem. Rev.* 112: 2100–2128.

113 Awad, W.H., Gilman, J.W., Nyden, M. et al. (2004). Thermal degradation studies of alkyl-imidazolium salts and their application in nanocomposites. *Thermochim. Acta* 409: 3–11.

114 Tokuda, H., Ishii, K., Susan, A.B.H. et al. (2006). Physicochemical properties and structures of room-temperature ionic liquids. 3. Variation of cationic structures. *J. Phys. Chem. B* 110: 2833–2839.

115 Zhang, S., Sun, N., He, X. et al. (2006). Physical properties of ionic liquids: database and evaluation. *J. Phys. Chem. Ref. Data* 35: 1475–1517.

116 Mackay, D., Shiu, W.-Y., Ma, K.-C., and Lee, S.C. (2006). *Handbook of Physical-Chemical Properties and Environmental Fate for Organic Chemicals*, 2nde. Boca Raton, FL: CRC Press/Taylor & Francis.

117 Freire, M.G., Carvalho, P.J., Fernandes, A.M. et al. (2007). Surface tensions of imidazolium based ionic liquids: anion, cation, temperature and water effect. *J. Colloid Interf. Sci.* 314: 621–630.

118 Ho, T.D., Canestraro, A.J., and Anderson, J.L. (2011). Ionic liquids in solid-phase microextraction: a review. *Anal. Chim. Acta* 695: 18.

119 Trujillo-Rodríguez, M.J., Rocío-Bautista, P., Pino, V., and Afonso, A.M. (2013). Ionic liquids in dispersive liquid-liquid microextraction. *TrAC - Trends Anal. Chem.* 51: 87.

120 Liu, L., Cheng, J., Matsadiq, G. et al. (2010). Application of DLLME to the determination of pyrethroids in aqueous samples. *Chromatographia* 72: 1017.

121 Aguilera-Herrador, E., Lucena, R., Cardenas, S., and Valcarcel, M. (2008). Determination of trihalomethanes in waters by ionic liquid-based single drop microextraction/gas chromatographic/mass spectrometry. *J. Chromatogr. A* 1209: 76.

122 Aguilera-Herrador, E., Lucena, R., Cárdenas, S., and Valcárcel, M. (2009). Ionic liquid-based single drop microextraction and room-temperature gas chromatography for on-site ion mobility spectrometric analysis. *J. Chromatogr. A* 1216: 5580.

123 Pena-Pereira, F., Lavilla, I., Bendicho, C. et al. (2009). Speciation of mercury by ionic liquid-based single-drop microextraction combined with high-performance liquid chromatography-photodiode array detection. *Talanta* 78: 537.

124 Vidal, L., Chisvert, A., Canals, A., and Salvador, A. (2010). Ionic liquid-based single-drop microextraction followed by liquid chromatography-ultraviolet spectrophotometry detection to determine typical UV filters in surface water samples. *Talanta* 81: 549.

125 Manzoori, J.L., Amjadi, M., and Abulhassani, J. (2009). Ultra-trace determination of lead in water and food samples by using ionic liquid-based single drop microextraction-electrothermal atomic absorption spectrometry. *Anal. Chim. Acta* 644: 48.

126 Martinis, E.M., Bertón, P., Altamirano, J.C. et al. (2010). Tetradecyl (trihexyl) phosphonium chloride ionic liquid single-drop microextraction for electrothermal atomic absorption spectrometric determination of lead in water samples. *Talanta* 80: 2034.

127 Martinis, E.M. and Wuilloud, R.G. (2010). Cold vapor ionic liquid-assisted headspace single-drop microextraction: a novel preconcentration technique for mercury species determination in complex matrix samples. *J. Anal. Atom. Spectrom.* 25: 1432.

128 Wang, Q., Qiu, H., Li, J. et al. (2010). On-line coupling of ionic liquid-based single-drop microextraction with capillary electrophoresis for sensitive detection of phenols. *J. Chromatogr. A* 1217: 5434.

129 Márquez-Sillero, I., Cárdenas, S., and Valcárcel, M. (2011). Direct determination of 2,4,6-tricholoroanisole in wines by single-drop ionic liquid microextraction coupled with multicapillary column separation and ion mobility spectrometry detection. *J. Chromatogr. A* 1218: 7574.

130 Wen, X., Deng, Q., and Guo, J. (2011). Ionic liquid-based single drop microextraction of ultra-trace copper in food and water samples before spectrophotometric determination. *Spectrochim. Acta, Part A* 79: 1941.

131 Carrillo-Carrión, C., Simonet, B.M., and Valcárcel, M. (2012). (CdSe/ZnS QDs)-ionic liquid-based headspace single drop microextraction for the fluorimetric determination of trimethylamine in fish. *Analyst* 137: 1152.

132 Vallecillos, L., Pocurull, E., and Borrull, F. (2012). Fully automated ionic liquid-based headspace single drop microextraction coupled to GC–MS/MS to determine musk fragrances in environmental water samples. *Talanta* 99: 824.

133 Wen, X., Deng, Q., Wang, J. et al. (2013). A new coupling of ionic liquid based-single drop microextraction with tungsten coil electrothermal atomic absorption spectrometry. *Spectrochim. Acta, Part A* 105: 320.

134 Jiang, C., Wei, S., Li, X. et al. (2013). Ultrasonic nebulization headspace ionic liquid-based single drop microextraction of flavour compounds in fruit juices. *Talanta* 106: 237.

135 Ruiz-Palomero, C., Soriano, M.L., and Valcárcel, M. (2014). Ternary composites of nanocellulose, carbonanotubes and ionic liquids as new extractants for direct immersion single drop microextraction. *Talanta* 125: 72.

136 An, J., Rahn, K.L., and Anderson, J.L. (2017). Headspace single drop microextraction versus dispersive liquid-liquid microextraction using magnetic ionic liquid extraction solvents. *Talanta* 167: 268.

137 Jahromi, Z., Mostafavi, A., Shamspur, T., and Mohamadim, M. (2017). Magnetic ionic liquid assisted single-drop microextraction of ascorbic acid before its voltammetric determination. *J. Sep. Sci.* 40: 4041.

138 Trujillo-Rodríguez, M.J., Pino, V., and Anderson, J.L. (2017). Magnetic ionic liquids as extraction solvents in vacuum headspace single-drop microextraction. *Talanta* 172: 86.

139 Fernández, E., Vidal, L., and Canals, A. (2018). Hydrophilic magnetic ionic liquid for magnetic headspace single-drop microextraction of chlorobenzenes prior to thermal desorption-gas chromatography-mass spectrometry. *Anal. Bioanal. Chem.* 410: 4679.

140 Li, M., Gu, C., Luo, L. et al. (2019). Determination of trace methanesulfonates in drug matrix using derivatization and headspace single drop microextraction followed by high-performance liquid chromatography with ultraviolet detection. *J. Chromatogr. A* 1591: 131.

141 Mafra, G., Vieira, A.A., Merib, J. et al. (2019). Single drop microextraction in a 96-well plate format: A step toward automated and high-throughput analysis. *Anal. Chim. Acta* 1063: 159.

142 Sarafraz-Yazdi, A. and Mofazzeli, F. (2010). Ionic liquid-based submerged single drop microextraction: a new method for the determination of aromatic amines in environmental water samples. *Chromatographia* 72: 867.

143 Yang, J., Wei, H., Yu, C. et al. (2012). Extraction of the volatile and semivolatile compounds in seeds of *Cuminum cyminum* L. using hydrodistillation followed by headspace-ionic liquid-based single-drop microextraction. *Chromatographia* 75: 1435.

144 Fu, J., Zhu, Y., Liu, B., and Tang, J. (2018). Fe_3O_4 sphere-assisted microwave distillation coupled with ionic liquid-based HS-SDME followed by GC-MS for the rapid analysis of essential oil in dried lavender. *Anal. Methods* 10: 652–659.

145 Márquez-Sillero, I., Aguilera-Herrador, E., Cárdenas, S., and Valcárcel, M. (2011). Determination of 2,4,6-tricholoroanisole in water and wine samples by ionic liquid-based single-drop microextraction and ion mobility spectrometry. *Anal. Chim. Acta* 702: 199.

146 Rahmani, M. and Kaykhaii, M. (2011). Determination of methylcyclopentadienyl-manganese tricarbonyl in gasoline and water via ionic-liquid headspace single drop microextraction and electrothermal atomic absorption spectrometry. *Microchim. Acta* 174: 413.

147 Amde, M., Tan, Z.-Q., Liu, R., and Liu, J.-F. (2015). Nanofluid of zinc oxide nanoparticles in ionic liquid for single drop liquid microextraction of fungicides in environmental waters prior to high performance liquid chromatographic analysis. *J. Chromatogr. A* 1395: 7.

148 Ahmad, F. and Wu, H.-F. (2011). Characterization of pathogenic bacteria using ionic liquid via single drop microextraction combined with MALDI-TOF MS. *Analyst* 136: 4020.

149 Akhtar, A., Kazi, T.G., Afridi, H.I. et al. (2020). Simultaneous preconcentration of toxic elements in eye makeup products through single drop ionic liquid based non-dispersive microextraction method using narrow glass column: multivariate application. *Microchem. J.* 157: 104963.

150 Manzoori, J.L., Amjadi, M., and Abulhassani, J. (2009). Ultra-trace determination of lead in water and food samples by using ionic liquid-based single drop microextraction-electrothermal atomic absorption spectrometry. *Anal. Chim. Acta* 644: 48–52.

151 Martinis, E.M., Berton, P., Altamirano, J.C. et al. (2010). Tetradecyl(trihexyl)phosphonium chloride ionic liquid single-drop microextraction for electrothermal atomic absorption spectrometric determination of lead in water samples. *Talanta* 80: 2034–2040.

152 Manzoori, J.L., Amjadi, M., and Abulhassani, J. (2009). Ionic liquid-based single drop microextraction combined with electrothermal atomic absorption spectrometry for the determination of manganese in water samples. *Talanta* 77: 1539–1544.

153 Amjadi, M., Manzoori, J.L., and Abulhassani, J. (2010). Ionic liquid-based, single-drop microextraction for preconcentration of cobalt before its determination by electrothermal atomic absorption spectrometry. *J. AOAC Int.* 93: 985–991.

154 Martinis, E.M. and Wuilloud, R.G. (2010). Cold vapor ionic liquid-assisted headspace single-drop microextraction: a novel preconcentration technique for mercury species determination in complex matrix samples. *J. Anal. Atom. Spectrom.* 25: 1432–1439.

155 Wen, X., Deng, Q., Wang, J. et al. (2013). A new coupling of ionic liquid based-single drop microextraction with tungsten coil electrothermal atomic absorption spectrometry. *Spectrochim. Acta A* 105: 320–325.

156 Wen, X., Deng, Q., and Guo, J. (2011). Ionic liquid-based single drop microextraction of ultra-trace copper in food and water samples before spectrophotometric determination. *Spectrochim. Acta A* 79: 1941–1945.

157 Xia, L., Li, X., Wu, Y. et al. (2008). Ionic liquids based single drop microextraction combined with electrothermal vaporization inductively coupled plasma mass spectrometry for determination of Co, Hg and Pb in biological and environmental samples. *Spectrochim. Acta B* 63: 1290–1296.

8

Functionalized 2D Nanomaterials for Miniaturized Analytical Devices

Thang P. Nguyen[1,2]

[1] *Hanoi University of Science and Technology, Department of Electronic Materials, School of Engineering Physics, No. 1, Dai Co Viet Street, 100000 Ha Noi, Vietnam*
[2] *Gachon University, Department of Chemical and Biological Engineering, Seongnam-si, Gyeonggido, 13120, South Korea*

8.1 Introduction

In modern life, nanotechnology is being a key solution to many issues. For example, nanomaterials are used in food packing as they improve stability, act as a gas barrier to polymers, remove toxicity or color in food processing and magnetic nanoparticles are used in cancer treatment, etc. [1, 2]. The great advantages come from the isolation of tiny materials from the bulk and the very high surface area to volume ratio which enhanced the physical/electrical/optical properties, catalytic properties, etc. The development of nanoscience has been tremendous, which is evident through numerous research publications in this field from the twentieth century until now [3, 4]. For analytical tools, the measurement of inorganic/organic materials with a size in micrometers or nanometers could not be easily resolved in traditional equipment in the microscopic level. The nanometer dimension is close to the wavelength of light and the structural unit size; therefore, a new phenomenon quantum confinement is observed that has never ever been measured in bulk size materials [5, 6]. Moreover, the smaller the size is, the larger the surface area, which leads to huge development in surface science, especially with catalytic reactions or optical applications [7]. The interaction of nanomaterials with matter or light will be stronger, increasing the ability of detecting a very low quantity (micro-, nano-, pico-, mol/liter, etc.). Therefore, the development of nanotechnology is going along with many new methods for detecting and measuring chemicals, biochemicals, and organic cells [8–10].

The achievement of nanotechnology is rapid introduction of miniaturized analytical devices, which are devices with a small size in millimeters or micrometers and which can be used in the synthesis, analysis, and detection of chemical compounds [11]. They are also called lab-on-a-chip (LOC) systems, which implies that laboratory functions are integrated on a single chip in tiny size. A complete LOC system requires many different optimizing steps in analytical procedures such as sample preparation,

Miniaturized Analytical Devices: Materials and Technology, First Edition.
Edited by Suresh Kumar Kailasa and Chaudhery Mustansar Hussain.

analytical separation, mechanism of detection, automation, reproducibility, portability [12].

This chapter provides an overview of various functionalized 2D nanomaterials (F2NMs) for miniaturized devices and their synthesis and characteristics such as graphene, transition metal oxides (TMOs), transition metal chalcogenides (TMCs), MXenes, and 2D metal–organic frameworks (MOFs). These F2NMs are promising candidates to be employed as the main reaction/detection part which provides the output signal of analytical devices.

8.2 2D Nanomaterials

8.2.1 Graphene

Graphene was discovered by cleaving graphite using the Scotch tape by the scientists at the University of Manchester. Since its discovery, graphene has been used in numerous applications such as photonic, electronic, energy, biosensors [13, 14]. Graphene nanosheets consist of a single layer of carbon atoms bound together by sp^2 bonding in a honeycomb structure. Graphene layers are stacked by weak Van der Waals forces, which form the layered structure of graphite. Based on the understanding of the structure of graphene, graphite can be easily cleaved into graphene nanosheets through physical or chemical methods. Due to its high conductivity, strength, and high transparency, graphene has been used in many applications including transparent electrodes, touch screen, membranes, solar cells, etc. [15–19]. Graphene has a significant impact on the production of flexible devices, wearable devices, and transparent displays. Therefore, graphene has received great attention and is considered an attractive material to be used in miniaturizing analytical devices. To further understand its potential applications, we discuss the characteristics in the following sections.

8.2.1.1 Synthesis of Graphene

Owing to the structure of graphene and the weak Van der Waals forces between the layers, the simplest method of fabrication is exfoliation from bulk graphite to a thin layer graphene. Figure 8.1 shows top-down and bottom-up methods for graphene preparation [20]. Graphene was discovered using the Scotch tape method, in which the tape is attached to both sides of graphite and pulled out to get a thinner layer for many times to get the single-carbon atom graphene. Later, many methods were developed for large-scale synthesis of graphene such as liquid sonication, liquid phase exfoliation by Li/Na ion intercalation, and oxidation, which is also known as Hummer's method. Liquid sonication method is a simple physical method that employs vibration to pull out the nanosheet layers. The Li/Na intercalation method is based on the interlayer of Li/Na after diffusing into the graphite, and then, the reaction of Li/Na with the solvent such as water breaks the layer out of the bulk material to single-layer nanosheets. Hummer developed a method based on the oxidation of graphite to produce a large amount of reduced graphene oxide

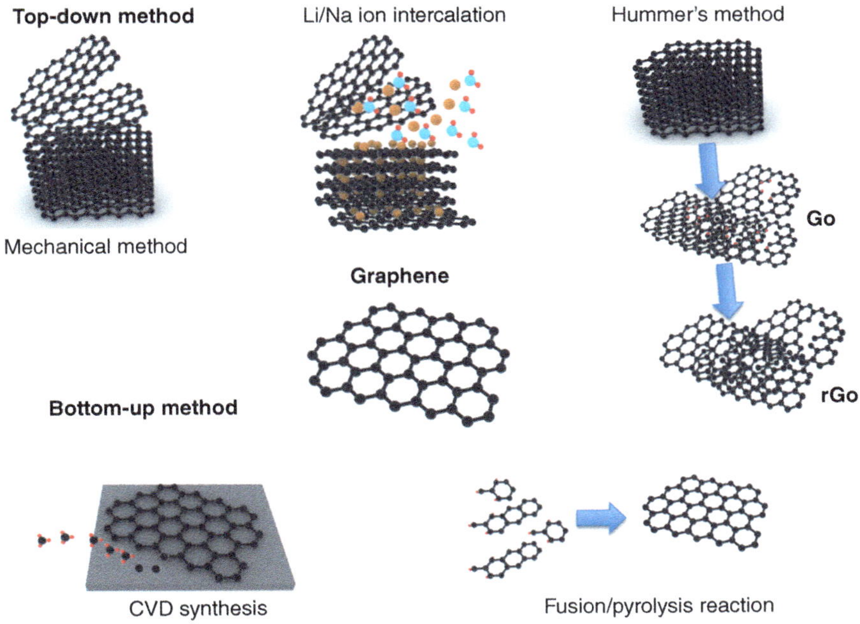

Figure 8.1 Top-down and bottom-up methods for synthesis of graphene.

(rGO), which is the imperfect graphene layer formed by reduction of GO. Typically, this method used strong oxidants such as $KMnO_4$ in high concentrated graphite with sodium nitrate and sulfuric acid. The strong reaction of this oxidant with water oxidizes and breaks the graphite into single-layer GO nanosheets; then, the dehydration with phosphorus pentoxide or the thermal treatment is used to form rGO nanosheets. After its success, many modified methods based on Hummer's procedure for highly efficient and safer synthesis were reported such as mixing of H_2SO_4 with H_3PO_4 or mixing of oxidant $KMnO_4$ with K_2FeO_4 [21, 22]. In contrast to top-down methods, bottom-up methods use the fusion reaction of carbon sources to form large graphene surface. For example, chemical vapor deposition (CVD) method uses the neutral argon gas mixed with H_2 and methane (CH_4) or ethylene (C_2H_4) gas as a C source and then reduced on the substrate at high temperatures (700–1000 °C), as described in Figure 8.1. The resulting graphene layers are deposited on a target substrate such as Cu, Si, or SiO_2, and then the substrate is etched by chemical etching in an acidic medium to obtain pure graphene layers [23]. Transferring techniques for large graphene surface were also developed such as using thermal release tape, poly(methyl methacrylate) (PMMA), or paraffin [24–26]. This method can synthesize large wafer-scale graphene layers by controlling the synthetic conditions for large-surface devices [27–29]. Graphene nanosheets were also fabricated by other bottom-up methods to make small flakes or graphene quantum dots. For example, to synthesize a single graphene layer, Neumann et al. used the self-assembled monolayer (SAM) method using organic chemicals [30].

For instance, 4-(1*H*-pyrrole-1-yl)thiophenol (PTP), 4-(4-thiophenyl)pyridine (TPP), and 4-(2,5-dimethyl-1*H*-pyrrole-1-yl)thiophenol (DPTP) have been used to make SAMs on copper substrates. The pyrolysis process was employed to reduce SAMs to graphene nanosheets.

8.2.1.2 Characteristics and Applications of Graphene

The single layer of carbon atoms of graphene is very thin that allows almost light pass through it. Transmittance of graphene is about 97.7% and reflectance is below 0.1% in the visible light spectrum [31]. Furthermore, graphene has a high charge carrier mobility of 200 000 cm^2/V via the field emission transistor structure [32]. Therefore, graphene can be used as a transparent electrode, which is the first brick for the screen touch and transparent display technology. Also, graphene has remarkable mechanical properties – it has 2D Young's modulus of about 340 N/m [33]. Graphene is the strongest material, which is 200 times stronger than steel. With very high surface area of about 2600 m^2/g, graphene is a promising candidate in supercapacitors or sensing applications such as electrochemical sensors or gas sensors [31, 32]. For instance, Feng et al. used graphene-based nanomaterials for detecting glucose at the low limit of detection about 2.7 μM [34]. The mechanism of this detection was based on the change in the charge carrier mobility of graphene, which led to the change in the electrical current and the resistance signal due to the absorption of glucose on graphene surface. Therefore, the concentration of glucose on the surface of graphene could be detectable with a tiny amount. Further, graphene derivatives such as rGO and graphene quantum dots are also promising materials for biosensors. Ng and Khor reported in detail the electrochemical properties of rGO in sensing of cytosine, adenine, thymine, etc. [35] Graphene doped with metal particles improves selectivity and stability of electrochemical sensors. For example, Zhou et al. reported gold-doped graphene for detection of enzymes [36]. Graphene/Au/chitosan composites were shown to produce H_2O_2 and O_2 in electrochemical reactions; thus, it showed great performance for glucose detection [37]. Gold or other metals act as a link to bind graphene to organic compounds; presence of these chemicals will affect the conductivity of the system. Thus, graphene has excellent characteristics to be used in many future applications and devices such as flexible touch screens, conductive ink, solar cells, batteries, supercapacitors, sensors.

8.2.2 Transition Metal Oxides

TMOs are conventional materials, which have been developed over many years, made of oxidation states of transition metal ions such as T, V, Mn, Fe, Co, Ni, Cu where the d orbital is partially filled [38]. The different electron number in transition metals creates various electron configurations, leads to the diversity of characteristics. For example, TMOs can act as an insulator such as $BaTiO_3$; or a metallic conductor as RuO_2; or a semiconductor as ZnO, CuO; or a magnetic substance such as Fe_3O_4. 2D TMOs introduce a strong ionic bonding of interlayer that causes strong surface polarization [38]. Different structures such as cubic, hexagonal, rutile, spinel, perovskite make different priorities on growth orientation in the synthesis of TMOs. Hence, the synthesis of 2D TMOs is a challenge for scientists.

8.2.2.1 Synthesis Method

As mentioned before, various structures of TMOs lead to different synthesis methods. The synthesis methods can be classified into layered TMO and nonlayered TMO synthesis. The layered TMOs such as materials having octahedral units (MO_6–M: metal atom) have the ionic bonding by alkaline cations, as shown in Figure 8.2b,c [39]. The layered TMOs are similar to graphene/graphene-like materials; therefore, they could be synthesized by physical/chemical exfoliation [38]. Exfoliation of these materials can be performed in liquid or surfactant-assisted method or by etching alkali ions in an acidic medium [38]. For example, Rui et al. demonstrated a liquid exfoliation method for synthesis of V_2O_5 nanosheets with few nanometer thickness [40]. Takagaki et al. exfoliated HNb_3O_8 by exfoliation and aggregation of layered HNb_3O_8 via solution processing with tetra(*n*-butylammonium)hydroxide [41]. The layered TMOs can also be prepared by bottom-up methods such as atomic layer deposition (ALD), hydrothermal method. For instance, Zhuiykov et al. used ALD method for deposition of single-layer WO_3 on SiO_2/Si substrate using bis(*tert*-butylimino)bis(dimethylamino)tungsten(VI) as W precursor [42]. In contrast, it is difficult to exfoliate nonlayered TMOs due to their complex structure; therefore, the best method is bottom-up synthesis such as physical vapor deposition (PVD), CVD, hydrothermal, solvothermal, microwave-assisted. The PVD and CVD methods for 2D TMOs involve deposition of the metal oxide layer on a pure metal substrate such as Pt, Pd, Ag. For example, Giordano and colleagues deposited FeO bilayer on Pt at elevated O_2 pressure [43]. The FeO(111) film was deposited on Pt(111) to avoid lattice constant mismatch. On the other hand, Addou et al. demonstrated CVD synthesis of Y_2O_3 on platinum-supported graphene [44]. Monolayer Y_2O_3(111) was grown on graphene/Pt(111) substrate by depositing Y in low vacuum O_2 atmosphere of 10^{-7} torr. Other methods were also developed to fabricate 2D TMOs such as self-assembly, template-assisted or liquid metal-based reaction methods [38]. Liu et al. fabricated CuO nanosheets by investigating the effect of pH via using ammonia solution or alkaline reactant (NaOH) and synthetic temperature [45]. Various morphologies could be obtained by adjusting the experimental condition, such as nanoribbons, nanosheets, nanoparticles, nanoflakes. Interestingly, CuO was used as a template for the synthesis of 2D α-Fe_2O_3 nanosheets by Cheng et al. [46] After the formation of the target material, the template was removed by washing with ammonia solution, deionized water, and ethanol. Therefore, the synthesis of 2D TMOs is a diverse process and challenging to have their full synthetic method.

8.2.2.2 Characteristics and Applications of TMOs

Since the discovery of metal oxides, many TMOs have exhibited characteristics such as magnetic, insulating, semiconducting, high-k dielectric properties. Semiconductor TMOs are well known as gas sensing materials such as WO_3, TiO_2, ZnO, SnO_2, Fe_2O_3, etc. TMOs have large active sites and high adsorption due to polarization on

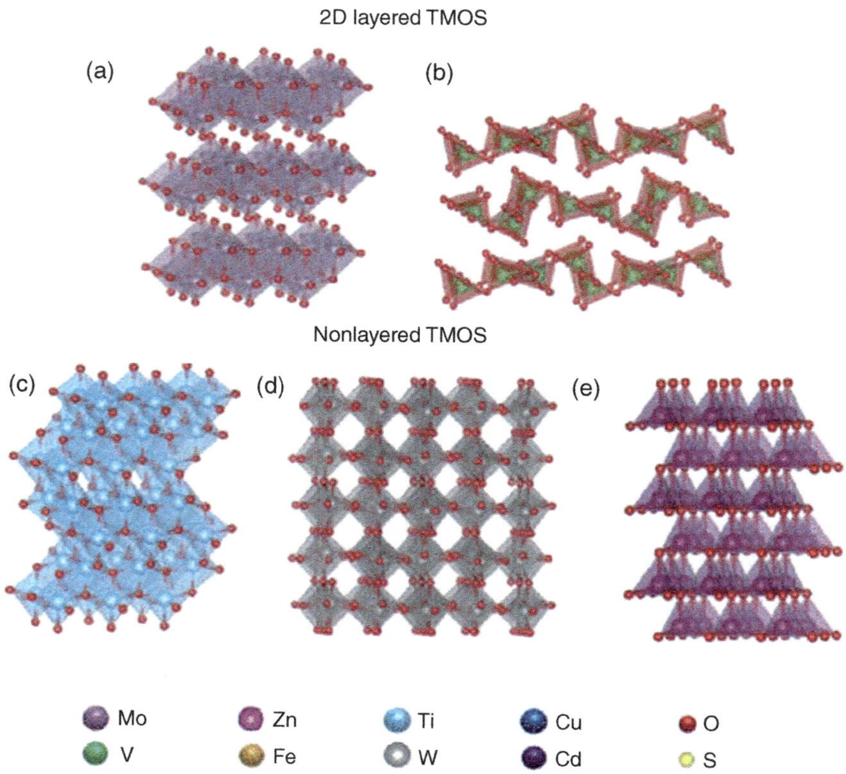

Figure 8.2 Structures of layered transition metal oxides: (a) orthorhombic MoO_3, (b) orthorhombic V_2O_5, and nonlayered transition metal oxide, (c) anatase TiO_2, (d) monoclinic WO_3, and (e) wurtzite ZnO. Source: Haque et al. [39]. Licensed under CC-BY-4.0.

their surface [38]. Thus, their selectivity, high responsibility, and low temperature operation are expected to enhance the performance of gas sensors. For example, Shendage et al. demonstrated an NO_2 gas sensor using WO_3 nanoplates operated in ambient air at a low temperature of 100 °C [47]. It is worth noting that the gas sensing temperature is an important factor in evaluating its performance; therefore, the operating temperature of 100 °C is a well achievement, while for most metal oxide gas sensors, the operating temperature is at about 200–300 °C and even higher in some conditions [48]. Besides, some insulator TMO materials such as $LaAlO_3$ and $SrTiO_3$ have the insulator–metal transition when the layer thickness is about 4-unit cells. Moreover, the magnetic ordering reconstruction was observed at the interface of $LaAlO_3$ and $SrTiO_3$. Zhou et al. reported the extraordinary interface conductivity of $LaAlO_3$ on $SrTiO_3$ and investigated the introduction of oxygen vacancies and lattice distortion at the interface [49]. $LaAlO_3$ and $SrTiO_3$ could be used as a superconducting material, while the interface of these two materials shows a superconductivity at low temperatures [50]. Most TMOs exhibit semiconducting properties, and they have been used in photocatalysts for years, such as TiO_2,

NiO, VO_2, V_2O_5, CuO [51–53]. The increasing surface area of 2D TMOs further enhance their catalytic properties as ideal catalyst materials. Also, the combination of TMOs for a higher conductivity with other 2D materials such as graphene or TMCs improves catalytic performance. Moreover, biosensing systems that are based on the photocatalytic properties of TMOs are also appropriate for detection of glucose, hemoglobin, etc. For example, Nallal et al. reported the TiO_2-based heterojunction nanohybrid for glucose and hemoglobin sensors without any enzyme auxiliary [54]. Gao et al. demonstrated a DNA sensor for the detection of polymerase chain reaction product from soybean based on TiO_2 nanorods/graphene composites [55]. On the other hand, with high-k dielectric 2D TMOs as Y_2O_3 or Al_2O_3, it is found that the dielectric constant of the Y_2O_3 (111) monolayer is 1.5 [56]. Therefore, it could be a promising material to replace SiO_2 as the conventional gate in MOS transistor. With all those promising applications and good properties, 2D TMOs are contributing to miniaturize analytical systems as handheld or wearable devices.

8.2.3 Transition Metal Chalcogenides

Along with the development of graphene, 2D TMCs, which have graphene-like or layered structures, have emerged as potential candidates for future semiconductor materials. TMCs have the formula MX_2, where M is a transition metal element and X is a chalcogenide element. Each layer of MX_2 consists of M atomic layers sandwiched between two layers of X atom, and each layer of MX_2 is bonded by a weak Van der Waals interaction. Therefore, TMCs can be easily cleaved into single or few layers, which are very thin and exhibiting excellent properties compared to bulk materials, such as high conductivity, high strength, and physical/chemical stability [8]. Interestingly, from bulk to single layer, the bandgap of TMCs changes from an indirect bandgap with a narrow band to a direct bandgap with higher bandgap energy, for example MoS_2 (from 1–1.2 to ~1.9 eV) and WS_2 (from ~1.2 to ~2.0 eV) [57–60]. These properties of thin-layer TMCs are ideal for many applications, including transistors, solar cells, photo detectors, light emitting diodes, gas sensors, photo/electrocatalyst, and biosensors [61–65]. Thus, TMCs will play key roles in miniaturized analytical devices as a support layer in solar cells, LED, transistors or an active layer in catalysts, biosensors, etc. [4].

8.2.3.1 Preparation of TMCs

TMC materials have graphene-like or layered structure in which the layers are bonded by the Van der Waals force bonds, which can be easily broken. Therefore, in a simple top-down method, a bulk material is cleaved into small particles or layers. Various methods have been used for preparation such as Scotch tapping method, sonication method, liquid exfoliation based on Li/Na ion (Li/Na intercalation method), and ball milling method, which is similar to graphene fabrication. The bulk materials of TMCs can be found in natural minerals or obtained from industrial fusion reactions. In the bulk state, materials are larger in size and thickness. The outstanding properties of 2D TMCs have led numerous researches to find efficient methods to fabricate these advanced materials in thin layers. The Scotch tapping

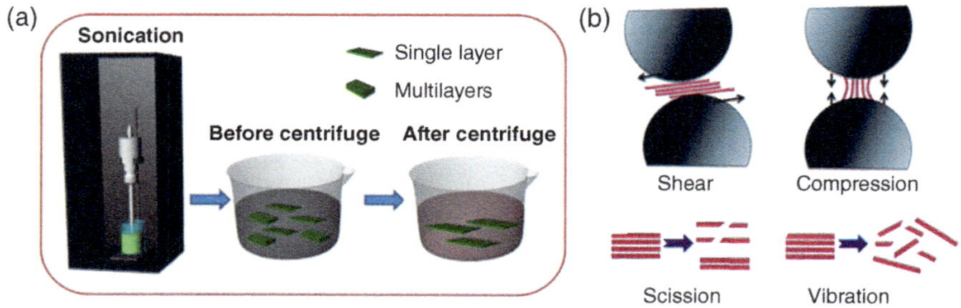

Figure 8.3 (a) Liquid sonication method. Source: Reproduced with permission Kim et al. [63]. (b) Mechanism of ball milling method. Source: Reproduced with permission Yao et al. [66].

method is a conventional method for cleaving large-sized single nanosheets. In this method, the tape is attached to the material surface and pulled out many times to get the single- or few-layer TMCs. For large-scale industrial production, ball milling, liquid sonication, and lithium intercalation method have been investigated, and these preparation methods can produce thin-layer TMCs of 1 to 10 nm in large amounts. Figure 8.3a and b shows the liquid sonication method using a probe tip and the ball milling mechanism, respectively. By using the sonication method, Coleman et al. investigated a list of solvents for TMCs and found some promising chemicals for TMC preparation such as *N*-methyl-2-pyrrolidone (NMP), dimethyl sulfoxide (DMSO), *N*-vinylpyrrolidone (NVP) [67]. Later, Nguyen et al. successfully fabricated TMC nanosheets such as MoS_2, WS_2, TiS_2 with NVP solvent; the method can produce few layer and single-layer TMDs with a lateral size from hundreds to few hundred nanometers [64]. However, the quality and efficiency of the sonication method are still limited. Therefore, ball milling is needed for efficiently grinding bulk materials into smaller size. Yao et al. demonstrated that ball milling can obtain uniform thin layers of TMCs [66]. The collision of rigid balls, during the milling process, creates shear and compression forces which effectively exfoliated the bulk materials into nanosheets, as shown in Figure 8.3b. Interestingly, by using lithium intercalation, the phase of TMCs changes from 2H to 1T due to the charge transfer from Li ions to the layer [68]. It has also been found that the 1T phase of TMCs has higher conductivity and high photoluminescence in comparison to the 2H phase. Therefore, using 1T phase TMCs will enhance the performance of optoelectronic devices. Compared to top-down methods, bottom-up methods have some advantages such as very large surface area of thin layers, uniformity, and compatibility in fabrication of electronic devices such as transistors, gas sensors, photoelectrocatalysts. There are some typical preparation methods such as CVD, PVD, hydrothermal, and hot injection, as shown in Figure 8.4a–c. In bottom-up methods, a TMC compound or a source consisting of transition metals and chalcogenide atoms is evaporated, hydrolyzed, or decomposed and then assembled in a typical medium or on a surface. CVD is a powerful method for the synthesis of 2D TMC with high quality and controllable thickness as described in Figure 8.4a [69]. In the synthesis procedure, the carrier gas will bring the

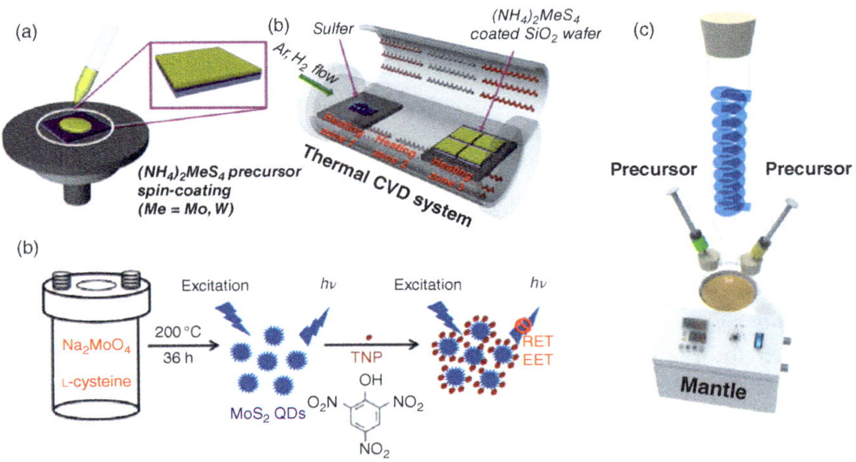

Figure 8.4 Bottom-up methods for synthesis of 2D-like TMC materials: (a) CVD method. Source: Reproduced with permission Kwon et al. [69]. (b) Hydrothermal method. Source: Wang and Ni [70]. (c) Hot injection method.

chalcogenide atom to the transition metal compound layer; on the surface, the reaction of combination occurs at very high temperatures of 700–1000 °C. The formation of 2D layer on the surface depends on the synthetic system settings such as source/substrate positions, thermal evaporation source temperature, deposition temperature, and reaction time. By controlling the reaction conditions and the substrate size, the CVD method can synthesize wafer-scale thin film layers in large scale as reported in the literature [71–73]. It has been shown TMC thin films are ideal candidates for energy conversion or as catalysts for hydrogen fuel generation. For example, Lee et al. successfully fabricated large-area MoS_2 by using CVD method, which showed the thin MoS_2 layer with a lateral size of micrometers to millimeters [74]. Kwon et al. successfully fabricated wafer-scale CVD-synthesized MoS_2/WS_2 thin films for a silicon p–n junction cell for the hydrogen evolution reaction [72, 75, 76]. However, the CVD method has limitations such as high temperature synthesis, low amount of sample, high cost. Therefore, other synthesis methods were also developed for industrial large-scale production such as the hot injection method or hydrothermal method [77]. These methods show uniformity, variation of morphologies, phases, and tunable bandgap. Liu et al. synthesized MoS_2 quantum dots and nanosheets using the hot injection method by injecting the sulfur source of *N,B′*-diphenyl thiourea in diphenyl ether into a mixture of $MoCl_5$ at 300 °C [78]. Van Le et al. successfully synthesized MoS_x nanodots for optoelectronic devices such as organic photovoltaic cells or organic light emitting diodes for replacing the traditional hole transport/injection layer (PEDOT : PSS) [79]. Wang et al. synthesized MoS_2 quantum dots via the hydrothermal method for optical sensor detection of 2,4,6-trinitrophenol [70]. Variations of TMC morphology are also reported such as nanoflowers, hierarchical structure, lamella structure, hollow spheres, nanorods [80–82].

8.2.3.2 Characteristics and Applications of TMCs

TMCs have trigonal, hexagonal, and rhombohedral crystal structures, which are named 1T, 2H, and 3R, respectively. As a natural mineral source, TMCs have 2H phase and almost selenide and sulfide compounds are semiconductors; however, telluride-based TMCs are metallic, paramagnetic, or superconducting. The most focused TMC semiconductors widely used in optoelectronic devices are compounds of Mo, W, or Ti with S or Se atom. In the 2H phase, these TMCs behave as a low bandgap semiconductor in the range of 0.3–2 eV [8]. The bulk form material has an indirect bandgap that prevents the direct transition of electrons from the valance to the conduction band and reversion; therefore, electrons will be trapped and lose energy as phonon vibration. However, their bandgap becomes larger when the thickness of nanosheet TMCs reduces from the bulk to single layer, which transforms to the direct bandgap and allows the direct transition of charge carriers. Therefore, electrons can easily transport between the valance and conduction band, improving the conductivity and optical properties, which can be applied for optoelectronic devices [61]. Moreover, the quantum dots of these materials such as MoS_2, WS_2 show the blue emission in the range 300–500 nm wavelength [83, 84]. These quantum dots can help materials act as active sites of catalytic reaction, biomarkers, or biosensors. For example, Anantharaj et al. used an electrochemical method to break the WS_2 nanosheets into WS_2 quantum dots for the water splitting reaction [85]. Liu et al. used MoS_2 quantum dots as the fluorescent biosensor for cancer detection [86]. The MoS_2 quantum dots showing an absorption of fluorophore-labeled ssDNA fluorescent probe (P_0-FAM), which causes the quenching of emission light. This binding will be lost when the MUC1 of cancer cell uptakes the ssDNA due to stronger affinity, thereby detecting the cancer cell. In contrast to semiconductors of 2H phase, 1T phase acts as a metallic conductor. Due to this characteristic, this phase is considered a potential candidate for the hydrogen evolution reaction as well as for other applications. For example, He et al. synthesized vertically aligned 1T-MoS_2 material via hydrothermal method [87]. The catalytic performance of 1T-MoS_2 was much improved with a low onset overpotential of 100 mV and a low Tafel slope of 43 mV/decade. Further, with the sulfur/selenide site on the surface and edges, TMCs can bind with, Au which is mostly used in biosensor systems for protein/virus/cancer detection. For example, Sarkar et al. reported MoS_2-based FET biosensor devices, which compromise sensitivity, label-free for single molecular detection [88]. This research opened the possibility of mini devices with a macrofluidic channel and an FET structure can be used for detection of biomolecules. Narang et al. also reported a $MoSe_2$-based sensing platform for detection of an *E. coli* toxin via electrochemical measurement [89]. Soni et al. demonstrated highly efficient polyaniline–MoS_2 hybrid nanostructures for cancer sensing applications [90]. The 3D structure of these composite materials showed a large active area, fast electron transportation, and high catalytic property. Baek et al. used Au-functionalized WS_2 to bind with a novel peptide as the electrochemical biosensor for the norovirus detection in oysters [91]. Other researches also showed that the catalytic properties of TMC-based materials can be used for glucose sensor via the redox reaction [92, 93]. It indicates that the 2D

TMCs in pure phase or composites or metal doped are the potential materials for optical/electronic devices and biosensing systems.

8.2.4 MXenes

The emergence of graphene and other 2D graphene-like materials encouraged the discovery of MXenes, which is a compound of metal with carbon or nitrogen (carbide or nitride compounds). MXenes also have the layered structure in which the metal layer is sandwiched between the carbide or nitride layer. The discovery and common fabrication of MXenes are related to the etching of MAX compounds, where M is metal, A is aluminum or silicon, X is carbon or nitrogen. MXene-based researches become wider in many applications such as energy storage, batteries, catalyst, or magnetic shielding [94]. MXenes inherit the advantages of other 2D graphene-like materials such as high transparent thin layer, high conductivity, strength, and physical/chemical stability. Therefore, MXenes can be a candidate for the current 2D material-based applications in optoelectronic devices, catalyst, or biosensors.

8.2.4.1 MXene Preparation

The structure of MXenes is named by the structure of bulk MAX compound, as shown in Figure 8.5a. The A atom in the intercalated compound MAX can be easily removed by acidic medium such as HF or basic medium such as NaOH. The etching of the A atom always brings the functional group on the surface of MXenes such as OH, O, and F. Therefore, the availability of MXenes depends on the existence of the alloy MAX, which is generally synthesized by the alloy melting process at high temperatures [97]. Besides, the MXenes such as molybdenum carbide or tungsten carbide were synthesized via carburization by the CVD method, which requires high temperatures of about 900 °C. Gong et al. reported the preparation of W_2C functionalized on carbon nanotubes for the hydrogen evolution reaction, as shown in Figure 8.5b [95]. This method is found to be promising for preparation of various MXene compounds. Recently, the low-temperature synthesis for the composition of TMC and MXene was made. By using a hydrothermal method, Nguyen et al. have found the formation of WS_2 with W_2C materials, as shown in Figure 8.5c [96, 98]. This alloy compound was employed as a promising catalyst for hydrogen evolution reaction.

8.2.4.2 Characteristics and Applications of MXenes

Like other graphene-like compounds, MXenes have the layered structure, which have excellent properties such as high conductivity, stability, and catalytic properties. Characteristics of MXenes are still to be explored. The DFT has shown that bandgap of MXenes is tunable by changing the terminal surface such as the bare MXenes are metallic conductors, but the functionalization of OH or F changes its bandgap to a semiconductor with a small bandgap from 0.05 to 2.87 eV [99]. With a variety of bandgap and superior properties, MXenes have been used in analytical purposes of H_2O_2, glucose, lactate, superoxide anion, urea, etc. [100] Zheng et al. demonstrated an inkjet-printed electrode using titanium carbide/graphene oxide for

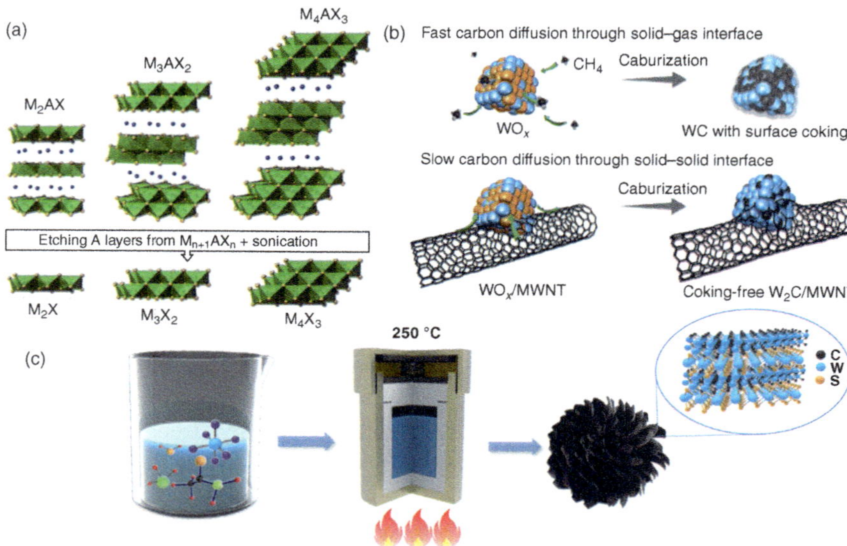

Figure 8.5 (a) Structures of MAX compounds and their corresponding MXenes. Source: Nguyen et al. [94]. (b) W_2C nanoparticles on multiwalled carbon nanotube fabrication via CVD method. Source: Gong et al. [95]. Licensed under CC-BY-4.0. (c) W_2C/WS_2 nanoflower alloys synthesis via hydrothermal method. Source: Nguyen et al. [96].

H_2O_2 sensing application [101]. W_2C nanoparticles have been reported as an efficient photo-/electrocatalyst when they were grown on carbon nanotubes for water splitting to produce hydrogen [95]. The catalytic properties of MXenes are not only valuable for hydrogen evolution but also for glucose detection. Lei et al. reported an MXene Ti_3C_2-based wearable biosensor system that can detect glucose and lactate in sweat from the human body [102]. Moreover, MXenes are also used in cancer biomarker sensors or detection of other biochemical compounds such as urea, uric acid, or creatinine [103, 104]. Notably, the remarkable application of MXenes is as a biosensor for pharmaceuticals due to their high surface area, biocompatibility, and hydrophilicity. Zhang et al. demonstrated an electrochemical sensor using titanium carbide for detection of acetaminophen and isoniazid, which are commonly present in drugs and can induce liver damage in certain circumstances [105]. Zheng et al. reported the use of MXene/DNa/Pd/Pt nanocomposites for a sensitive dopamine sensor [106]. Thus, from these reports, it is evident that MXenes and their superior properties are widely used for various applications in analytical measurements and contribute to miniaturization of analytical devices.

8.2.5 2D Metal–Organic Frameworks

2D MOFs are polymeric hybrid materials composed of metal ions or metal clusters and organic linkers that are assembled in a specific structure. MOFs were reported in the early 1990s. Later, Omar Yaghi and coworkers developed and created many MOF materials by designing reticular chemistry structure [1]. This combination brings them unique characteristics of the organic–inorganic hybrid nature such as large surface area, ultrathin thickness, tunable structure, high conductivity, facile functionalization on surface [107, 108]. Due to these superior properties, MOFs have great potential for many applications, such as energy storage, energy conversion, sensors, catalysts, biomedicines, gas storage and separation.

8.2.5.1 Synthesis of 2D MOFs

For layered materials, 2D MOFs have also been prepared by many methods such as sonication exfoliation, lithium intercalation, surfactant-assisted synthesis, sonication synthesis, which can be classified into top-down and bottom-up methods [108]. MOF layers are bonded by weak van der Waals forces and hydrogen bonding, so these weak interlayer interactions can be easily broken via sonication along with liquid solvents or lithium intercalation, etc. Thus, the top-down method is quite similar to synthesis of TMC materials where liquid sonication and lithium intercalation are useful for cleaving layer by layer of these materials [109]. For instance, Li et al. reported the sonication exfoliation of MOF-2(Zn(TPA)·(H_2O)·DMF, TPA = terephthalate) in acetone [110]. Song et al. used the lithium intercalation method to synthesize 2D MnDMS for microRNA sensing systems [111]. More recently, Ding et al. synthesized 2D MOFs via the organic ligand–assisted sonication method. In this work, 4,4′-dipyridyl disulfide (DPDS) was employed to break the bonding of the layered structure and to passivate the surface of 2D materials [109]. Therefore, the MOF can be synthesized with a thickness of ~1 nm and the production yield up to 57%. In contrast to top-down methods, bottom-up synthesis methods of MOF materials can be a short-cut from raw inorganic/organic materials to 2D MOFs. Typically, this method for hybrid materials is based on the interaction at the interface and modulation of the structure; therefore, various methods were developed base on the specific functional groups or surfactants such as interfacial synthesis, three-layer synthesis, surfactant-assisted synthesis, sonication synthesis [108]. Makiura used the Langmuir–Blodgett method to synthesize MOF CoTCPP-py-Cu layer by layer by using the interfacial of liquid/air, as shown in Figure 8.6a,b [113]. Thus, the number of layers or thickness can be controlled by the number of dipping process. In another interface synthesis method, Huang et al. reported the formation of Cu–BHT (BHT = benzene hexathiol) at the water/dichloromethane interface. While water can dissolve the metal ions, dichloromethane can dissolve organic chemicals; therefore, at the interface, the materials combine and form the MOF structure, as shown in Figure 8.6c [114]. The three-layer synthesis is an upgradation of the surficial synthesis method, in which the interface is formed by the partial dissolution of two solvents, creating the three layers separately. For example, Rodenas et al. used this method to fabricate the 3D crystalline structure of Cu-BDC MOF (BDC = 1,4-benzenedicarboxylate) for CO_2/CH_4 gas separation [115]. Due to high surface energy, 2D MOFs are easily aggregated, thereby limiting the stability for a long period. The surfactant synthesis is used to form the MOF structure and passivate its surface, thereby preventing the precipitation of 2D MOFs in solvent. Zhao et al. demonstrated the advantages of surfactant-assisted synthesis over traditional synthesis methods with sub-10-nm thick 2D MOFs [116]. They employed

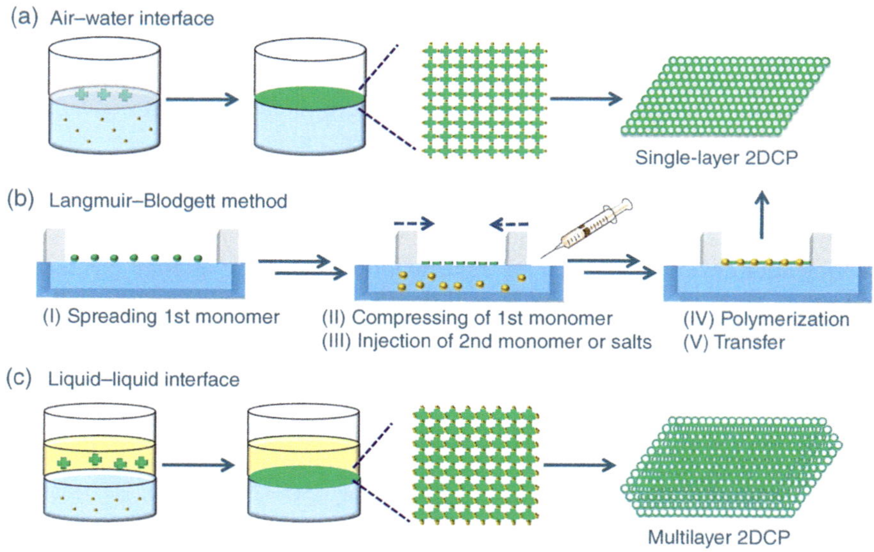

Figure 8.6 Interfacial synthesis of 2D MOF: (a) air–water interface; (b) Langmuir–Blodgett method; and (c) liquid–liquid interface. Source: Wang et al. [112]. Licensed under CC-BY-4.0.

polyvinyl pyrrolidone as the surfactant for preparation of 2D M-TCPP (M = Cu, Cd, Co, Zn, TPP = tetrakis(4-carboxyphenyl)porphyrin) ultrathin nanosheets. To control the morphology, the modulated synthesis of 2D MOFs uses small-molecule modulators such as acetic acid, pyridine, formic acid. The modulators act as the organic linker that can coordinate with the metal, thus affect the growth of MOFs, and then define certain crystal planes for orientating the growth direction. With the functional groups from modulators, MOFs can be grown with various morphologies such as cubic, nanorods, nanosheets. For example, Hu et al. used a mixture of acetic acid, formic acid, and trifluoroacetic acid on water as the key solvent for crystallization of Zn/Hf-MOF [117]. Jiang et al. successfully synthesized 2D-ZIP-8-NS-C (zeolitic imidazolate frameworks eight nanosheets carbon denoted) embedded in hierarchical pores for high-capacity lithium sulfur batteries [118].

8.2.5.2 Characteristics and Applications of MOFs

Each designated MOF has its own structure and functional groups on the surface; therefore, they have various types and utilities for numerous applications. It is worth to note that MOFs have abundant active sites and high porosity, which make them suitable catalysts for reactions such as hydrogen evolution, oxygen evolution, or carbon dioxide reduction. Ran et al. reported on 2D MOF nanosheets for efficient visible-light-induced hydrogen production [119]. 2D MOF [Ni(phen) (oba)]$_n$·0.5nH$_2$O (phen = 1,10-phenanthroline, oba = 4,4′-oxybis(benzoate)) was prepared by the hydrothermal method and the liquid exfoliation method combined with CdS quantum dots to prove light absorption and conversion. Kornienko et al. also demonstrated an MOF as electrocatalyst for CO_2 reduction [107]. The cobalt–porphyrin MOF, $Al_2(OH)_2$TCPP–Co (TCPP–H_2 = 4,4′,4″,4‴-(porphyrin-5,10,15,20-tetrayl)tetrabenzoate) showed high selectivity for CO production and stability of over 7 h. Moreover, MOFs are combined from metal ions with an organic frame; therefore, they are very compatible with human cells if a suitable organic compound is selected. This characteristic makes MOFs applicable in biosensors, drug delivery, or biomedicines. For instance, in biosensing applications, Meng et al. used MOF ZIP-67 for detection of glucose [120]. Glucose is easily oxidized to gluconolactone by the catalysis of MOFs in NaOH solution; therefore, the amount of glucose can be found with high sensitivity of 0.379 $\mu A/\mu M/cm^2$ and a low detection limit of 0.66 μM. Zhao et al. demonstrated synthesis of ultrathin 2D MOF nanosheets by the surfactant-assisted method for DNA sensing system with a low detection limit of 20×10^{-12} M [116]. Remarkably, the MOFs have high compatibility to human cells; therefore, they can be used in medicinal applications. For instance, Lin et al. revealed the Universitetet i Oslo (UiO) MOF nanosheets for cisplatin delivery and pool small interfering RNAs to improve the therapeutic efficiency [121]. This report showed great potential of MOFs in drug delivery and efficient treatment of drug-resistant cancers. These all recent reports indicate MOFs to be a key material in various applications in optical/electronic/biological devices; therefore, they are potential candidates for many miniaturized analytical devices in the future.

8.3 Functionalization Methodologies

Although 2D nanomaterials have many excellent properties such as high conductivity, large surface area, physical and chemical stability, they have limitations that should be solved before applied in commercial production. These limitations include the weak binding of 2D nanomaterials with organic materials, the not matching of the material redox potentials, the low performance due to the high redox potential, or the toxicity of materials, etc. They could be resolved by surface passivation, modification, doping method, and co-activated function groups. Thus, the functionalization of 2D materials also plays an important role in their applicability and should be investigated in detail.

8.3.1 Inorganic Doping Method

Foreign atoms that contribute a small amount to the atomic ratio or weight ratio (%) in pure materials are called doping agents. Doping enhances the properties of materials, for example, metal doping of novel metals (Pt, Au, Ag, etc.) increases charge carriers in the materials, thereby improving conductivity of the resulting materials. Moreover, the introduction of some metals such as Ag and Au could be used as a linkage for organic material binding such as peptides, which later can be conjugated to DNA or protein molecules [122]. Doping methods can be divided into two types: lattice doping and surface doping. Lattice doping is the presence of foreign atoms in substitution of original atoms, causing lattice mismatch. This doping is mostly

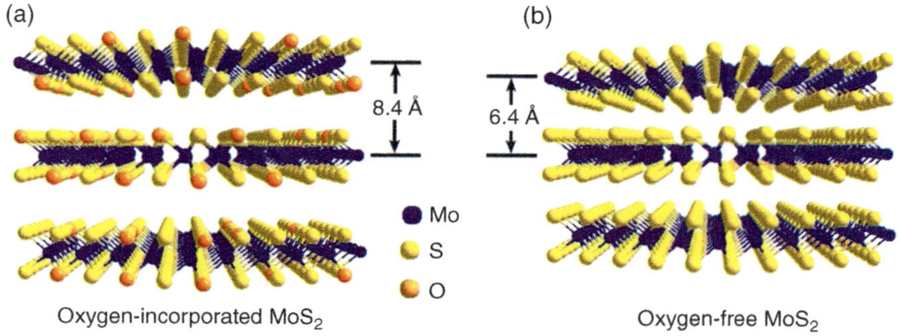

Figure 8.7 Structural models of (a) oxygen-incorporated MoS_2 and (b) bare MoS_2 nanosheets. Source: Reproduced with permission Zhou et al. [123].

formed in fabrication of materials, but sometimes due to the implantation of other ions via ion beam treatment. For instance, oxygen atoms incorporated into TMCs such as MoS_2, WS_2 nanosheets were reported as efficient catalysts for hydrogen production [123–125]. By sulfurizing the TMO or its compound but imperfectly, the oxygen-incorporated MoS_2 was formed, as shown in Figure 8.7. The introduction of oxygen atoms in the lattice bonded to Mo atoms creates a 2D structure with an enlarged interlayer spacing of 0.84 Å, which is larger than 6.5 Å of bare-MoS_2 material. Therefore, this material has potential to be used in metal ion storage applications such as Li or Na ion storage. Moreover, this defect-rich structure of MoS_2 exhibits high intrinsic conductivity and consists of more active sites for HER but also retains long-term stability of catalytic performance. In contrast, surface doping has received significant attention due to its ability to introduce other atoms/molecules or functional materials into 2D nanomaterials. For instance, Wang et al. decorated graphene by gold nanoclusters, which was later used for detection of human immunodeficiency virus [126]. Due to the high conductivity of graphene, gold ions can easily chemically or physically absorb on its surface, and can then be reduced to gold metal nanoclusters through thermal reduction, chemical reduction, photochemical reduction, etc. [127, 128] Another method for metal doping is using a functional carboxyl or hydroxyl group as a target for metal ion absorption. For instance, Zhou et al. demonstrated a method for decorating the TMC MoS_2 by silver metal using mercaptopropionic acid, L-cysteine, or 1-thioglycerol, as shown in Figure 8.8 [129, 130]. The uniform size and distribution of Ag nanoparticles on MoS_2 nanosheets indicate this method's high efficiency.

8.3.2 Functionalized Organic Functional Groups

Most of inorganic nanomaterials are harmful to organic cells or not available for binding with organic chemicals; therefore, functionalization of these materials is a prerequisite to various wearable or bio applications. Materials rich of oxygen atoms such as TMOs can easily attach to organic binders such as chitosan, cysteine. However, for other materials such as TMC, graphene, MXenes end with sulfide,

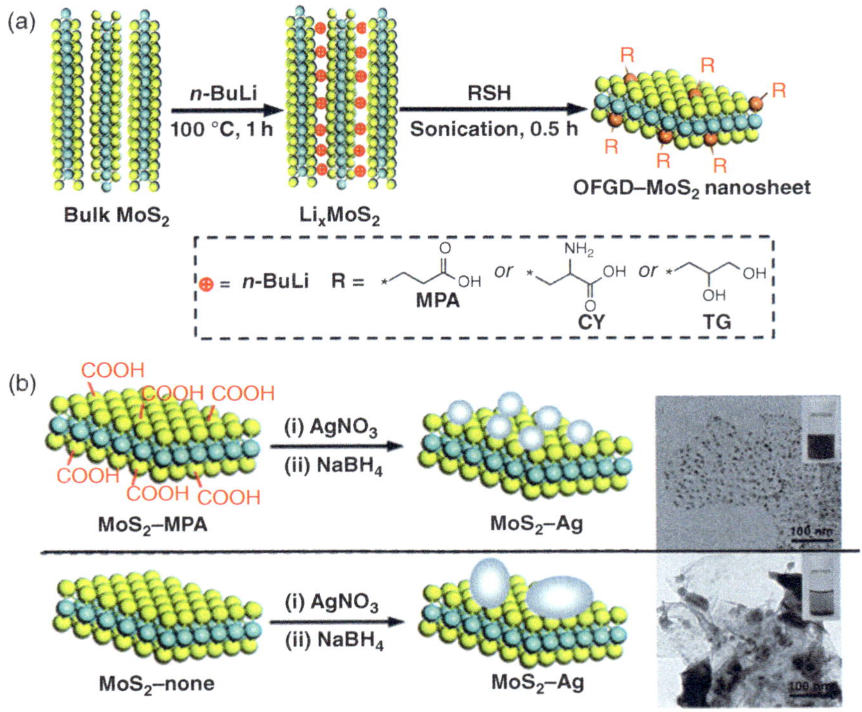

Figure 8.8 (a) Schematic synthesis of organic functional groups on MoS_2 nanosheets and (b) schematic synthesis and TEM images of Ag-decorated MoS_2 nanosheets with and without MPA functionalization. Source: Reproduced with permission Zhou et al. [129].

carbide, nitride bonding, finding a suitable organic binder is not easy. Thus, these materials should be designed with functional groups by various ways such as using a metal doping of Au, Ag, which easily binds with organic binders such as peptide, for biosensing applications. On the other hand, the surface of material can be modified by carboxyl or hydroxyl groups, which help the surface to be bind with organic chemicals. For example, MXenes could have functional groups of –F or –OH depending on the etching medium condition; therefore, MXene functional groups can be designed by choosing an acid or base medium. Graphene oxide, a derivative of graphene, with carboxyl and hydroxyl groups can be easily used in biological applications. The use of MOF functional groups was designed by the synthetic process by choosing a suitable organic material as the organic framework.

8.4 Outlook

Analytical systems play an important role in diagnosis of human diseases. In modern life, miniaturization of analytical devices plays an important role to complete the system with fast detection, high accuracy, wearable and biomedical supports.

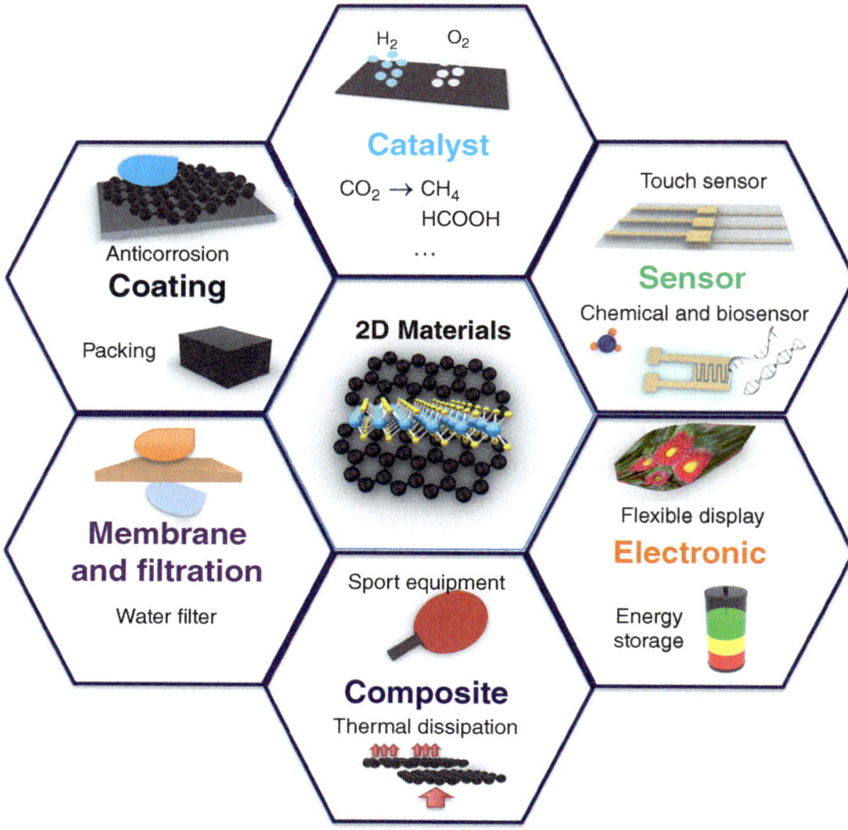

Figure 8.9 Applications of 2D materials.

Figure 8.9 shows the common real-time and promising applications of 2D materials in electronics/optical devices, sensing, thermal dissipation, biosensor applications, etc. Numerous researches in this field will continuously contribute to gradually fulfill the requirements of the system. With excellent material properties and selective functional groups, 2D functionalized materials in nano- and microsizes are found to be promising to create devices with high response, high selectivity, low detection limit, and stability, thereby, minimum size and energy source requirements and low production cost.

However, to obtain good performance and meeting all requirements of materials, there are many challenges for scientists. For example, production of 2D thin single layer is still difficult for industrial production at low cost. The CVD and PVD methods require high temperatures in the range of 700–1000 °C and in neutral gas environments such as N_2 or Ar_2 gas, which are difficult for large-scale synthesis and well being controlled. Materials and their properties are not yet fully elucidated; therefore, the basic research on electronic, spin, and momentum properties need many theories and practices to be done. Furthermore, bare materials do not naturally match the device requirement; therefore, search for functional groups, doping behavior, or other modifications that improve the characteristics of materials is necessary. For example, recent research has found that 1T phase TMCs have better conductivity than 2H phase; however, stability is a limitation in these materials, while chemical exfoliation always contains the damage on the surfaces. Voiry et al. reported the treatment of iodine to recover the edge of 1T MoS_2 nanosheets for the hydrogen evolution reaction [68]. The defect engineering on TMC allows the functionalized inorganic/organic materials desorbing/absorbing the metal atoms and other functional groups as the terminal surface, therefore, enhancing the electrical/optical properties or creating a binding bridge with protein molecules. For traditional 2D materials such as graphene or TMOs, the power tools for deep discovery the characteristic of materials is limited due to the number of the powerful analyzer or the expensive of that measurement. For example, the wide-angle X-ray scattering or small-angle X-ray scattering are some types of powerful measurement for crystalline structure of thin layer materials, but, they are expensive and new to researchers. There is more scope for research with new materials such as TMCs, MXenes, and MOFs, and exploration of their synthetic procedures, characteristics, and practical applications is under way. The F2NM-based miniaturized analytical devices will bring a bright future with wearable, health-care, and convenience devices.

References

1 O'Keeffe, M., Peskov, M.A., Ramsden, S.J., and Yaghi, O.M. (2008). The Reticular Chemistry Structure Resource (RCSR) database of, and symbols for, crystal nets. *Acc. Chem. Res.* 41 (12): 1782–1789.

2 Giustini, A.J., Petryk, A.A., Cassim, S.M. et al. (2010). Magnetic nanoparticle hyperthermia in cancer treatment. *Nano Life* 1 (1n02) https://doi.org/10.1142/S1793984410000067.

3 Choi, W., Lahiri, I., Seelaboyina, R., and Kang, Y.S. (2010). Synthesis of graphene and its applications: a review. *Crit. Rev. Solid State Mater. Sci.* 35 (1): 52–71.

4 Chhowalla, M., Shin, H.S., Eda, G. et al. (2013). The chemistry of two-dimensional layered transition metal dichalcogenide nanosheets. *Nat. Chem.* 5 (4): 263–275.

5 Reimann, S.M. (2002). Electronic structure of quantum dots. *Rev. Mod. Phys.* 74: 1283–1342.

6 Yoffe, A.D. (1993). Low-dimensional systems: quantum size effects and electronic properties of semiconductor microcrystallites (zero-dimensional systems) and some quasi-two-dimensional systems. *Adv. Phys.* 42.

7 Kudo, A. and Miseki, Y. (2009). Heterogeneous photocatalyst materials for water splitting. *Chem. Soc. Rev.* 38 (1): 253–278.

8 Xu, M., Liang, T., Shi, M., and Chen, H. (2013). Graphene-like two-dimensional materials. *Chem. Rev.* 113 (5): 3766–3798.

9 Chen, Y., Tan, C., Zhang, H., and Wang, L. (2015). Two-dimensional graphene analogues for biomedical applications. *Chem. Soc. Rev.* 44 (9): 2681–2701.

10 Tan, C. and Zhang, H. (2015). Two-dimensional transition metal dichalcogenide nanosheet-based composites. *Chem. Soc. Rev.* 44 (9): 2713–2731.
11 Zhang, B. and Radisic, M. (2017). Organ-on-a-chip devices advance to market. *Lab Chip* 17 (14): 2395–2420.
12 Rios, A., Escarpa, A. and Simonet, B. (2009). *Miniaturization in Analytical Chemistry*. John Wiley & Sons.
13 Geim, A.K. and Novoselov, K.S. (2007). The rise of graphene. *Nat. Mater.* 6 (3): 183–191.
14 Geim, A.K.G. (2009). Status and prospects. *Science* 324 (5934): 1530–1534.
15 Du, X., Skachko, I., Barker, A., and Andrei, E.Y. (2008). Approaching ballistic transport in suspended graphene. *Nat. Nanotechnol.* 3 (8): 491–495.
16 Novoselov, K.S., Jiang, Z., Zhang, Y. et al. (2007). Room-temperature quantum hall effect in graphene. *Science* 315 (5817): 1379–1379.
17 Li, X., Zhu, Y., Cai, W. et al. (2009). Transfer of large-area graphene films for high-performance transparent conductive electrodes. *Nano Lett.* 9 (12): 4359–4363.
18 Choi, K.S., Park, Y., and Kim, S.Y. (2013). Comparison of graphene oxide with reduced graphene oxide as hole extraction layer in organic photovoltaic cells. *J. Nanosci. Nanotechnol.* 13 (5): 3282–3287.
19 Lee, C.Y., Le, Q.V., Kim, C., and Kim, S.Y. (2015). Use of silane-functionalized graphene oxide in organic photovoltaic cells and organic light-emitting diodes. *Phys. Chem. Chem. Phys.* 17 (14): 9369–9374.
20 Wang, X.-Y., Narita, A., and Müllen, K. (2017). Precision synthesis versus bulk-scale fabrication of graphenes. *Nat. Rev. Chem.* 2 (1): 0100.
21 Park, J., Kim, Y.S., Sung, S.J. et al. (2017). Highly dispersible edge-selectively oxidized graphene with improved electrical performance. *Nanoscale* 9 (4): 1699–1708.
22 Yu, H., Zhang, B., Bulin, C. et al. (2016). High-efficient synthesis of graphene oxide based on improved hummers method. *Sci. Rep.* 6 (1): 36143.
23 Kwon, K.C., Ham, J., Kim, S. et al. (2014). Eco-friendly graphene synthesis on Cu foil electroplated by reusing Cu etchants. *Sci. Rep.* 4: 4830.
24 Polsen, E.S., McNerny, D.Q., Viswanath, B. et al. (2015). High-speed roll-to-roll manufacturing of graphene using a concentric tube CVD reactor. *Sci. Rep.* 5 (1): 10257.
25 Leong, W.S., Wang, H., Yeo, J. et al. (2019). Paraffin-enabled graphene transfer. *Nat. Commun.* 10 (1): 867.
26 Plutnar, J., Pumera, M., and Sofer, Z. (2018). The chemistry of CVD graphene. *J. Mater. Chem. C* 6 (23): 6082–6101.
27 Gao, L., Ni, G.-X., Liu, Y. et al. (2014). Face-to-face transfer of wafer-scale graphene films. *Nature* 505 (7482): 190–194.
28 Kong, W., Kum, H., Bae, S.-H. et al. (2019). Path towards graphene commercialization from lab to market. *Nat. Nanotechnol.* 14 (10): 927–938.
29 Bae, S., Kim, H., Lee, Y. et al. (2010). Roll-to-roll production of 30-inch graphene films for transparent electrodes. *Nat. Nanotechnol.* 5 (8): 574–578.

30 Neumann, C., Kaiser, D., Mohn, M.J. et al. (2019). Bottom-up synthesis of graphene monolayers with tunable crystallinity and porosity. *ACS Nano* 13 (6): 7310–7322.

31 Dong, H.S. and Qi, S.J. (2015). Realising the potential of graphene-based materials for biosurfaces – a future perspective. *Biosurf. Biotribol.* 1 (4): 229–248.

32 Morozov, S.V., Novoselov, K.S., Katsnelson, M.I. et al. (2008). Giant intrinsic carrier mobilities in graphene and its bilayer. *Phys. Rev. Lett.* 100 (1): 016602.

33 Papageorgiou, D.G., Kinloch, I.A., and Young, R.J. (2017). Mechanical properties of graphene and graphene-based nanocomposites. *Prog. Mater Sci.* 90: 75–127.

34 Feng, X., Cheng, H., Pan, Y., and Zheng, H. (2015). Development of glucose biosensors based on nanostructured graphene-conducting polyaniline composite. *Biosens. Bioelectron.* 70: 411–417.

35 Ng, K.L. and Khor, S.M. (2017). Graphite-based nanocomposite electrochemical sensor for multiplex detection of adenine, guanine, thymine, and cytosine: a biomedical prospect for studying DNA damage. *Anal. Chem.* 89 (18): 10004–10012.

36 Zhou, K., Zhu, Y., Yang, X., and Li, C. (2010). Electrocatalytic oxidation of glucose by the glucose oxidase immobilized in graphene-Au-Nafion biocomposite. *Electroanalysis* 22 (3): 259–264.

37 Shan, C., Yang, H., Han, D. et al. (2010). Graphene/AuNPs/chitosan nanocomposites film for glucose biosensing. *Biosens. Bioelectron.* 25 (5): 1070–1074.

38 Yang, T., Song, T.T., Callsen, M. et al. (2019). Atomically thin 2D transition metal oxides: structural reconstruction, interaction with substrates, and potential applications. *Adv. Mater. Interf.* 6 (1): 1801160.

39 Haque, F., Daeneke, T., Kalantar-zadeh, K., and Ou, J.Z. (2017). Two-dimensional transition metal oxide and chalcogenide-based photocatalysts. *Nano-Micro Lett.* 10 (2): 23.

40 Rui, X., Lu, Z., Yu, H. et al. (2013). Ultrathin V_2O_5 nanosheet cathodes: realizing ultrafast reversible lithium storage. *Nanoscale* 5 (2): 556–560.

41 Takagaki, A., Lu, D., Kondo, J.N. et al. (2005). Exfoliated HNb_3O_8 nanosheets as a strong protonic solid acid. *Chem. Mater.* 17 (10): 2487–2489.

42 Zhuiykov, S., Hyde, L., Hai, Z. et al. (2017). Atomic layer deposition-enabled single layer of tungsten trioxide across a large area. *Appl. Mater. Today* 6: 44–53.

43 Giordano, L., Lewandowski, M., Groot, I.M.N. et al. (2010). Oxygen-induced transformations of an FeO(111) film on Pt(111): a combined DFT and STM study. *J. Phys. Chem. C* 114 (49): 21504–21509.

44 Addou, R., Dahal, A., and Batzill, M. (2013). Growth of a two-dimensional dielectric monolayer on quasi-freestanding graphene. *Nat. Nanotechnol.* 8 (1): 41–45.

45 Liu, J., Huang, X., Li, Y. et al. (2006). Self-assembled CuO monocrystalline nanoarchitectures with controlled dimensionality and morphology. *Cryst. Growth Des.* 6 (7): 1690–1696.

46 Cheng, W., He, J., Yao, T. et al. (2014). Half-unit-cell α-Fe_2O_3 semiconductor nanosheets with intrinsic and robust ferromagnetism. *J. Am. Chem. Soc.* 136 (29): 10393–10398.

47 Shendage, S.S., Patil, V.L., Vanalakar, S.A. et al. (2017). Sensitive and selective NO_2 gas sensor based on WO_3 nanoplates. *Sens. Actuator B-Chem.* 240: 426–433.

48 Dey, A. (2018). Semiconductor metal oxide gas sensors: a review. *Mater. Sci. Eng. B* 229: 206–217.

49 Zhou, J., Asmara, T.C., Yang, M. et al. (2015). Interplay of electronic reconstructions, surface oxygen vacancies, and lattice distortions in insulator-metal transition of $LaAlO_3/SrTiO_3$. *Phys. Rev. B* 92 (12): 125423.

50 Reyren, N., Thiel, S., Caviglia, A.D. et al. (2007). Superconducting interfaces between insulating oxides. *Science* 317 (5842): 1196.

51 Salam, Z., Vijayakumar, E., Subramania, A. et al. (2015). Graphene quantum dots decorated electrospun TiO_2 nanofibers as an effective photoanode for dye sensitized solar cells. *Sol. Energy Mater. Sol. Cells* 143: 250–259.

52 Lee, K., Yoon, H., Ahn, C. et al. (2019). Strategies to improve the photocatalytic activity of TiO_2: 3D nanostructuring and heterostructuring with graphitic carbon nanomaterials. *Nanoscale* 11 (15): 7025–7040.

53 Védrine, J.C. (2019). Metal oxides in heterogeneous oxidation catalysis: state of the art and challenges for a more sustainable world. *ChemSusChem* 12 (3): 577–588.

54 Nallal, M., Anantha Iyengar, G., and Pill-Lee, K. (2017). New titanium dioxide-based heterojunction nanohybrid for highly selective photoelectrochemical–electrochemical dual-mode sensors. *ACS Appl. Mater. Interf.* 9 (42): 37166–37183.

55 Gao, H., Sun, M., Lin, C., and Wang, S. (2012). Electrochemical DNA biosensor based on graphene and TiO_2 nanorods composite film for the detection of transgenic soybean gene sequence of MON89788. *Electroanalysis* 24 (12): 2283–2290.

56 Song, T.T., Yang, M., Callsen, M. et al. (2015). Graphene stabilized high-κ dielectric Y_2O_3 (111) monolayers and their interfacial properties. *RSC Adv.* 5 (102): 83588–83593.

57 Kolobov, A.V. and Tominaga, J. (2016). *Two-dimensional Transition-Metal Dichalcogenides*. Springer International Publishing.

58 Newaz, A.K.M., Prasai, D., Ziegler, J.I. et al. (2013). Electrical control of optical properties of monolayer MoS_2. *Solid State Commun.* 155: 49–52.

59 Kadantsev, E.S. and Hawrylak, P. (2012). Electronic structure of a single MoS_2 monolayer. *Solid State Commun.* 152 (10): 909–913.

60 Mak, K.F., Lee, C., Hone, J. et al. (2010). Atomically thin MoS_2: a new direct-gap semiconductor. *Phys. Rev. Lett.* 105 (13): 136805.

61 Radisavljevic, B., Radenovic, A., Brivio, J. et al. (2011). Single-layer MoS_2 transistors. *Nat. Nanotechnol.* 6 (3): 147–150.

62 Le, Q.V., Nguyen, T.P., Jang, H.W., and Kim, S.Y. (2014). The use of UV/ozone-treated MoS_2 nanosheets for extended air stability in organic photovoltaic cells. *Phys. Chem. Chem. Phys.* 16 (26): 13123–13128.

63 Kim, C., Nguyen, T.P., Le, Q.V. et al. (2015). Performances of liquid-exfoliated transition metal dichalcogenides as hole injection layers in organic light-emitting diodes. *Adv. Funct. Mater.* 25 (28): 4512–4519.

64 Nguyen, T.P., Choi, S., Jeon, J.M. et al. (2016). Transition metal disulfide nanosheets synthesized by facile sonication method for the hydrogen evolution reaction. *J. Phys. Chem. C* 120 (7): 3929–3935.

65 Kim, T.H., Hasani, A., Quyet, L.V. et al. (2019). NO_2 sensing properties of porous Au-incorporated tungsten oxide thin films prepared by solution process. *Sens. Actuator B-Chem.* 286: 512–520.

66 Yao, Y., Lin, Z., Li, Z. et al. (2012). Large-scale production of two-dimensional nanosheets. *J. Mater. Chem.* 22 (27): 13494–13499.

67 Coleman, J.N., Lotya, M., O'Neill, A. et al. (2011). Two-dimensional nanosheets produced by liquid exfoliation of layered materials. *Science* 331 (6017): 568–571.

68 Voiry, D., Salehi, M., Silva, R. et al. (2013). Conducting MoS_2 nanosheets as catalysts for hydrogen evolution reaction. *Nano Lett.* 13 (12): 6222–6227.

69 Kwon, K.C., Kim, C., Le, Q.V. et al. (2015). Synthesis of atomically thin transition metal disulfides for charge transport layers in optoelectronic devices. *ACS Nano* 9 (4): 4146–4155.

70 Wang, Y. and Ni, Y. (2014). Molybdenum disulfide quantum dots as a photoluminescence sensing platform for 2,4,6-trinitrophenol detection. *Anal. Chem.* 86 (15): 7463–7470.

71 Hu, C., Jiang, Z., Zhou, W. et al. (2019). Wafer-scale sulfur vacancy-rich monolayer MoS_2 for massive hydrogen production. *J. Phys. Chem. Lett.* 10 (16): 4763–4768.

72 Kwon, K.C., Choi, S., Hong, K. et al. (2016). Wafer-scale transferable molybdenum disulfide thin-film catalysts for photoelectrochemical hydrogen production. *Energy Environ. Sci.* 9 (7): 2240–2248.

73 Xu, H., Zhang, H., Guo, Z. et al. (2018). High-performance wafer-scale MoS_2 transistors toward practical application. *Small* 14 (48): 1803465.

74 Lee, Y.H., Zhang, X.Q., Zhang, W. et al. (2012). Synthesis of large-area MoS_2 atomic layers with chemical vapor deposition. *Adv. Mater.* 24 (17): 2320–2325.

75 Chang Kwon, K., Choi, S., Lee, J. et al. (2017). Drastically enhanced hydrogen evolution activity by 2D to 3D structural transition in anion-engineered molybdenum disulfide thin films for efficient Si-based water splitting photocathodes. *J. Mater. Chem. A* 5 (30): 15534–15542.

76 Kwon, K.C., Choi, S., Hong, K. et al. (2017). Tungsten disulfide thin film/p-type Si heterojunction photocathode for efficient photochemical hydrogen production. *MRS Commun.* 7 (2): 272–279.

77 Altavilla, C., Sarno, M., and Ciambelli, P.A. (2011). Novel wet chemistry approach for the synthesis of hybrid 2D free-floating single or multilayer nanosheets of MS2@oleylamine (M = Mo, W). *Chem. Mater.* 23 (17): 3879–3885.

78 Liu, Y., Zhong, Q., Chen, K. et al. (2017). Morphologies controllable synthesis of MoS_2 by hot-injection method: from quantum dots to nanosheets. *J. Mater. Sci. Mater. Electron.* 28 (18): 13633–13637.

79 Van Le, Q., Nguyen, T.P., Park, M. et al. (2016). Bottom-up synthesis of MeS_x nanodots for optoelectronic device applications. *Adv. Opt. Mater.* 4 (11): 1796–1804.

80 Hasani, A., Nguyen, T.P., Tekalgne, M. et al. (2018). The role of metal dopants in WS_2 nanoflowers in enhancing the hydrogen evolution reaction. *Appl. Catal. A.* 567: 73–79.

81 Li, Y.D., Li, X.L., He, R.R. et al. (2002). Artificial lamellar mesostructures to WS_2 nanotubes. *J. Am. Chem. Soc.* 124 (7): 1411–1416.

82 Nguyen, T.P., Nguyen, D.L.T., Nguyen, V.-H. et al. (2020). Facile synthesis of WS_2 hollow spheres and their hydrogen evolution reaction performance. *Appl. Surf. Sci.* 505: 144574.

83 Tekalgne, M., Hasani, A., Le, Q.V. et al. (2019). CdSe quantum dots doped WS_2 nanoflowers for enhanced solar hydrogen production. *Phys. Status Solidi A* https://doi.org/10.1002/pssa.201800853.

84 Nguyen, T.P., Sohn, W., Oh, J.H. et al. (2016). Size-dependent properties of two-dimensional MoS_2 and WS_2. *J. Phys. Chem. C* 120 (18): 10078–10085.

85 Anantharaj, S., Valappil, M.O., Karthick, K. et al. (2019). Electrochemically chopped WS_2 quantum dots as an efficient and stable electrocatalyst for water reduction. *Catal. Sci. Technol.* 9 (1): 223–231.

86 Liu, Y., Zhang, J., Shen, Y. et al. (2017). MoS_2 quantum dots featured fluorescent biosensor for multiple detection of cancer. *RSC Adv.* 7 (86): 54638–54643.

87 He, Q., Wang, L., Yin, K., and Luo, S. (2018). Vertically aligned ultrathin 1T-WS2 nanosheets enhanced the electrocatalytic hydrogen evolution. *Nanoscale Res. Lett.* 13 (1): 167.

88 Sarkar, D., Liu, W., Xie, X. et al. (2014). MoS_2 Field-effect transistor for next-generation label-free biosensors. *ACS Nano* 8 (4): 3992–4003.

89 Narang, J., Mishra, A., Pilloton, R. et al. (2018). Development of $MoSe_2$ nano-urchins as a sensing platform for a selective bio-capturing of *Escherichia coli* shiga toxin DNA. *Biosensors* 8 (3): 77.

90 Soni, A., Pandey, C.M., Pandey, M.K., and Sumana, G. (2019). Highly efficient polyaniline-MoS_2 hybrid nanostructures based biosensor for cancer biomarker detection. *Anal. Chim. Acta* 1055: 26–35.

91 Baek, S.H., Park, C.Y., Nguyen, T.P. et al. (2020). Novel peptides functionalized gold nanoparticles decorated tungsten disulfide nanoflowers as the electrochemical sensing platforms for the norovirus in an oyster. *Food Control* 114: 107225.

92 Jeong, J.-M., Yang, M., Kim, D.S. et al. (2017). High performance electrochemical glucose sensor based on three-dimensional MoS_2/graphene aerogel. *J. Colloid Interf. Sci.* 506: 379–385.

93 Huang, J., He, Y., Jin, J. et al. (2014). A novel glucose sensor based on MoS_2 nanosheet functionalized with Ni nanoparticles. *Electrochim. Acta* 136: 41–46.

94 Nguyen, T.P., Tuan Nguyen, D.M., Tran, D.L. et al. (2020). MXenes: applications in electrocatalytic, photocatalytic hydrogen evolution reaction and CO_2 reduction. *Molec, Catal.* 486: 110850.

95 Gong, Q., Wang, Y., Hu, Q. et al. (2016). Ultrasmall and phase-pure W_2C nanoparticles for efficient electrocatalytic and photoelectrochemical hydrogen evolution. *Nat. Commun.* 7: 13216.

96 Nguyen, T.P., Kim, S.Y., Lee, T.H. et al. (2020). Facile synthesis of $W_2C@WS_2$ alloy nanoflowers and their hydrogen generation performance. *Appl. Surf. Sci.* 504: 144389.

97 Deysher, G., Shuck, C.E., Hantanasirisakul, K. et al. (2020). Synthesis of Mo_4VAlC_4 MAX phase and two-dimensional Mo_4VC_4 MXene with five atomic layers of transition metals. *ACS Nano* 14 (1): 204–217.

98 Nguyen, T.P., Choi, K.S., Kim, S.Y. et al. (2020). Strategy for controlling the morphology and work function of W_2C/WS_2 nanoflowers. *J. Alloys Compd.* 829: 154582.

99 Zhang, C., Ma, Y., Zhang, X. et al. (2020). Two-dimensional transition metal carbides and nitrides (MXenes): synthesis, properties, and electrochemical energy storage applications. *Energy Environ. Mater.* 3 (1): 29–55.

100 Kalambate, P.K., Gadhari, N.S., Li, X. et al. (2019). Recent advances in MXene–based electrochemical sensors and biosensors. *TrAC - Trends Anal. Chem.* 120: 115643.

101 Zheng, J., Diao, J., Jin, Y. et al. (2018). An inkjet printed Ti_3C_2-GO electrode for the electrochemical sensing of hydrogen peroxide. *J. Electrochem. Soc.* 165 (5): B227–B231.

102 Lei, Y., Zhao, W., Zhang, Y. et al. (2019). MXene-based wearable biosensor system for high-performance in vitro perspiration analysis. *Small* 15 (19): 1901190.

103 Kumar, S., Lei, Y., Alshareef, N.H. et al. (2018). Biofunctionalized two-dimensional Ti_3C_2 MXenes for ultrasensitive detection of cancer biomarker. *Biosens. Bioelectron.* 121: 243–249.

104 Liu, J., Jiang, X., Zhang, R. et al. (2019). MXene-enabled electrochemical microfluidic biosensor: applications toward multicomponent continuous monitoring in whole blood. *Adv. Funct. Mater.* 29 (6): 1807326.

105 Zhang, Y., Jiang, X., Zhang, J. et al. (2019). Simultaneous voltammetric determination of acetaminophen and isoniazid using MXene modified screen-printed electrode. *Biosens. Bioelectron.* 130: 315–321.

106 Zheng, J., Wang, B., Ding, A. et al. (2018). Synthesis of MXene/DNA/Pd/Pt nanocomposite for sensitive detection of dopamine. *J. Electroanal. Chem.* 816: 189–194.

107 Kornienko, N., Zhao, Y., Kley, C.S. et al. (2015). Metal–organic frameworks for electrocatalytic reduction of carbon dioxide. *J. Am. Chem. Soc.* 137 (44): 14129–14135.

108 Zhao, M., Huang, Y., Peng, Y. et al. (2018). Two-dimensional metal-organic framework nanosheets: synthesis and applications. *Chem. Soc. Rev.* 47 (16): 6267–6295.

109 Ding, Y., Chen, Y.-P., Zhang, X. et al. (2017). Controlled intercalation and chemical exfoliation of layered metal–organic frameworks using a chemically labile intercalating agent. *J. Am. Chem. Soc.* 139 (27): 9136–9139.

110 Li, P.-Z., Maeda, Y., and Xu, Q. (2011). Top-down fabrication of crystalline metal–organic framework nanosheets. *Chem. Commun.* 47 (29): 8436–8438.

111 Song, W.-J. (2017). Intracellular DNA and microRNA sensing based on metal-organic framework nanosheets with enzyme-free signal amplification. *Talanta* 170: 74–80.

112 Wang, L., Sahabudeen, H., Zhang, T., and Dong, R. (2018). Liquid-interface-assisted synthesis of covalent-organic and metal-organic two-dimensional crystalline polymers. *NPJ 2D Mater. Appl.* 2 (1): 26.

113 Makiura, R., Motoyama, S., Umemura, Y. et al. (2010). Surface nano-architecture of a metal–organic framework. *Nat. Mater.* 9 (7): 565–571.

114 Huang, X., Sheng, P., Tu, Z. et al. (2015). A two-dimensional π–d conjugated coordination polymer with extremely high electrical conductivity and ambipolar transport behaviour. *Nat. Commun.* 6 (1): 7408.

115 Rodenas, T., Luz, I., Prieto, G. et al. (2015). Metal–organic framework nanosheets in polymer composite materials for gas separation. *Nat. Mater.* 14 (1): 48–55.

116 Zhao, M., Wang, Y., Ma, Q. et al. (2015). Ultrathin 2D metal–organic framework nanosheets. *Adv. Mater.* 27 (45): 7372–7378.

117 Hu, Z., Castano, I., Wang, S. et al. (2016). Modulator effects on the water-based synthesis of Zr/Hf metal–organic frameworks: quantitative relationship studies between modulator, synthetic condition, and performance. *Cryst. Growth Des.* 16 (4): 2295–2301.

118 Jiang, Y., Liu, H., Tan, X. et al. (2017). Monoclinic ZIF-8 nanosheet-derived 2D carbon nanosheets as sulfur immobilizer for high-performance lithium sulfur batteries. *ACS Appl. Mater. Interf.* 9 (30): 25239–25249.

119 Ran, J., Qu, J., Zhang, H. et al. (2019). 2D metal organic framework nanosheet: a universal platform promoting highly efficient visible-light-induced hydrogen production. *Adv. Energy Mater.* 9 (11): 1803402.

120 Meng, W., Wen, Y., Dai, L. et al. (2018). novel electrochemical sensor for glucose detection based on Ag@ZIF-67 nanocomposite. *Sens. Actuator B-Chem.* 260: 852–860.

121 He, C., Lu, K., Liu, D., and Lin, W. (2014). Nanoscale metal–organic frameworks for the co-delivery of cisplatin and pooled siRNAs to enhance therapeutic efficacy in drug-resistant ovarian cancer cells. *J. Am. Chem. Soc.* 136 (14): 5181–5184.

122 Zong, J., Cobb, S.L., and Cameron, N.R. (2017). Peptide-functionalized gold nanoparticles: versatile biomaterials for diagnostic and therapeutic applications. *Biomater. Sci.* 5 (5): 872–886.

123 Zhou, J., Fang, G., Pan, A., and Liang, S. (2016). Oxygen-incorporated MoS_2 nanosheets with expanded interlayers for hydrogen evolution reaction and pseudocapacitor applications. *ACS Appl. Mater. Interf.* 8 (49): 33681–33689.

124 Huang, H., Feng, X., Du, C. et al. (2015). Incorporated oxygen in MoS_2 ultrathin nanosheets for efficient ORR catalysis. *J. Mater. Chem. A* 3 (31): 16050–16056.

125 Sarma, P.V., Tiwary, C.S., Radhakrishnan, S. et al. (2018). Oxygen incorporated WS_2 nanoclusters with superior electrocatalytic properties for hydrogen evolution reaction. *Nanoscale* 10 (20): 9516–9524.

126 Wang, Y., Bai, X., Wen, W. et al. (2015). Ultrasensitive electrochemical biosensor for HIV gene detection based on graphene stabilized gold nanoclusters with exonuclease amplification. *ACS Appl. Mater. Interf.* 7 (33): 18872–18879.

127 Yin, H., Tang, H., Wang, D. et al. (2012). Facile synthesis of surfactant-free Au cluster/graphene hybrids for high-performance oxygen reduction reaction. *ACS Nano* 6 (9): 8288–8297.

128 Darabdhara, G., Das, M.R., Singh, S.P. et al. (2019). Ag and Au nanoparticles/reduced graphene oxide composite materials: Synthesis and application in diagnostics and therapeutics. *Adv. Colloid Interface Sci.* 271: 101991.

129 Zhou, L., He, B., Yang, Y., and He, Y. (2014). Facile approach to surface functionalized MoS_2 nanosheets. *RSC Adv.* 4 (61): 32570–32578.

130 Presolski, S. and Pumera, M. (2016). Covalent functionalization of MoS_2. *Mater. Today* 19 (3): 140–145.

9

Functionalized Materials for Miniaturized Analytical Devices

Hani Nasser Abdelhamid

Assiut University, Department of Chemistry, Advanced Multifunctional Materials Laboratory, Assiut 71516, Egypt

9.1 Introduction

Modern analytical chemistry aims at simplification ("easy to implement or do"), automation ("electromechanically self-operation"), and miniaturization ("small scale" or "make to a very small scale") [1–4]. Miniaturized analytical systems, micro-total analysis systems (μ-TAS) and "lab-on-a-chip," are used to represent small-scale analytical processes. Miniaturization process implies not only decreases in scale but also that other forces and phenomena are present in microsize environments [5, 6]. These processes require advanced materials with properties such as high sensitivity, flexibility, and easy for fabrication and implementation into electronic devices.

Nanotechnology is the term used to describe the technology that uses materials with at least one dimension in the nanometer range (1 nm = 10^{-9} m, nanoparticles). Nanoparticles have unique properties that significantly distinguish them from bulk materials. The new properties of nanoparticles have developed several applications and open new venues for modern technologies as well as new products and devices. Nanoparticles have been applied in analytical chemistry [7–16], proteomics [17], biosensing [18–20], biotechnology [21–30], nanomedicines [31–38], antibacterials [39], drug delivery [40–42], gene delivery [43, 44], wound healing [45], energy-based applications [46–48], adsorption [49], and environmental applications [50–59]. Nanoparticles offer high performance [60–68]. Nanoparticles can be easily implemented for miniaturized devices including wearable electronics [69]. Thus, they offer new opportunities to improve the quality of life via the prevention of fatal and chronic diseases by enabling very early detection of a disease.

The advanced technology offers processing of materials including nanoparticles to ensure simple usages. Several processing methods were reported including three-dimensional (3D) printing [70], wax printing technology [71], electrospinning [72], screen-printing [73], femtosecond laser direct writing [74], metal organic chemical

Miniaturized Analytical Devices: Materials and Technology, First Edition.
Edited by Suresh Kumar Kailasa and Chaudhery Mustansar Hussain.

vapor deposition (MOCVD) [75], and electrowetting-on-dielectric (EWOD) [76]. These methods offer several advantages such as the preferential design of electrodes [73]. They represent the links between laboratory work and commercial products.

Among the wide applications, nanoparticles have advanced analytical methods such as polymerase chain reactions (PCR), electrochemical (EC) based methods, mass spectrometry (MS), chromatography, electrophoresis, and surface plasmon-based methods. Electrochemical techniques are interference-less compared to other analytical methods such as spectroscopy including chemiluminescence or fluorescence. Nanoparticles involve simple sample pretreatments and are easier to miniaturize. Consequently, nanoparticle-based analytic methods have become the method of choice for point-of-care analysis. Nanoparticle-based methods have paved the way for the innovative analysis of several target analytes with a low limit of detection (LOD) compared to the conventional methods. These methods decrease the required time for analysis and are simple to use and require no skilled personnel for authenticated results. Nanoparticles make the analysis simple even for complex samples. Furthermore, nanoparticles can be used for miniaturized analytical devices and implemented for small and portable devices.

9.2 Miniaturized Devices

Several miniaturized devices have been reported. Miniaturization has been applied for energy storage devices [77] such as supercapacitors [78], tetherless medical tools [79], flow cytometry [80], capillary gas chromatography (CGC), micro-gas chromatography (μGC) and microcapillary electrophoresis (μCE), and μTAS (also called lab-on-a-chip, LOC) [81].

Miniaturization of analytical instruments has led to advanced health care and lifestyle. Analytical devices are miniaturized in three scales: (i) miniscale (mm–μl), (ii) microscale (size of mm–50 μm and sample volume of μl and 10 nl), and (iii) nanoscale (50 μm–1 μm, and sample volume below 10 nl). It is highly required for bulky analytical methods. Analytical methods such as chromatography and mass spectrometry are bulky instruments and are difficult to miniaturize on a millimeter scale. Nanomaterial-based devices were reported for point-of-care diagnostic applications [82], including mycotoxin detection [83]. Nowadays, portable devices that can monitor and diagnose target analytes are commercially available. Well-known examples are glucose meters, miniaturized portable (handheld) Raman spectrometer (TacticID), and the immunoassay system "POCKET" [84]. These devices require a few microliters of samples and can determine the concentration of the analyte in the sample within a few minutes. The World Health Organization (WHO) surveyed some of the automated portable platforms for the detection of the human immunodeficiency virus (HIV) [85]. Several methods such as GeneXpert® HIV-1 Viral Load Test (Cepheid), cobas® Liat™ System (Roche), Alere™ q system (Alere™), and EOSCAPE-HIV™ (Wave 80 Biosciences) were listed.

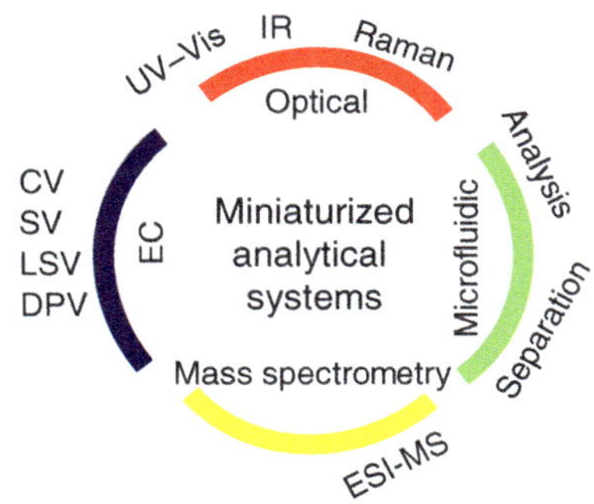

Figure 9.1 Classes of miniaturized analytical systems covered in this chapter.

9.3 Miniaturized Devices for Analysis

Portable devices can be classified into two main categories: body fluid–based devices and volatile biomarker-based devices. They can be also classified into contact approaches (for analysis of samples such as tears, sweat, and saliva) and contactless technologies (for analysis of samples such as breath and perspiration). These devices may require chemical interactions or physical signals such as light and pressure. They can be further categorized as optical, electrochemical, microfluidic devices, and mass spectrometry (Figure 9.1).

9.3.1 Optical Devices

Several optical devices have been reported for point-of-care analysis. Lateral flow biosensors (LFBs) are fabricated on paper substrates, the best-known example of which is the common pregnancy test [86]. It is based on thin-layer immunoaffinity chromatography.

Miniaturized Raman instruments were also reported for qualitative analysis [87] such as analysis of cancer [88, 89], West Nile virus [90], and diabetes [91].

Several review articles were published on the applications of functional nanomaterials for metal ions detection [92–97]. The devices for metal ion detection should include the features of miniaturization and portability [98]. The design of detection devices includes (i) recognition unit of the target analyte, (ii) generation unit for signal, (iii) signal transduction or amplification unit, (iv) signal reception unit, and (v) signal processing unit. Several materials were functionalized to miniaturize these units. Generally, nanoparticles can be integrated to miniaturize the first three units by exploration of suitable functional nanomaterials, such as metallic nanoparticles, metal oxide nanoparticles, quantum dots, carbon nanomaterials.

Microfluidic devices, integrated into the analysis process on a piece of chip, can be used for the separation of target analytes in a complex sample. The analytes can be detected using mobile phones, photometers, intelligence chips, and other smart devices. These methods replace the conventional signal detectors. Nanomaterials and these convenient instruments can be integrated to explore various kinds of metal ions detection.

9.3.2 Electrochemical Methods

Electrochemical analytical methods such as amperometry (current measurements), impedimetric (charge transfer resistance measurements), potentiometry (potential measurements), and voltammetry (potentiodynamic measurements) were investigated for the detection of several analytes such as proteins, biomolecules, metals, and biomarkers [99]. Electrochemical methods are based on changes in measurable signals. Voltammetric techniques such as cyclic voltammetry (CV), linear sweep voltammetry (LSV), differential pulse voltammetry (DPV), square-wave voltammetry (SWV), and stripping voltammetry (SV) are used for biosensing.

Electrochemical methods such as nanogap electrodes were also reported. Nanogap electrodes are a pair of electrodes with a nanometer gap that fits the size of the target analyte. The electrical signals (resistance/impedance, capacitance/dielectric, or field effect) of the two electrodes after trapping the target analytes were recorded [100]. The change in the electrical signals before and after encapsulation can be used for the quantitative analysis of the target analyte.

A wearable tattoo-based iontophoretic sensing system was reported for detection of alcohol in sweat [101]. Detecting the sweat alcohol content is an alternative method for measuring blood alcohol concentration (BAC). The method is a noninvasive source of alcohol monitoring [101]. Wearable health-care devices such as transducers have been used for biosensing applications. Printed organic transistors were reported for biosensing of glucose [102]. The method is an enzyme-based sensor using glucose oxidase (GOx)–Prussian blue (PB)-modified extended-gate electrodes [102]. A sensor belt comprised of sweat-wicking materials and ion-selective electrodes (ISE) was used for monitoring the sodium content in human sweat [103]. Amperometry-based sensing performance of porous WO_3 nanofibers is evaluated toward low concentrations (1.8–12.5 ppm) of acetone [72]. The analysis can be further improved via light excitation (365 nm UV) and applied bias voltage (3–7 V) [72].

Several materials including graphene (G) [104], poly(3,4-ethylenedioxythiophene) : poly(styrenesulfonate) (PEDOT : PSS) [105] were investigated for sensing and biosensing [106]. Inorganic [106] and organic electrochemical transistors (OECTs) [105] were reported. Different analytes such as ammonium (NH_4^+)\ calcium (Ca^{2+}) [105] and glucose [102] were reported.

9.3.3 Magnetic Relaxation Switches (MRSw) Assays

Portable nuclear magnetic resonance (NMR) technique combined with new surface-functionalized magnetic nanoparticles (MNPs), known as magnetic relaxation switch (MRSw), was investigated for *in vitro* diagnostics [107–109]. This technique provides ultrasensitive, selective, immune, and molecular diagnostics in point-of-care settings [110]. The assay is based on the difference in the magnetic particles' transverse relaxation times (T_2) between dispersed and aggregated states. Two situations were observed: in nanoparticle-based systems (also called Type I systems), T_2 decreases with aggregation, whereas in larger magnetic particles (also

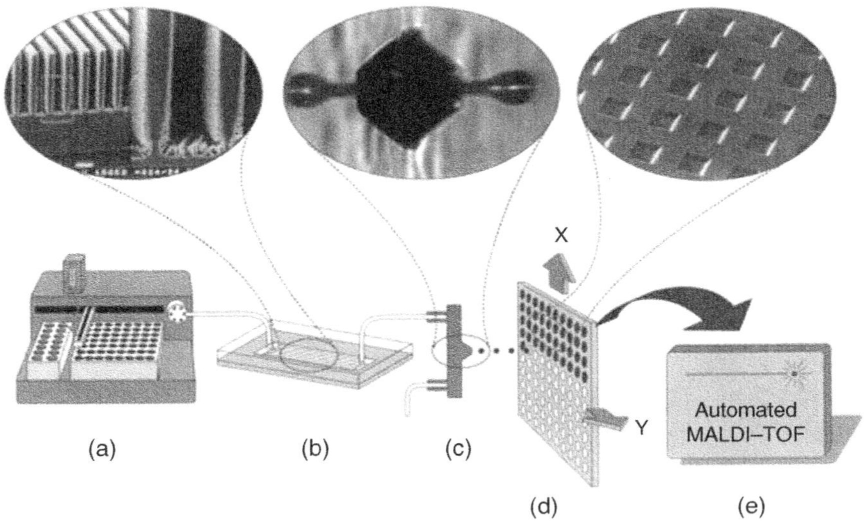

Figure 9.2 Integrated microfluidic system for proteins analysis: (a) sample injection and pretreatment, (b) enzyme digestion, and (c) sample dispensing into (d) microvials for detection, and (e) MALDI–TOF–MS. Source: Reproduced with permission Ekström et al. [112]. © 2000, American Chemical Society.

called Type II systems), T_2 increases with aggregation [111]. The technique has several advantages such as the use of biocompatibility of MNPs, high sensitivity, good specificity, and can be used to detect analytes in opaque samples (unresponsive to light-based interferences) [107].

A miniaturized MR device, called diagnostic magnetic resonance (DRM), has also been reported (Figure 9.2) [108]. DRM is capable of measuring T_2 changes in Type I MRSw sensors, has high sensitivity, and requires small volumes (<5 μl). The technique is based on miniaturized NMR electronics using a portable permanent magnet that generates a polarizing magnetic field [108]. μNMR-2, the second-generation DRM device [108], provides a high signal-to-noise ratio and requires a small sample volume (1 μl). μNMR was reported for the analysis of *Staphylococcus aureus* in samples with as little as 10 colony-forming units (CFU) [108].

9.3.4 Microfluidic Technology

Microfluidics is a chip-based technology with small-scale properties of a tiny dimensional channel. It can be used to achieve an efficient and controlled fluids movement. Thus, it can be applied for various fields such as physics, chemistry, biochemistry, biotechnology. Several materials such as cyclo-olefin copolymer (COC, suitable for fluorescence-based applications because of low background), poly(methyl methacrylate) (PMMA, biocompatible for biological applications), polystyrene (for flexible microfluidic devices) [113], fluorinated ethylene propylene (FEP, suitable for solvent resistance applications), polycarbonate (PC), and polydimethylsiloxane (PDMS) were used for microfluidic chips. Microfluidic technology resolves real-time practical problems. It provides high throughput and can be integrated into devices facilitating various applications. Microfluidic systems were coupled with techniques such as colorimetry [114], fluorescence [115], chemiluminescence [116], micro-PCR [117], and electrochemical methods [118]. They can be integrated with poly(vinyl alcohol) (PVA) nanofiber mats for the separation of negatively charged nanoparticles [119].

Several technologies for microfluidic systems were reported including microfluidic paper-based analytical devices (μPADs) [120, 121], droplet microfluidics [122], and μTAS (lab-on-a-chip) [123, 124]. These technologies are simple, can be easily used, are cheap, are disposable, and can be easily integrated with detection devices such as NMR [125], surface-enhanced Raman scattering (SERS) [126], mass spectrometry [127, 128], protein immunoblotting [129], organic electrochemical transistors [130], polymerase chain reaction (PCR) devices [131], Fourier-transform infrared (FT-IR) spectroscopy [132], and laser-controlled microactuators [133]. They can be used for on-site total analysis with high-throughput analysis. They require only tiny amounts of samples and reagents.

Microfluidic devices can be used for sample pretreatment and enzyme microreactors for proteolysis [112]. Silicon micromachined analytical tools (microchip-immobilized enzyme reactor, μ-chip IMER) were used for the on-line enzymatic digestion of protein samples (1 μl) within 1–3 min (Figure 9.2) [112]. The device offered on-line picoliter sample preparation in a high-density format [112]. It was integrated with automated matrix-assisted laser desorption ionization–time-of-flight–mass spectrometry (MALDI–TOF–MS) that analyze 100 protein samples in 3.5 hours (Figure 9.2).

Microfluidic devices were reported for rapid and ultrasensitive detection of analytes such as bioapplications [134],foodborne pathogens [135], HIV [136], pathogens [137], and pollution analysis [138]. The performance of microfluidic devices can be improved using liposomes [139]. A microfluidic flow cell based on gold interdigitated array microelectrode (IDAM) and magnetic NP–antibody conjugates (MNACs) were used to detect pathogenic bacteria in ground-beef samples [140]. The biosensor can detect as low as 1.6×10^2 and 1.2×10^3 cells of *Escherichia coli* O157: H7 cells present in pure cultures and ground beef samples, respectively, in just 35 min [140].

9.3.5 Mass Spectrometry

Mass spectrometry is a sensitive analytical method with great potential for miniaturization [141, 142]. Nanoparticles have advanced mass spectrometry [143–152] offering high sensitivity [153–158] and can be easily used for small and portable devices. The miniaturization of mass analyzers such as quadrupole ion trap [159], time of flight [160], magnetic sector [161], and linear quadrupole [162] was investigated. Several miniaturized MS including microfluidic electrospray ionization mass spectrometry (ESI-MS) [163] and microfluidic MALDI–MS [164] were reported for proteomics. Single-probe MS was used for metabolomic analysis of individual living cells [165].

9.4 Applications of Nanomaterials in Miniaturized Separation Techniques

Nanomaterials are used for miniaturization of separation and extraction techniques [166] such as solid-phase extraction [167], porous membrane, dispersive solid-phase microextraction, capillary electrophoresis [168], and continuous dielectrophoretic cell sorter [169]. Several forms such as microwave-assisted μ-SPE accelerated solvent extraction (ASE), and stir-bar-supported micro-solid-phase extraction (SB-μ-SPE) were also reported. These techniques are relatively fast techniques and require a very small amount of sorbent and volume of desorption solvent. They perform extraction, cleanup, and preconcentration of target analytes in a single step. They can be used for the analysis of complex samples directly without any pretreatment. They are reusable several times, requiring only washing. These techniques are robust, durable, and portable. Furthermore, they can be easily manipulated into other devices.

Various plasma separation microdevices such as capillary imbibition [170], blood cell sedimentation [171], cross-flow filtration [172], microfluidic free-flow isoelectric focusing glass chip [173], liquid-phase chromatography [174], capillary electrophoresis (isotachophoresis) [175], superhydrophobic plasma separator [176], and on-chip centrifugation [177] have been reported. These techniques can be used for analysis of analytes that are not suitable for centrifugation. Membrane-based, sedimentation-assisted plasma separators can be used for extracting 275 ± 33.5 μl of plasma from 1.8 ml of undiluted whole blood within less than 7 min (Figure 9.3) [178].

Free-flow electrophoresis (FFE) is generally used to separate charged proteins and cells [179]. The separation process is achieved by flushing the solution through a separation compartment under the effect of an electrical field perpendicular to the hydrodynamic flow [180]. The analyte becomes under two orthogonal velocity vectors, which deflect the species for the flow direction depending on electrophoretic mobility. The analytes are then separated at different outlets based on the time difference. The first attempt to miniaturize FFE was μFFE [181]. Miniaturization of FFE requires a small amount of sample and buffer solution and consequently offers faster separations. The substrate of FFE devices can be made from glass [182], indium tin oxide (ITO)-coated glass [183], and PDMS [184].

9.5 Advantages, Disadvantages, and Challenges

Miniaturized analytical devices are attractive for various reasons (Table 9.1). They require smaller volumes of (environmentally friendly) reagents; are cheaper, are

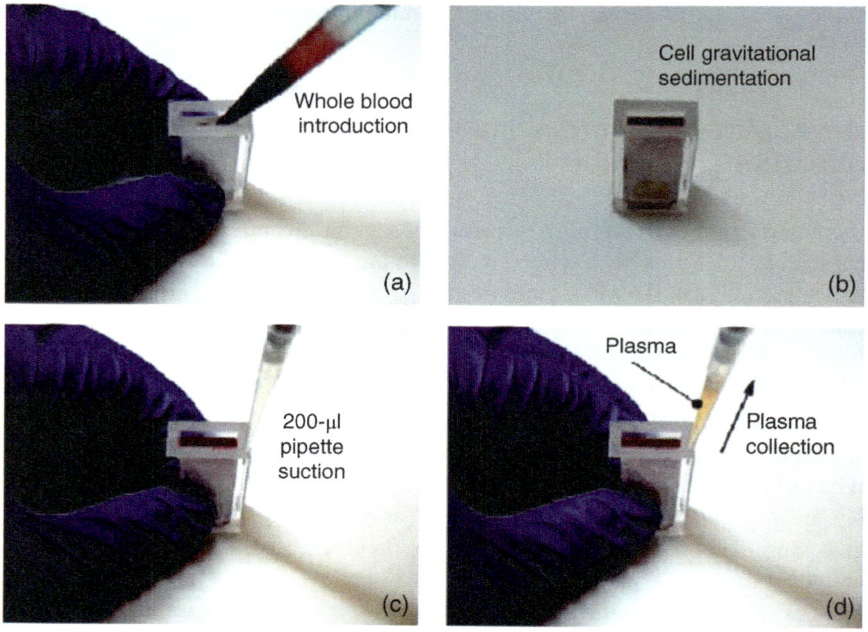

Figure 9.3 Sequence for plasma separation from blood samples using sedimentation-assisted plasma separator. Source: Figure reprinted with permission from Liu et al. [178]. Copyright belongs to the American Chemical Society.

quicker, enable on-site analysis; and are suitable for small analyses such as blood sample analysis.

Small devices require a tiny volume of sample species (Table 9.1), so it will be challenging to analyze biological samples such as fluid-containing cells. A small drop of the cell liquid sometimes contains no cell because most of these species are heterogeneous solutions. Thus, one should typically have 10^6 cells/ml at the start for a droplet size of ≈100 μl [185].

It is not easy to miniaturize all electronics and mechanical parts of an analytical system. Miniaturization of analytical techniques that require high pressure is difficult. For example, high-performance liquid chromatography requires a pumping power of up to 400 bar, while it will be up to 1000 bar for miniaturization devices. The detection part in miniaturized analytical systems should be very sensitive for ultra-small sample volumes.

9.6 Conclusions

Functionalized materials have led to advanced miniaturized analytical methods. Microfabrication on polymers is cheaper and can be used for mass production than on glass. On the other side, glass substrates offer electroosmotic flow (EOF) and can be easily modified. Advances in detection systems improve the sensitivity

Table 9.1 Advantages, disadvantages, and challenges of miniaturized analytical systems.

Advantages	Disadvantages	Challenges
Compatibility with micro-/nanosensors	Low level of automation/integration	Reproducibility in the fabrication and analysis
Small size and portability	Lack of representativeness of results	Low sensitivity of current detection methods
No skilled personnel	Require controlled experimental conditions	
Offer real-time, in situ analysis		
Require very low sample and reagent/solvent consumption		
Require a short time of analysis		
Good resolution and selectivity		

of miniaturized analytical systems. The scientific literature covers relatively few examples of chip-based chromatographic instruments.

Acknowledgments

The author would like to thank the Ministry of Higher Education and Scientific Research (MHESR) and Institutional Review Board (IRB) of the Faculty of Science at Assiut University, Egypt, for the support.

References

1 Quake, S.R. (2000). *Science* 290: 1536–1540.
2 Derkus, B. (2016). *Biosens. Bioelectron.* 79: 901–913.
3 Malcos, J.L. and Hancock, W.O. (2011). *Appl. Microbiol. Biotechnol.* 90: 1–10.
4 A. RiosEscarpa, A. and Simonet, B. (2009). *Miniaturization of Analytical Systems: Principles, Designs and Applications.* Wiley.
5 Greenwood, P.A. and Greenway, G.M. (2002). *TrAC Trends Anal. Chem.* 21: 726–740.
6 Geschke, O., Klank, H., and Telleman, P. (2004). *Microsystems Engineering of Lab-on-a-Chip Devices.* Weinheim: Wiley-VCH.
7 Wu, H.-F., Gopal, J., Abdelhamid, H.N., and Hasan, N. (2012). *Proteomics* 12: 2949–2961.
8 Abdelhamid, H.N. and Wu, H.-F. (2012). *Anal. Chim. Acta* 751: 94–104.

9 Abdelhamid, H.N. and Wu, H.-F. (2015). *TrAC Trends Anal. Chem.* 65: 30–46.
10 Abdelhamid, H.N. (2016). *TrAC Trends Anal. Chem.* 77: 122–138.
11 Abdelhamid, H.N., Gopal, J., and Wu, H.F. (2013). *Anal. Chim. Acta* 767: 104–111.
12 Yang, Y., Shen, K., Lin, J. et al. (2016). *RSC Adv.* 6: 45475–45481.
13 Nasser Abdelhamid, H., Wu, B.-S., and Wu, H.-F. (2014). *Talanta* 126: 27–37.
14 Abdelhamid, H.N. (2017). *TrAC Trends Anal. Chem.* 89: 68–98.
15 Abdelhamid, H.N. (2018). *Microchim. Acta* 185: 200.
16 Abdelhamid, H.N., Bermejo-Gómez, A., Martín-Matute, B., and Zou, X. (2017). *Microchim. Acta* 184: 3363–3371.
17 Abdelhamid, H.N., Kumaran, S., and Wu, H.-F. (2016). *RSC Adv.* 6: 97629–97635.
18 Abdelhamid, H.N. and Wu, H.-F. (2018). *Spectrochim. Acta Part A Mol. Biomol. Spectrosc.*: 188.
19 Abdelhamid, H.N. (2017). Lanthanide Metal-Organic Frameworks and Hierarchical Porous Zeolitic Imidazolate Frameworks: Synthesis, Properties, and Applications, Stockholm University, Faculty of Science.
20 Abdelhamid, H.N. and Wu, H.-F. (2015). *RSC Adv.*: 5.
21 Abdelhamid, H.N. and Wu, H.-F. (2013). *J. Mater. Chem. B* 1: 3950–3961.
22 Abdelhamid, H.N. and Wu, H.-F. (2013). *J. Mater. Chem. B* 1: 6094–6106.
23 Manikandan, M., Nasser Abdelhamid, H., Talib, A., and Wu, H.-F. (2014). *Biosens. Bioelectron.* 55: 180–186.
24 Gopal, J., Abdelhamid, H.N., Hua, P.-Y., and Wu, H.-F. (2013). *J. Mater. Chem. B* 1: 2463–2475.
25 Abdelhamid, H.N. and Wu, H.-F. (2016). *Anal. Bioanal. Chem.* 408: 4485–4502.
26 Bhaisare, M.L., Abdelhamid, H.N., Wu, B.-S., and Wu, H.-F. (2014). *J. Mater. Chem. B* 2: 4671–4683.
27 Abdelhamid, H.N. and Wu, H.-F. (2014). *Mater. Sci. Eng. C* 45: 438–445.
28 Hua, P.-Y., Manikandan, M., Abdelhamid, H.N., and Wu, H.-F. (2014). *J. Mater. Chem. B* 2: 7334–7343.
29 Abdelhamid, H.N. and Wu, H.-F. (2013). *Colloids Surf. B: Biointerf.* 115C: 51–60.
30 Sekar, R., Kailasa, S.K., Abdelhamid, H.N. et al. (2013). *Int. J. Mass Spectrom.* 338: 23–29.
31 Shahnawaz Khan, M., Abdelhamid, H.N., and Wu, H.-F. (2015). *Colloids Surf. B: Biointerf.* 127C: 281–291.
32 Abdelhamid, H.N., Khan, M.S., and Wu, H.F. (2014). *RSC Adv.* 4: 50035–50046.
33 Wu, B.-S., Abdelhamid, H.N., and Wu, H.-F. (2014). *RSC Adv.* 4: 3722.
34 Abdelhamid, H.N., Talib, A., and Wu, H.-F. (2015). *RSC Adv.* 5: 34594–34602.
35 Dowaidar, M., Abdelhamid, H.N., Hällbrink, M. et al. (2017). *Biochim. Biophys. Acta - Gen. Subj.* 1861: 2334–2341.
36 Dowaidar, M., Abdelhamid, H.N., Hällbrink, M. et al. (2017). *Sci. Rep.* 7: 9159.
37 Abdelhamid, H.N., Talib, A., and Wu, H.F. (2017). *Talanta* 166: 357–363.
38 Abdelhamid, H.N. (2016). *Mater. Focus* 5: 305–323.
39 Abdelhamid, H.N., Talib, A., and Wu, H.-F. (2015). *RSC Adv.* 5: 39952–39953.

40 Abdelhamid, H.N., El-Bery, H.M., Metwally, A.A. et al. (2019). *Carbohydr. Polym.* 214: 90–99.
41 Keservani, R.K. and Sharma, A.K. (2019). Nanoparticulate Drug Delivery Systems. In: *Apple Academic Press.*
42 Abdelhamid, H.N. and Wu, H.-F. (2019). *Nanoparticle drug delivery systems* (eds. R.K. Keservani and A.K. Sharma), 121–150. Apple Academic Press.
43 Dowaidar, M., Nasser Abdelhamid, H., Hällbrink, M. et al. (2018). *J. Biomater. Appl.* 33: 392–401.
44 Abdelhamid, H.N., Dowaidar, M., Hällbrink, M., and Langel, Ü. (2019). *SSRN Electron. J.* 3435895.
45 Hussein, K.H., Abdelhamid, H.N., Zou, X., and Woo, H.-M. (2019). *Mater. Sci. Eng. C* 94: 484–492.
46 Abdelhamid, H.N., El-Zohry, A.M., Cong, J. et al. (2019). *R. Soc. Open Sci.* 6: 190723.
47 Goda, M.N., Abdelhamid, H.N., and Said, A.E.-A.A. (2020). *ACS Appl. Mater. Interf.* 12: 646–653.
48 Kassem, A.A., Abdelhamid, H.N., Fouad, D.M., and Ibrahim, S.A. (2019). *Int. J. Hydrogen Energy* 44: 31230–31238.
49 Abdelhamid, H.N. (2020). *Mater. Today Chem.* 15: 100222.
50 Yao, Q., Bermejo Gómez, A., Su, J. et al. (2015). *Chem. Mater.* 27: 5332–5339.
51 Abdelhamid, H.N., Huang, Z., El-Zohry, A.M. et al. (2017). *Inorg. Chem.* 56: 9139–9146.
52 Abdelhamid, H.N. and Wu, H.-F. (2014). *J. Am. Soc. Mass Spectrom.* 25: 861–868.
53 Emam, H.E., Abdelhamid, H.N., and Abdelhameed, R.M. (2018). *Dye. Pigment.* 159: 491–498.
54 Abdelhamid, H.N., Bhaisare, M.L., and Wu, H.-F. (2014). *Talanta* 120: 208–217.
55 Etman, A.S., Abdelhamid, H.N., Yuan, Y. et al. (2018). *ACS Omega* 3: 2193–2201.
56 Ashour, R.M., Abdelhamid, H.N., Abdel-Magied, A.F. et al. (2017). *Solvent Extr. Ion Exch.* 35: 91–103.
57 Abdelhamid, H.N. and Zou, X. (2018). *Green Chem.* 20: 1074–1084.
58 Abdelhamid, H.N. and Wu, H.-F. (2015). *Analyst* 140: 1555–1565.
59 Abdelhamid, H.N. (2015). *Mater. Sci. Forum.* 832: 28–53.
60 Nasser Abdelhamid, H. and Wu, H.F. (2013). *Talanta* 115: 442–450.
61 Abdelhamid, H.N., Khan, M.S., and Wu, H.F. (2014). *Anal. Chim. Acta* 823: 51–60.
62 Gopal, J., Abdelhamid, H.N., Huang, J.H., and Wu, H.F. (2016). *Sens. Actuators B Chem.* 224: 413–424.
63 Valencia, L. and Abdelhamid, H.N. (2019). *Carbohydr. Polym.* 213: 338–345.
64 Abdel-Magied, A.F., Abdelhamid, H.N., Ashour, R.M. et al. (2019). *Micropor. Mesopor. Mater.* 278: 175–184.
65 Abdelhamid, H.N. and Spectrom, M. (2015). *Purif. Tech.* 01: 109–119.
66 Iqbal, M.N., Abdel-Magied, A.F., Abdelhamid, H.N. et al. (2017). *ACS Sustain. Chem. Eng.* 5: 9651–9656.

67 Abdelhamid, H.N., Wilk-Kozubek, M., El-Zohry, A.M. et al. (2019). *Micropor. Mesopor. Mater.* 279: 400–406.
68 Abdelhamid, H.N. (2016). *J. Data Mining Genomics Proteomics* 7: 2153–0602. 1000.
69 Tricoli, A., Nasiri, N., and De, S. (2017). *Adv. Funct. Mater.* 27: 1605271.
70 Sultan, S., Abdelhamid, H.N., Zou, X., and Mathew, A.P. (2018). *Adv. Funct. Mater.* 29: 1805372.
71 Zhang, Z., Azizi, M., Lee, M. et al. (2019). *Lab Chip* 19: 3448–3460.
72 Imran, M., Rashid, S.S.A.A.H., Sabri, Y. et al. (2019). *J. Mater. Chem. C* 7: 2961–2970.
73 Silveira, C., Monteiro, T., and Almeida, M. (2016). *Biosensors* 6: 51.
74 Xu, B.-B., Zhang, Y.-L., Xia, H. et al. (2013). *Lab Chip* 13: 1677.
75 Pla, D., Jimenez, C., and Burriel, M. (2017). *Adv. Mater. Interf.* 4: 1600974.
76 Barber, R.W. and Emerson, D.R. (2012). *Microdroplet Technology*, 77–116. New York, NY: Springer New York.
77 Zhang, P., Wang, F., Yu, M. et al. (2018). *Chem. Soc. Rev.* 47: 7426–7451.
78 Soavi, F., Bettini, L.G., Piseri, P. et al. (2016). *J. Power Sources* 326: 717–725.
79 Fernandes, R. and Gracias, D.H. (2009). *Mater. Today* 12: 14–20.
80 Rosenauer, M., Buchegger, W., Finoulst, I. et al. (2011). *Microfluid. Nanofluidics* 10: 761–771.
81 Manz, A., Graber, N., and Widmer, H.M. (1990). *Sens. Actuators B Chem.* 1: 244–248.
82 Quesada-González, D. and Merkoçi, A. (2018). *Chem. Soc. Rev.* 47: 4697–4709.
83 Malhotra, B.D., Srivastava, S., Ali, M.A., and Singh, C. (2014). *Appl. Biochem. Biotechnol.* 174: 880–896.
84 Sia, S.K., Linder, V., Parviz, B.A. et al. (2004). *Angew. Chem. Int. Ed.* 43: 498–502.
85 https://www.who.int/medical_devices/diagnostics/selection_in-vitro/selection_in-vitro-meetings/00007_01_WHO_Laboratory_Manual_STIs.pdf.
86 Quesada-González, D. and Merkoçi, A. (2015). *Biosens. Bioelectron.* 73: 47–63.
87 Carron, K. and Cox, R. (2010). *Anal. Chem.* 82: 3419–3425.
88 Qian, X., Peng, X.-H., Ansari, D.O. et al. (2008). *Nat. Biotechnol.* 26: 83–90.
89 Sha, M.Y., Xu, H., Penn, S.G., and Cromer, R. (2007). *Nanomedicine* 2: 725–734.
90 Harpster, M.H., Zhang, H., Sankara-Warrier, A.K. et al. (2009). *Biosens. Bioelectron.* 25: 674–681.
91 Lyandres, O., Yuen, J.M., Shah, N.C. et al. (2008). *Diabetes Technol. Ther.* 10: 257–265.
92 Chansuvarn, W., Tuntulani, T., and Imyim, A. (2015). *TrAC Trends Anal. Chem.* 65: 83–96.
93 Duarte, K., Justino, C.I.L., Freitas, A.C. et al. (2015). *TrAC Trends Anal. Chem.* 64: 183–190.
94 Bahadır, E.B. and Sezgintürk, M.K. (2016). *TrAC Trends Anal. Chem.* 82: 286–306.
95 Ding, Y., Wang, S., Li, J., and Chen, L. (2016). *TrAC Trends Anal. Chem.* 82: 175–190.

96 Huber, J. and Leopold, K. (2016). *TrAC Trends Anal. Chem.* 80: 280–292.
97 Leopold, K., Philippe, A., Wörle, K., and Schaumann, G.E. (2016). *TrAC Trends Anal. Chem.* 84: 107–120.
98 Li, Y., Chen, Y., Yu, H. et al. (2018). *TrAC Trends Anal. Chem.* 98: 190–200.
99 Mahato, K., Kumar, A., Maurya, P.K., and Chandra, P. (2018). *Biosens. Bioelectron.* 100: 411–428.
100 Chen, X., Guo, Z., Yang, G.-M. et al. (2010). *Mater. Today* 13: 28–41.
101 Kim, J., Jeerapan, I., Imani, S. et al. (2016). *ACS Sensors* 1: 1011–1019.
102 Mano, T., Nagamine, K., Ichimura, Y. et al. (2018). *ChemElectroChem* 5: 3881–3886.
103 Schazmann, B., Morris, D., Slater, C. et al. (2010). *Anal. Methods* 2: 342.
104 Valentini, F., Carbone, M., and Palleschi, G. (2013). *Anal. Bioanal. Chem.* 405: 3449–3474.
105 Keene, S.T., Fogarty, D., Cooke, R. et al. (2019). *Adv. Healthcare Mater.* 8: 1901321.
106 Siangproh, W., Dungchai, W., Rattanarat, P., and Chailapakul, O. (2011). *Anal. Chim. Acta* 690: 10–25.
107 Alcantara, D., Lopez, S., García-Martin, M.L., and Pozo, D. (2016). *Nanomed. Nanotechnol. Biol. Med.* 12: 1253–1262.
108 Lee, H., Sun, E., Ham, D., and Weissleder, R. (2008). *Nat. Med.* 14: 869–874.
109 Haun, J.B., Yoon, T.J., Lee, H., and Weissleder, R. (2010). *Wiley Interdiscip. Rev. Nanomed. Nanobiotechnol.* 2: 291–304.
110 Koh, I. and Josephson, L. (2009). *Sensors* 9: 8130–8145.
111 Josephson, L., Perez, J.M., and Weissleder, R. (2001). *Angew. Chem. Int. Ed.* 40: 3204–3206.
112 Ekström, S., Önnerfjord, P., Nilsson, J. et al. (2000). *Anal. Chem.* 72: 286–293.
113 Yuen, P.K. and DeRosa, M.E. (2011). *Lab Chip* 11: 3249.
114 Yu, L., Li, C.M., Liu, Y. et al. (2009). *Lab Chip* 9: 1243.
115 Hu, M., Yan, J., He, Y. et al. (2010). *ACS Nano* 4: 488–494.
116 Wang, S., Ge, L., Song, X. et al. (2012). *Biosens. Bioelectron.* 31: 212–218.
117 Ahmad, F. and Hashsham, S.A. (2012). *Anal. Chim. Acta* 733: 1–15.
118 Su, M., Ge, L., Ge, S. et al. (2014). *Anal. Chim. Acta* 847: 1–9.
119 Matlock-Colangelo, L., Cho, D., Pitner, C.L. et al. (2012). *Lab Chip* 12: 1696.
120 Mahato, K., Srivastava, A., and Chandra, P. (2017). *Biosens. Bioelectron.* 96: 246–259.
121 Yamada, K., Henares, T.G., Suzuki, K., and Citterio, D. (2015). *Angew. Chem. Int. Ed.* 54: 5294–5310.
122 Mashaghi, S., Abbaspourrad, A., Weitz, D.A., and van Oijen, A.M. (2016). *TrAC Trends Anal. Chem.* 82: 118–125.
123 Janasek, D., Franzke, J., and Manz, A. (2006). *Nature* 442: 374–380.
124 Ríos, Á., Zougagh, M., and Avila, M. (2012). *Anal. Chim. Acta* 740: 1–11.
125 Liong, M., Hoang, A.N., Chung, J. et al. (2013). *Nat. Commun.* 4: 1752.
126 Chen, L. and Choo, J. (2008). *Electrophoresis* 29: 1815–1828.
127 Gao, D., Liu, H., Jiang, Y., and Lin, J.-M. (2013). *Lab Chip* 13: 3309.
128 Feng, X., Liu, B.-F., Li, J., and Liu, X. (2015). *Mass Spectrom. Rev.* 34: 535–557.

129 He, M. and Herr, A.E. (2010). *Nat. Protoc.* 5: 1844–1856.

130 Yang, S.Y., DeFranco, J.A., Sylvester, Y.A. et al. (2009). *Lab Chip* 9: 704–708.

131 Zhang, Y. and Jiang, H.-R. (2016). *Anal. Chim. Acta* 914: 7–16.

132 Kazarian, S.G. (2007). *Anal. Bioanal. Chem.* 388: 529–532.

133 Jung, J.H., Han, C., Lee, S.A. et al. (2014). *Lab Chip* 14: 3781.

134 Yeo, L.Y., Chang, H.-C., Chan, P.P.Y., and Friend, J.R. (2011). *Small* 7: 12–48.

135 Jiang, Y., Zou, S., and Cao, X. (2016). *Anal. Methods* 8: 6668–6681.

136 Mauk, M., Song, J., Bau, H.H. et al. (2017). *Lab Chip* 17: 382–394.

137 Nasseri, B., Soleimani, N., Rabiee, N. et al. (2018). *Biosens. Bioelectron.* 117: 112–128.

138 Kudr, J., Zitka, O., Klimanek, M. et al. (2017). *Sens. Actuators B Chem.* 246: 578–590.

139 Edwards, K.A., Bolduc, O.R., and Baeumner, A.J. (2012). *Curr. Opin. Chem. Biol.* 16: 444–452.

140 Varshney, M., Li, Y., Srinivasan, B., and Tung, S. (2007). *Sens. Actuators B Chem.* 128: 99–107.

141 Chen, Y.-C., Abdelhamid, H.N., and Wu, H.-F. (2017). *Mass Spectrom. Lett.*: 8.

142 Abdelhamid, H.N. and Wu, H.-F. (2019). *J. Am. Soc. Mass Spectrom.* 30: 1174.

143 Abdelhamid, H.N. and Wu, H.-F. (2014). *RSC Adv.* 4: 53768–53776.

144 Gedda, G., Abdelhamid, H.N., Khan, M.S., and Wu, H.-F. (2014). *RSC Adv.* 4: 45973–45983.

145 Abdelhamid, H.N., Lin, Y.C., and Wu, H.F. (2017). *Microchim. Acta* 184: 1517–1527.

146 Chen, Z.-Y., Abdelhamid, H.N., and Wu, H.-F. (2016). *Rapid Commun. Mass Spectrom.* 30: 1403–1412.

147 Shastri, L., Abdelhamid, H.N., Nawaz, M., and Wu, H.-F. (2015). *RSC Adv.* 5: 41595–41603.

148 Abdelhamid, H.N. (2013). *Applications of Nanomaterials and Organic Semiconductors for Bacteria &Biomolecules Analysis/ Biosensing Using Laser Analytical Spectroscopy*. National Sun-Yat Sen University.

149 Abdelhamid, H.N., Chen, Z.-Y., and Wu, H.-F. (2017). *Anal. Bioanal. Chem.* 409: 4943–4950.

150 Abdelhamid, H.N. and Wu, H.-F. (2015). *RSC Adv.* 5: 76107–76115.

151 Khan, N., Abdelhamid, H.N., Yan, J.-Y. et al. (2015). *Anal. Chem. Res.* 3: 89–97.

152 Abdelhamid, H.N., Lin, Y.C., and Wu, H.-F. (2017). *RSC Adv.* 7: 41585–41592.

153 Abdelhamid, H.N. (2019). *Handbook of Smart Materials in Analytical Chemistry*, 729–755. Chichester, UK: Wiley.

154 Abdelhamid, H.N. and Wu, H.-F. (2015). *Mass Spectrom. Lett.* 6: 43–47.

155 Kumaran, S., Abdelhamid, H.N., and Wu, H.-F. (2017). *RSC Adv.* 7: 30289–30294.

156 Abdelhamid, H.N. (2019). *Microchim. Acta* 186: 682.

157 Abdelhamid, H.N. and Wu, H.-F. (2019). *Handbook of Graphene 2*, Physics, Chem. Biol., vol. 2 (ed. T. Stauber), 273–290. Scrivener Publishing.

158 Abdelhamid, H.N. (2018). *Meth. Protoc.* 1: 23.

159 Riter, L.S., Laughlin, B.C., Nikolaev, E., and Cooks, R.G. (2002). *Rapid Commun. Mass Spectrom.* 16: 2370–2373.
160 Berkout, V.D., Cotter, R.J., and Segers, D.P. (2001). *J. Am. Soc. Mass Spectrom.* 12: 641–647.
161 Diaz, J.A., Giese, C.F., and Gentry, W.R. (2001). *J. Am. Soc. Mass Spectrom.* 12: 619–632.
162 Boumsellek, S. and Ferran, R.J. (2001). *J. Am. Soc. Mass Spectrom.* 12: 633–640.
163 Oleschuk, R.D. and Harrison, D.J. (2000). *TrAC Trends Anal. Chem.* 19: 379–388.
164 Brivio, M., Fokkens, R.H., Verboom, W. et al. (2002). *Anal. Chem.* 74: 3972–3976.
165 Pan, N., Rao, W., Kothapalli, N.R. et al. (2014). *Anal. Chem.* 86: 9376–9380.
166 Kailasa, S.K. and Wu, H.F. (2015). *TrAC - Trends Anal. Chem.* 65: 54–72.
167 Fritz, J.S. and Masso, J.J. (2001). *J. Chromatogr. A* 909: 79–85.
168 Harrison, D.J., Fluri, K., Seiler, K. et al. (1993). *Science* 261: 895–897.
169 Valero, A., Braschler, T., Demierre, N., and Renaud, P. (2010). *Biomicrofluidics* 4: 022807.
170 Lee, K.K. and Ahn, C.H. (2013). *Lab Chip* 13: 3261.
171 Zhang, X.-B., Wu, Z.-Q., Wang, K. et al. (2012). *Anal. Chem.* 84: 3780–3786.
172 CHEN, X., CUI, D., LIU, C., and LI, H. (2008). *Sens. Actuators B Chem.* 130: 216–221.
173 Kohlheyer, D., Eijkel, J.C.T., Schlautmann, S. et al. (2007). *Anal. Chem.* 79: 8190–8198.
174 Kutter, J.P. (2012). *J. Chromatogr. A* 1221: 72–82.
175 Tetala, K.K.R. and Vijayalakshmi, M.A. (2016). *Anal. Chim. Acta* 906: 7–21.
176 Liu, C., Liao, S.-C., Song, J. et al. (2016). *Lab Chip* 16: 553–560.
177 Amasia, M. and Madou, M. (2010). *Bioanalysis* 2: 1701–1710.
178 Liu, C., Mauk, M., Gross, R. et al. (2013). *Anal. Chem.* 85: 10463–10470.
179 Kohlheyer, D., Eijkel, J.C.T., van den Berg, A., and Schasfoort, R.B.M. (2008). *Electrophoresis* 29: 977–993.
180 Becker, M., Mansouri, A., Beilein, C., and Janasek, D. (2009). *Electrophoresis* 30: 4206–4212.
181 Raymond, D.E., Manz, A., and Widmer, H.M. (1994). *Anal. Chem.* 66: 2858–2865.
182 Kohlheyer, D., Besselink, G.A.J., Schlautmann, S., and Schasfoort, R.B.M. (2006). *Lab Chip* 6: 374.
183 Lao, A.I.K., Trau, D., and Hsing, I.-M. (2002). *Anal. Chem.* 74: 5364–5369.
184 Zhang, C.-X. and Manz, A. (2003). *Anal. Chem.* 75: 5759–5766.
185 Golberg, A., Linshiz, G., Kravets, I. et al. (2014). *PLoS One* 9: e86341.

10

Microvolume UV–Visible Spectrometry for Assaying of Pesticides

Jigneshkumar V. Rohit[1] *and Vaibhavkumar N. Mehta*[2]

[1] *Department of Chemistry, National Institute of Technology, Srinagar 190006, Jammu and Kashmir, India*
[2] *ASPEE Shakilam Biotechnology Institute, Navsari Agricultural University, Ghod-Dod Road, Surat 395007, Gujarat, India*

10.1 Introduction

To keep up with the increasing global population, more food production is required. It has been estimated that 34% of the global crop production is lost annually due to various pests including insects, fungi, and weeds [1–3]. To resolve pest problems, large quantities of pesticides (insecticides, fungicides, herbicide, etc.) are applied on pests (insect, fungi, weed, etc.) to prevent crop losses and increase yield (Figure 10.1). Enormous use of pesticides over the years has caused a risk to human health as pesticide residues can enter the food chain through air, water, and soil [4]. When pesticides enter the human body, they can damage the central nervous system, cause skin and eye irritation, lead to hormonal imbalance, and reproductive effects; in some cases, high dosages cause death. Also, when pesticides destroy non-pest things, this can radically alter the normal work function of the ecosystem [5]. At present, more than nine hundred active ingredients are available for pest control, which causes a significant and growing concern about soil, water, and food quality, especially where nonbiodegradable pesticide residues remain in the environment for years [6]. Thus, methods for determining pesticide residues in agricultural products are constantly receiving attention. Available instrumental analytical techniques (HPLC, GC–MS, LC–MS) are sophisticated, expensive, require tedious sample pretreatment procedures, and are time-consuming, which limit their on-site screening of pesticides in environmental and food samples. Undoubtedly, it is of great importance to develop a simple, selective, sensitive, and real-time detection method for analysis of pesticides by using simple analytical tools.

Metal nanoparticles (Ag/Au NPs) are used for sensing due to their higher extinction coefficients, narrow photoluminescence bandwidth, sharper extinction bands, strong field enhancements, higher ratio of scattering to extinction, and large surface area for functionalization. These properties of NPs can be readily tuned by varying their size, shape, and the surrounding chemical environment. For example, the binding experience between the recognition constituent and the analyte can

Miniaturized Analytical Devices: Materials and Technology, First Edition.
Edited by Suresh Kumar Kailasa and Chaudhery Mustansar Hussain.

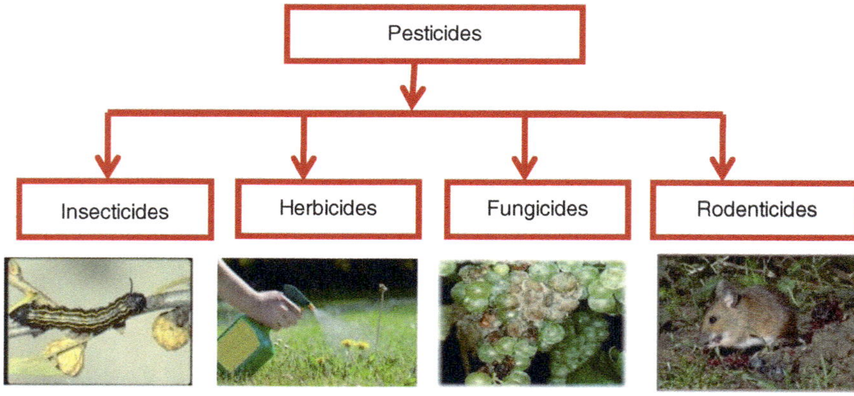

Figure 10.1 Use of pesticides according to their mode of action.

alter physicochemical properties of NPs, such as plasmon resonance, conductivity, and fluorescence behavior, which produce recognizable signals [7]. More importantly, NPs absorb and scatter light intensely at its surface plasmon resonance (SPR) wavelength region, which can be tuned by changing the size while strong absorbance at the visible light region makes them easily observable sensing tool for determination of target analytes [8]. Moreover, surface functionalization on metallic NPs with an appropriate capping ligand is the deciding factor for their application in quantifying the trace targets from complex sample matrices [9]. Many interactions (charge–transfer, ion–pair, π–π, and covalent) taking place between organic molecules on NP surfaces and target analytes lead to aggregations of NPs with color changes observable to the naked eyes and a red shift in the SPR peak, which are recorded by simple UV–visible spectrometer with a very less sample volume [10].

This chapter is a comprehensive review of all research reports that deal with development of new methods for pesticide analysis using microvolume UV–visible spectroscopy. Also, this chapter is a big data bank for readers providing all reported data in the form of summary tables like type of pesticides analyzed, functionalization on NPs, sensitivity of methods, and real samples (water, soil, and food) analyzed. Further, this chapter discusses working methodology of reported methods such as specific interactions between pesticides and molecules on NP surfaces, response of NPs at binding events, volume of sample, and calculations used for determining particular pesticides.

10.2 Ag NP–Based Microvolume UV–Visible Spectrometry for Analysis of Pesticides

Ag NPs show some advantages over Au NPs in noble metal NP–based UV–visible spectrometric analysis, as they have higher extinction coefficients than Au NPs of similar sizes [11]. Also, Ag NPs exhibit excellent optical properties and a simple discrimination between their dispersed and aggregation states they are used for analysis of many chemical species [12]. Many authors reported on the use of Ag NPs in microvolume UV-visible spectrometry for analysis of pesticides (fungicides, herbicides, and insecticides), which are listed in Table 10.1.

10.2.1 Analysis of Fungicides

Fungicides are one of the important classes of pesticides that reduce crop production loss due to fungal diseases. However, fungicides are hazardous chemical compounds and remain on treated crops for long time; therefore they will enter the human body via the food chain and cause risks to human health. Therefore, analysis of fungicides in food, water, and soil samples is an important task for researchers, and many efforts have been put into developing simple methods for analysis of fungicides using Ag NP-based UV-visible spectrometry. In this connection, a simple and selective method was developed for analysis of tricyclazole fungicides using 5-sulfo anthranilic acid dithiocarbamate (SAADTC) Ag NPs [13]. In this method, tricyclazole was successfully extracted from rice samples and detected using SAADTC-Ag NPs, with observable color changes (yellow to pink) and a red shift in absorption spectra of SAADTC-Ag NPs with a very less total sample volume (900 μl). This method was able to detect tricyclazole in rice samples up to a concentration of 1.8×10^{-7} M based on aggregation of SAADTC-Ag NPs using simple UV-visible spectrometry, and the results were comparable with many sophisticated instrumental techniques. Similarly, Rohit's group prepared a selective sensor based on dopamine dithiocarbamate (DDTC)-functionalized Ag NPs for analysis of mancozeb fungicides [14]. As the concentration of mancozeb increased, color of DDTC-Ag NPs changed from brownish orange to bluish and spectral changes were in situ monitored by UV-visible spectrometry. Also, the authors demonstrated practicability of the proposed method for the detection of mancozeb in natural water (tap, canal, and river) samples and fruit juice (apple, grape, and tomato) samples. The same group developed a method for analysis of thiram fungicides and paraquat herbicides via the "host-guest" chemistry mechanism by using cyclen dithiocarbamate (CN-DTC)-capped Ag NPs as a dual sensing probe [15]. This method demonstrated the use of simple UV-visible spectrometer for sensitive determination with a limit of detection of 2.81×10^{-6} M for thiram, and the method was able to detect thiram in water, potato, and wheat samples.

In another report, one of the widely used fungicides carbendazim was analyzed using 4-aminobenzenethiol (ABT)-capped Ag NP-based microvolume (0.6 ml) UV-visible spectrometry method [16]. This method was highly selective and able to detect carbendazim among 14 pesticides, and as shown in Figure 10.2, only carbendazim showed a different change in color from yellow to orange, which resulted in a long shift in UV-visible absorption spectra of ABT-Ag NPs from 397 to 510 nm. Similarly, thiophanate-methyl fungicide was detected by using microvolume UV-visible spectroscopy with citrate-Ag NP-based detection method [17]. Thiophanate-methyl was determined quantitatively by recording the absorbance ratio at A_{525}nm/A_{394}nm with addition of thiophanate-methyl in the range of 2-100 μM. As a result, the SPR peak intensity of citrate-Ag NPs at

Table 10.1 Ag NP–based microvolume UV–visible spectrometry for assaying pesticides.

Type of pesticide	Analyte	Surface functionalization of Ag NPs	Real samples	Limit of detection	References
Fungicide	Tricyclazole	SAADTC–Ag NPs	Rice	1.8×10^{-7} M	[13]
	Mancozeb	DDTC–Ag NPs	Tap water, canal water, river water, apple, grapes, tomato	21.1×10^{-6} M	[14]
	Thiram	CN–DTC–Ag NPs	Tap water, canal water, river water, potato, wheat	2.81×10^{-6} M	[15]
	Carbendazim	ABT–Ag NPs	Tap water, canal water, river water, apple, carrot	1.04×10^{-6} M	[16]
	Thiophanate-methyl	Citrate–Ag NPs	River water, tap water, tomato	0.12 μM	[17]
	Cymoxanil	Citrate–Ag NPs	River water, soil, plant epidermis	2 nM	[18]
Herbicide	Paraquat	CN–DTC–Ag NPs	Tap water, canal water, river water, potato, wheat	7.21×10^{-6} M	[15]
	Paraquat	Citrate–Ag NPs	Ground water, canal water, Chinese cabbage, green apple	0.05 mg/l	[19]
	Glyphosate	MNBZ–Ag NPS	Tap water, canal water, river water, potato, corn	17.1 nM	[20]
Insecticide	Trichlorfon and malathion	AChE–Ag NPs	Agricultural runoff water, apple, cabbage	0.078 and 2.402 nM	[21]
	Malathion	Aptamer–Ag NPs	Tap water, lake water	0.5 pM	[22]
	Triazophos	MPA–GAA–Ag NPs	Tap water, canal water, river water, rice, apple	0.5 μM	[23]
	Triazophos	Citrate–Ag NPs	River water, tap water, apple	5 nM	[24]
	Dimethoate	pSC_4R–Ag NPs	Industrial wastewater	80 nM	[25]
	Dipterex	Citrate–Ag NPs	Water	0.18 ng/ml	[26]
	Lambda-cyhalothrin	$Ag@SiO_2–NH_2$ NPs	—	1.0×10^{-6} M	[27]
	Endrin	Sucrose–Ag NPs	River water, pond water, rice, wheat, potato	0.015 mg/ml	[28]
Other	6-Benzylaminopurine	ABT–Ag NPs	Water, bean sprouts	0.1 μM	[29]

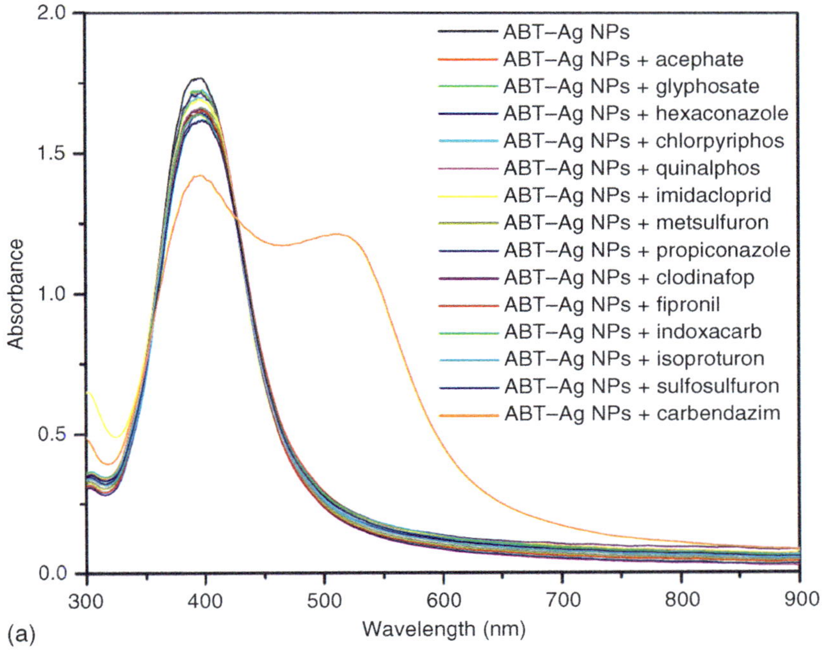

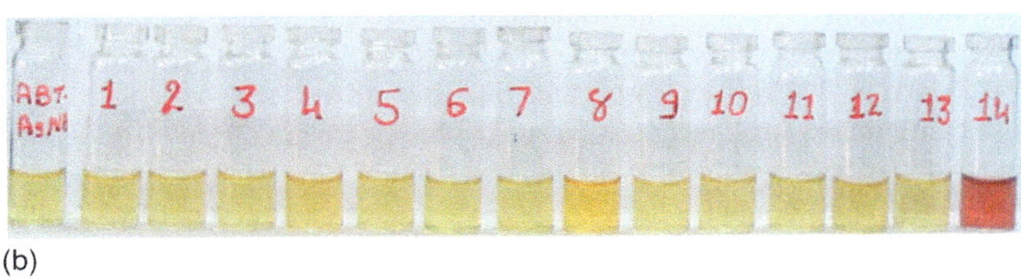

Figure 10.2 (a) UV–visible spectra of ABT–Ag NPs in the presence of different pesticides: 1, acephate; 2, glyphosate; 3, hexaconazole; 4, chlorpyriphos; 5, quinalphos; 6, imidacloprid; 7, metsulfuron; 8, propiconazole; 9, clodinafop; 10, fipronil; 11, indoxacarb; 12, isoproturon; 13, sulfosulfuron; and 14, carbendazim (75 μL and 1mM) at sodium acetate buffer pH 5 and (b) the corresponding photo images of ABT–Ag NPs in the presence of the listed pesticides. Source: Reproduced from Patel et al. [16].

394 nm decreased, and a new SPR peak at 525 nm appeared and gradually increased with a gradual color change of citrate–Ag NPs from yellow to cherry red, and the limit of detection was found to be 0.12 μM. Jiang's group synthesized a dual colorimetric and fluorometric probe for determination of cymoxanil fungicides based on aggregation of citrate–Ag NPs and carbon dots [18]. To investigate the performance of this dual colorimetric and fluorometric cymoxanil detection, river water, soil, and plant epidermis samples were analyzed with or without the addition of cymoxanil, and the sensor provided accurate visual and spectral determination of cymoxanil fungicides. This method showed a limit of detection of 2 nM, which indicates superior performance of simple UV–visible spectrometer.

10.2.2 Analysis of Herbicides

Paraquat is a nonselective and fast-acting contact herbicide and has been widely used to control weeds in agricultural fields for many years. However, it exhibit harmful effects on DNA, cell membrane, and many biomolecules in humans and animals. As discussed in subpart 2.1, Rohit's group developed a method for detection of thiram herbicides and paraquat fungicides by using CN–DTC–Ag NPs as a dual sensing probe, and the reported method was very sensitive and able to detect paraquat up to a concentration of 7.21×10^{-6} M [15]. In this method, UV–visible spectroscopy proved to be a promising approach to detect paraquat as the SPR peak of CN–DTC–Ag NPs at 396 nm gradually decreased and new peaks appeared at 510 nm with increasing concentrations of paraquat. Also, Siangproh's group prepared a sensor for detection of the same herbicide paraquat using negatively charged Ag NPs. In this method, analysis of paraquat was based on the mechanism of aggregation of negatively charged Ag NPs, which was induced by positively charged paraquat. Color changed from pale yellow to deep greenish with increasing concentrations of paraquat. This method was able to determine paraquat from water, cabbage, and apple samples using UV–visible spectrometer with microvolume sample sizes (100 μl) [19]. Another widely used fungicide glyphosate was detected by Rawat's group by addition of glyphosate solution (750 μl) to the solution of 2-mercapto-5-nitrobenzimidazole (MNBZ)-capped Ag NPs (750 μl) in the presence of Mg^{2+} (7.5 μl), and response of MNBZ–Ag NPs in the presence of glyphosate was observed by using UV–visible spectrometry [20]. The SPR peak of MNBZ–NPs at 399 nm decreased and intensity of the new peak at 517 nm increased with increasing concentrations of glyphosate ranging from 0.1 to 1.2 μM and showed high sensitivity with a limit of detection of 17.1 nM. This method was highly selective and was able to detect glyphosate among eight pesticides of similar structures and to efficiently detect glyphosate from water, potato, and corn samples.

10.2.3 Analysis of Insecticides

Among various chemical classes of pesticides, organophosphorus (OP) insecticides are the most widely used one. They inhibit the function of acetylcholinesterase (AChE), an essential enzyme for functioning of the central nervous system, thereby resulting in paralysis and even death of insects. Through this mechanism, OP insecticides effectively kill the insects and increase the yield of crop production, but due to their severe toxicity, they cause high risks to human health because they can easily contaminate soil, water, and food from agriculture sites. In this connection, Nanda Kumar's group developed a method for analysis of malathion and trichlorfon organophosphorus insecticides [21]. This method was based on the catalytic reaction of acetylcholine (ATCh) and AChE during the synthesis of Ag NPs, and formation of Ag NPs was controlled by malathion and trichlorfon insecticides, which was monitored by UV–visible spectrophotometer. The reported method showed a high sensitivity of 0.078 and 2.402 nM for trichlorfon and malathion, respectively, and the method was effectively applied for the detection of trichlorfon

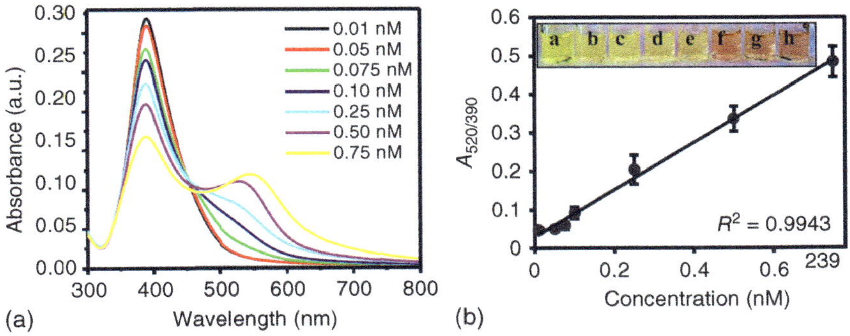

Figure 10.3 (a) Absorbance spectra of silver nanoparticles (2 nM) with 10 μM peptide, 500 nM aptamer, and varying malathion concentrations. (b) Linear calibration plot obtained with different concentrations of malathion. The inset depicts the visual color changes of Ag NPs with increasing malathion concentrations. (a) Ag NPs and (b) to (h) 0.01 nM to 0.75 nM malathion. Each point represents an average of three individual measurements and error bars indicate standard deviation. Source: Reproduced from Bala et al. [22].

and malathion in spiked water and food samples with good recovery percentage. Similarly, Bala's group reported a highly sensitive method for determination of malathion insecticides using aptamer–peptide–Ag NPs [22]. In this method, interaction of Ag NPs with aptamer and peptides yielded different optical responses in the absence and presence of malathion, which were recorded by simple UV–visible spectrometry with a very microvolume sample (0.7 ml). As the concentration of malathion increased from 0.01 to 0.75 nM, the SPR band at around 390 nm belonging to well-dispersed Ag NPs decreased, whereas a new band at around 520 nm appeared, indicating the aggregation of Ag NPs, which was also visually confirmed by a color change of Ag NPs from yellow to orange (Figures 10.3a,b). Furthermore, for quantitative determination of malathion, a standard calibration curve was plotted by using the absorption ratio of Ag NPs at A_{520} nm/A_{390} nm vs. various concentrations of malathion (Figure 10.3b), and the limit of detection was 0.5 pM, which was considerably lower compared to that of many other instrumental techniques, which proves the significance of simple UV–visible spectroscopy in analytical science.

Triazophos, another broad spectrum organophosphorus insecticide, was detected using UV–visible spectrometry by Bhamore's group [23]. In this method, 3-mercaptopropionic acid (MPA) and guanidineacetic acid (GAA) were capped on Ag NP surfaces, and bifunctionalized Ag NPs were used as a colorimetric sensor for the selective detection of triazophos among other organophosphorus insecticides. Here, hydrogen bonding interaction between triazophos and MPA–GAA on Ag NP surfaces led to aggregation of MPA-GAA–Ag NPs, and responses were detected using UV–visible spectroscopy with a very less sample size. Triazophos insecticide was also detected by using Cit–Ag NPs from water and fruit samples [24]. This reported method was highly sensitive and able to detect triazophos up to a concentration of 5 nM using simple UV–visible spectroscopy, which was comparable with many sophisticated instrumental techniques. Another organophosphorus insecticide dimethoate was detected by using *p*-sulfonato-calix[4]resorcinarene (pSC_4R)-capped Ag NPs via the host–guest interaction between dimethoate and pSC_4R [25]. This interaction led to aggregation of Ag NPs, which was confirmed by recorded absorption spectra using UV–visible spectrometer, and the SPR peak of pSC_4R–Ag NPs at 420 nm was shifted to a longer wavelength at 519 nm, with a color change from yellow to red. The limit of detection was 80 nM, and the method demonstrated its practicability by detecting dimethoate from industrial wastewater samples. Dipterex (Trichlorfan), another organophosphorus insecticide, was analyzed by citrate–Ag NPs, in which AChE catalyzing acetylthiocholine (ATCh) to form thiocholine (TCh) led to aggregation of Ag NPs [26]. This UV–visible spectrophotometric method showed linearity at the absorbance ratio A_{396} nm/A_{520} nm with the Dipterex concentration in the range of 0.25–37.5 ng/ml and a limit of detection 0.18 ng/ml.

Pyrethroids are a class of insecticides that are alternative to highly toxic organophosphorus insecticides and have been used on a large scale due to their high insecticidal activity and moderate toxicity. λ-Cyhalothrin, one of the widely used pyrethroids, was analyzed using amine-modified silver–silica core–shell nanoparticles ($Ag@SiO_2$–NH_2 NPs) based on the strong hydrogen bond interactions between the trifluoromethyl fluorine functional groups of λ-cyhalothrin and amine groups on the surface of $Ag@SiO_2$–NH_2 NPs [27]. Only upon the addition of λ-cyhalothrin, color of $Ag@SiO_2$–NH_2 NPs changed from yellow to pink with an increase in absorbance intensity at 525 nm, while other pyrethroids did not exhibit any interference to absorption spectra of $Ag@SiO_2$–NH_2 NPs recorded on UV–visible spectrophotometer. Similarly, endrin, an organochlorine insecticide, was determined by Shrivas's group using sucrose-capped Ag NPs [28]. As the concentration of endrin increased from 0.05 to 5.00 mg/ml, degree of aggregation of Ag NPs increased, which was recorded by decreasing intensity of absorption spectra of Ag NPs at 400 nm and increasing intensity at 580 nm. By plotting a standard calibration curve, limit of detection was found to be 0.015 μg/ml using UV–visible spectroscopy with very low sample volumes.

10.2.4 Analysis of Other Pesticides

Pesticide 6-benzylaminopurine is widely used as a plant growth regulator for many agriculture crops but has disadvantages such as damage to the eyes, skin, upper respiratory tract, and mucous membranes upon contact and contamination. Ma's group developed a simple method for analyzing 6-benzylaminopurine using *p*-aminobenzenethiol (ABT)-functionalized Ag NP–based UV–visible spectrometry [29]. This method was based on the interaction between 6-benzylaminopurine and ABT–Ag NPs via hydrogen bonding, π–π stacking, and electrostatic interactions, resulting in a visible color change from yellow to reddish orange of ABT–Ag NPs. This method was able to determine 6-benzylaminopurine from water and bean sprouts with a limit of detection of 0.1 μM in a microvolume sample (1 ml) using simple UV–visible spectrometer.

10.3 Au NP-based Microvolume UV-Visible Spectrometry for Analysis of Pesticides

Au NPs have attracted great attention over many years as their significant physico-chemical properties make them a reliable platform for preparation of highly selective and sensitive sensors for analysis of target analytes [30]. SPR is one of the most interesting optical properties of Au NPs, which depends on their shape, size, and the surrounding medium. Due to SPR, Au NPs display strong absorption in the visible region, which makes them a valuable analytical tool to examine chemical and biological species using simple UV-visible spectrophotometer [31]. Many authors reported on the use of Au NPs in microvolume UV-visible spectrometry for analysis of pesticides (fungicides, herbicides, and insecticides), which are listed in Table 10.2.

10.3.1 Analysis of Fungicides

Dithiocarbamate is one of the widely used fungicides for controlling fungal diseases in agricultural crops for many years. They cause eye inflammation, skin allergy, and respiratory tract problems in humans. Thiram, a largely used fungicide from the dithiocarbamate group, was detected by Rastegarzadeh's group by using the formation of Au NPs and with UV-visible spectroscopy [32]. The intensity of the absorption peak of ascorbic acid-Au NPs decreased in the presence of thiram and the responses were analyzed using UV-visible spectrometer with a limit of detection of 1.7×10^{-7} mol/l. Another dithiocarbamate group fungicide ziram was detected using in situ formation of Au NPs, and the trace levels of ziram affected the formation of Au NPs and the process was monitored by UV-visible spectroscopy [33]. The reported method demonstrated its practicability to determine ziram in water, soil, and food with good accuracy and precision.

Iprobenfos and edifenphos are toxic organophosphate fungicides used in many agricultural crops such as fruits, vegetables, and grains for managing fungal diseases. These fungicides were detected by Kwon's group by using the aptamer-Au NP-based colorimetric method [34]. Aptamer selectively binds only to iprobenfos and edifenphos, which is confirmed by Au NP assays by visual color changes and long shifting of the SPR peak with UV-visible spectroscopy. In this method, very less amounts of analytes and sensors were used (a total sample volume of only 400 µl), making this method an exact microvolume UV-visible analysis of pesticides. Another fungicide tebuconazole was determined using very low sample volume (1.5 ml) using citrate-Au NPs in the presence of imidazole [35]. This reported method was very sensitive with a limit of detection 52.0 ppb using simple UV-visible spectroscopy, and the results were well comparable with those of HPLC.

10.3.2 Analysis of Herbicides

Glyphosate is a widely used herbicide that helps in controlling weed and vegetation for many decades. It affects fertility and disturbs cell cycle regulation in humans. Zheng's group developed a method for determination of glyphosate herbicide by

Table 10.2 Au NP–based microvolume UV–visible spectrometry for assaying pesticides.

Type of pesticide	Analyte	Surface functionalization of Au NPs	Real samples	Limit of detection	References
Fungicide	Thiram	Ascorbic acid–Au NPs	Tap and river water, tomato, cucumber, water melon	1.7×10^{-7} M	[32]
	Ziram	Ascorbic acid–Au NPs	Well water, river water, soil, potato, carrot, wheat	0.06 ng/ml	[33]
	Iprobenfos and Edifenphos	Aptamer–Au NPs	Paddy rice, polished rice	10 and 5 nM	[34]
	Tebuconazole	Citrate–Au NPs	Strawberry	52.0 ppb	[35]
Herbicide	Glyphosate	CS–Au NPs	Water	5.88×10^{-8} M	[36]
	Atrazine	Melamine–Au NPs	Tap water	0.0165 μM	[37]
	Atrazine	Cysteamine–Au NPs	Rice	0165 μg/g	[38]
	Ametryn	Citrate–Au NPs	River water, sewage sludge water, industrial wastewater	0.15 μg/l	[39]
	Metsulfuron-methyl	DTC–PTBCA–Au NPs	Tap water, canal water, river water, rice, wheat	1.9×10^{-7} M	[40]
	Pendimethalin	RAC–DTC–Au NPs	Tap water, canal water, river water, rice, wheat, tomato	0.22 μM	[41]
	Terbuthylazine	Citrate–Au NPs	Tap water, green tea, apple juice	0.3 μM	[42]
Insecticide	Malathion	Aptamer–Au NPs	Lake water, apple	0.06 pM	[43]
	Methyl-parathion	Lanthanum–Au NPs	River water, soil	0.1 nM	[44]
	Parathion	Au^{3+}–CTAB–Au NPs	Tap water, sea water, and apple	35 ppb	[45]
	Ethyl-parathion	Cysteine–Au NPs	Tap water, lake water, mineral water	0.081 ng/ml	[46]
	Phorate	Aptamer–Au NPs	Apple	0.01 nM	[47]
	Phorate	HA–Tyr–Au NPs	Tomato	0.005 μg/ml	[48]

(Continued)

Table 10.2 (Continued)

Type of pesticide	Analyte	Surface functionalization of Au NPs	Real samples	Limit of detection	References
	Mathamidophos	Citrate–Au NPs	Chinese cabbage, cowpea, Chinese leek, cucumber	1.40 ng/ml	[49]
	Paraoxon	Alkyne–Au NPs	Apple	10^{-6} g/l	[50]
	Dichlorvos	AA–Au NPs	Tap water, wheat, apple juice	42.94 μM	[51]
	Diazinon, iprobenfos and edifenphos	Citrate–Au NPs	Deionized water and river water	53.3, 53.6 and 27.9 ppb	[52]
	Quinalphos	p-NA–DTC–Au NPs	Tap water, river water, canal water, tomato, rice, wheat	3.2×10^{-6} M	[53]
	Terbufos and thiacloprid	NBI–DTC–Au NPs and HBI–DTC–Au NPs	Tap water, river water, canal water, corn, sorghum, apple, tomato	7.3×10^{-8} M and 6.0×10^{-7} M	[54]
	Dimethoate	Citrate–Au NPs	Tomato, cucumber, cabbage	4.7 μg/l	[55]
	Imidacloprid	Imidazole ionic liquid–Au NPs	River water, lake water	5×10^{-7} M	[56]
	Imidacloprid	Piperidine-calix[4]arene–Au NPs	Tap water	5.0×10^{-6} M	[57]
	Cyromazine	Citrate–Au NPs	Cucumber	252 ng/g	[58]
	Cyromazine	Citrate–Au NPs	River water	0.1 mg/l	[59]
	Cyromazine	Tween 20–Au NPs	Cucumber, milk	0.04 mg/kg	[60]
	Acetamiprid	Citrate–Au NPs	Green vegetables, eggplant, cucumber	4.4×10^{-8} M	[61]
	Acetamiprid	Citrate–Au NPs	Soil	5 nM	[62]
	Acetamiprid	Aptamer–Au NPs	—	0.5 mM	[63]
	Acetamiprid	Aptamer–Au NPs	Tomato, wastewater	1.02 μg/l	[64]
	Acetamiprid	Aptamer–Au NPs	Wastewater, soil, cucumber	7.5×10^{-8} M	[65]

(Continued)

Table 10.2 (Continued)

Type of pesticide	Analyte	Surface functionalization of Au NPs	Real samples	Limit of detection	References
	Cartap	Citrate–Au NPs	Tea, cabbage	0.04 mg/kg	[66]
	Cyhalothrin	4-Amino-3-mercaptobenzoic acid–Au NPs	River water, lake water	0.75 μM	[67]
	Diafenthiuron	4-Hydroxy-6-methyl-3-nitro-2-pyridone-dithiocarbamate–Au NPs	Tap water, river water, apple, tomato, cabbage	7.1 nM	[68]
	Deltamethrin	2-Mercapto-6-nitrobenzothiazole–Au NPs	Cherry, mini tomato	0.25 μM	[69]
	Pymetrozine	Melamine–Au NPs	Tap water, lake water, green tea, apple juice	10 nM	[70]

using cysteamine (CS)-capped Au NP–based UV–visible spectroscopy [36]. The electrostatic interaction between glyphosate and CS on Au NP surfaces resulted in a shift in the absorption band, with a color change of CS–Au NPs from red to blue, which was monitored by UV–visible spectroscopy. The reported method was highly sensitive with a detection limit of 5.88×10^{-8} M and was able to determine glyphosate from environmental water samples.

Triazines, an important class of herbicides, are used on a large scale to control grassy weeds. Triazines exhibit carcinogenic effects. Atrazine, a widely used triazine herbicide, was detected by using melamine-modified Au NPs *via* hydrogen bonding between atrazine and melamine on surfaces of Au NPs [37]. This reported method was highly sensitive with a detection limit 0.0165 μM, which was less than that of other sophisticated instrumental techniques used for analysis of atrazine, which shows the advantage of simple UV–visible spectroscopic method. Similarly, Liu's group prepared cysteamine–Au NPs and analyzed atrazine [38]. As shown in Figure 10.4, atrazine was extracted from rice samples and detected using cysteamine–Au NPs as a colorimetric probe. The $-NH_2$ groups on the surface of the cysteamine–Au NPs were positively charged in acidic pH, and it formed hydrogen bond with atrazine which induced aggregation of cysteamine–Au NPs. This aggregation led to a color change of cysteamine–Au NPs from wine red to blue and a decrease in intensity of the SPR peak at 523 nm and appearance of a new peak at 640 nm in the presence of atrazine, which was monitored by UV–visible spectroscopy. In this method, the total volume of sample was only 0.6 ml and the detection limit was 0.0165 μg/g, which shows the practical use of UV–visible spectroscopic analysis for microvolume samples. Very recently, another triazine

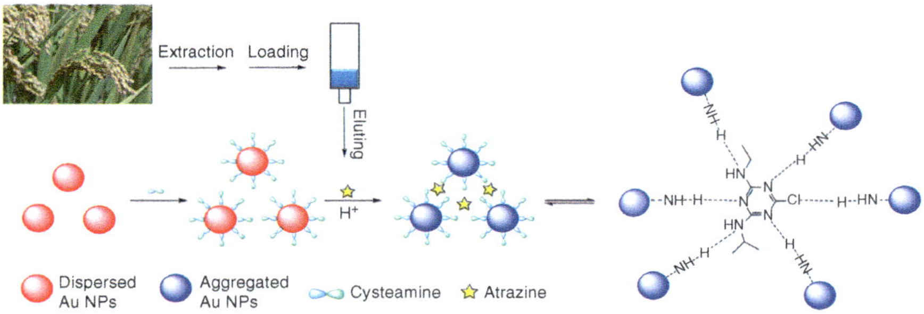

Figure 10.4 Colorimetric sensing mechanism for atrazine. Source: Reproduced from Liu et al. [38], with permission from the Royal Society of Chemistry Publication.

herbicide ametryn was detected by Qu's group by using aggregation mechanism of citrate–Au NPs in the presence of ametryn [39]. This aggregation resulted in a color change of citrate–Au NPs with a decrease in intensity of the absorption peak at 530 nm and appearance of a new peak at 690 nm, which was determined by simple UV–visible spectroscopy (Figure 10.5A). Furthermore, this ametryn-induced aggregation of citrate–Au NPs was confirmed by transmission electron microscope (TEM) images, which clearly indicated that citrate–Au NPs were well dispersed in solution due to repulsion between negatively charged citrate ions on Au NP surfaces. But after addition of ametryn to citrate–Au NPs, interparticle distance between citrate–Au NPs was reduced due to the electrostatic interaction between the negatively charged citrate on Au NP surfaces and the three-nitrogen hybrid ring of ametryn (Figure 10.5B).

Metsulfuron-methyl, a sulfonylurea derivative and widely used herbicide, was detected by Rohit's group by using dithiocarbamate-*p-tert*-butylcalix[4]arene (DTC–PTBCA)-capped Au NPs [40]. In this method, guest molecule metsulfuron-methyl interacted with host DTC–PTBCA–Au NPs via host–guest chemistry, which resulted in color changes and a red shift in the absorption band, which was monitored by UV–visible spectroscopy. This reported method was highly selective and was able to detect metsulfuron-methyl among 12 other herbicides using a microvolume sample (0.7 ml). The same group developed a method for analysis of pendimethalin herbicide using ractopamine–dithiocarbamate (RAC–DTC) Au NPs [41]. This method was highly selective and detected pendimethalin among other 11 herbicides using a very low sample volume (0.4 ml). Also, Chen's group developed a method for dual analysis of terbuthylazine herbicide and dimethoate insecticide using citrate–Au NPs based on UV–visible spectrometry [42]. This reported method was highly sensitive and determined terbuthylazine concentration up to 0.3 μM in water and food samples.

10.3.3 Analysis of Insecticides

Organophosphorus is a class of largely used insecticides in agriculture, but classified as class-I (highly toxic) or class-II (moderately toxic) pesticides by the Environmental Protection Agency (EPA). Therefore, development of simple methods for analysis

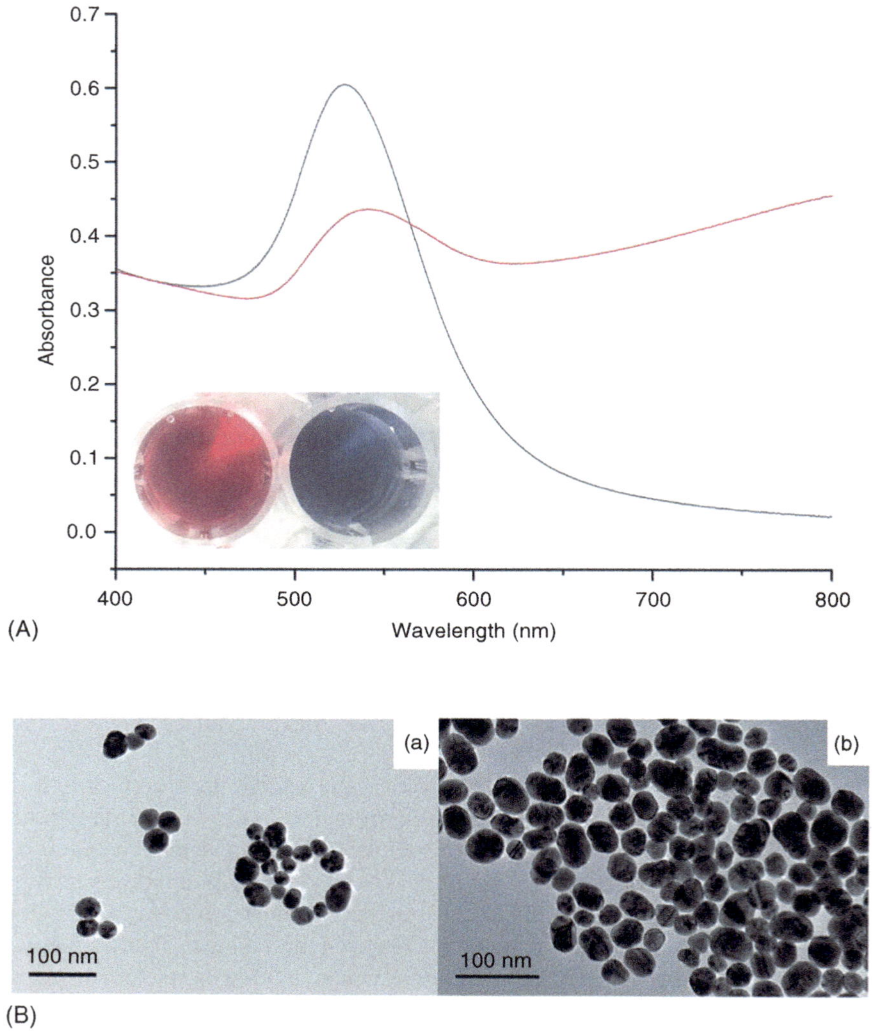

Figure 10.5 The visible absorption spectra (A) and TEM images (B) of Au NP-based colorimetric sensor in the (a) absence and (b) presence of 1.0 μg/l ametryn. Source: Reproduced from Qu et al. [39], with permission from the Royal Society of Chemistry Publication.

of organophosphorus insecticides is a very important task. In this regard, Bala's group developed a method for analysis of organophosphorus insecticide malathion using Au NP–based simple UV–visible spectroscopy [43]. This reported method was highly sensitive and was able to detect malathion up to 0.06 pM concentration. The method demonstrated its practicability to detect malathion from lake water and apple samples, and the results were more accurate than those obtained from HPLC. Another organophosphorus insecticide methyl-parathion was analyzed by Wang's group by using lanthanum-functionalized Au NPs [44]. Coordination-bonding interactions between methyl-parathion and lanthanum (La^{3+}) induced aggregation of lanthanum–Au NP probes and resulted in a change in solution color from red to blue, which was in situ monitored by UV–visible spectrometer for the quantitative determination of methyl-parathion. Similarly, Au^{3+}-cetyltrimethylammonium bromide (Au^{3+}-CTAB) Au NPs were used for analysis of parathion using simple UV–visible spectroscopy [45]. Practical sample analyses of parathion in tap water, sea water, and apple samples indicated on-site use of this method. Also, a method for detection of ethyl-parathion was developed using aggregation mechanism of cysteine–Au NPs [46]. This reported method was highly sensitive and a limit of detection was 0.081 ng/ml with very less sample volume (1 ml) using UV–visible spectroscopy. Another organophosphate insecticide phorate was detected by Bala's group in which phorate-induced aggregation of aptamer–Au NPs led to color and spectral changes of aptamer–Au NPs, which were analyzed using UV–visible spectrophotometer [47]. Same insecticide phorate was determined by Yang's group using hyaluronan–tyrosine (HA–Tyr) Au NPs with very less sample size [48]. This reported method was highly sensitive with a detection limit of 0.005 μg/ml, which was lower than the maximum residue limit (MRL) for phorate prescribed by the European Union.

Another organophosphorus insecticide methamidophos was detected using simple UV–visible spectroscopy based on aggregation of citrate–Au NPs [49]. The concentration of methamidophos in the range of 0.02–1.42 μg/ml was examined with a detection limit of 1.40 ng/ml and the method was able to determine methamidophos from vegetable samples. Paraoxon, a widely used insecticide, was determined based on Cu(I)-catalyzed click chemistry with alkyne-capped Au NP–based UV–visible spectrometry [50]. Insecticide dichlorvos was analyzed by D'souza's group using ascorbic acid (AA)-functionalized Au NPs [51]. Hydrogen bonding interaction between dichlorvos and AA–Au NPs led to aggregation of AA–Au NPs, which was confirmed by a color change from red to purple and a shift in UV–visible absorption peak from 525 to 620 nm. Three organophosphorus insecticides diazinon, iprobenfos, and edifenphos were determined by single-probe citrate–Au NPs and the method was able to determine the aforementioned insecticides from water samples [52]. Also, quinalphos insecticide was detected using *p*-nitroaniline dithiocarbamate (*p*-NA-DTC)-capped Au NPs using microvolume UV–vis spectrometric method with a total sample volume of only 250 μl by Rohit's group [53]. The same group developed a method for dual analysis of insecticides terbufos and thiacloprid by using nitro and hydroxy benzylindole–dithiocarbamate (NBI- and HBI–DTC) Au NPs [54]. The reported method was highly sensitive with detection limits 7.3×10^{-8} M and 6.0×10^{-7} M for terbufos and thiacloprid, respectively. Recently, Hu's group reported a method for determination of dimethoate insecticide by mimicking catalytic activity of Au NPs [55]. In this UV–visible spectroscopic method, very less sample volume (500 μl) was used to determine dimethoate.

Many other classes of insecticides, which are widely used in agriculture field as an alternative to highly toxic organophosphorus insecticides, were determined by simple UV–visible spectroscopy with minimum sample volume. Similarly, imidacloprid, a novel nitromethylene class insecticide used to control sucking mouthparts insects, was determined by Au NP–based UV–visible spectrometric methods [56, 57]. These methods are good alternatives to conventional instrumental techniques for analysis of imidacloprid. Another insecticide cyromazine belongs to triazine group of pesticides used to control growth of insects, but is toxic to the environment and human health. In this connection, many author groups put efforts and developed simple methods to determine cyromazine from water and food samples using Au NP–based UV–visible spectroscopy [58–60]. Acetamiprid, a new and most efficient neonicotinoid class of insecticide, was determined by Au NP–based UV–visible spectroscopic method. As it is carcinogenic to humans, determination of acetamiprid concentration in agriculture crops and food samples will be very useful to avoid its use above acceptable limits. Many research groups worked hard and developed simple and on-site detection method for analysis of acetamiprid using simple UV–visible spectrometric methods based on aggregation mechanism of Au NPs [61–65]. All these reported methods are highly selective and sensitive, which are able to determine acetamiprid among other pesticides at very low concentrations using very less sample size. Similarly, many other researchers have successfully developed Au NP–based simple UV–visible spectrometry methods for analysis of various insecticides (cartap, cyhalothrin, diafenthiuron, deltamethrin, and pymetrozine) [66–70]. These methods are selective toward specific insecticides and are able to determine reported insecticides from water, soil, and food samples. Also, all these reported methods are highly sensitive and are capable of detecting particular insecticides below their MRL values defined by various agencies.

10.4 Summary

This chapter illustrated the use of simple UV–visible spectroscopy for analysis of various pesticides based on NPs sensing methods using microvolume samples. The chapter concluded that all methods discussed here were selective toward particular pesticides and able to determine among other pesticides by easily observable color changes and red shift in absorption spectra of NPs. Furthermore, NP-based UV–visible spectroscopic methods were highly sensitive and capable of determining trace-level pesticides from water, soil, and food samples. Also, these methods are simple and can analyze microvolume samples, which make them ideal for real-time and on-site monitoring of pesticides in agricultural products. Finally, results of pesticide analysis obtained from NP-based simple UV–visible spectroscopy were comparable with or better than those of many other sophisticated instrumental techniques, which shows the importance of UV–visible spectroscopy in analytical science.

References

1 Oerke, E.C. (2006). Crop losses to pests. *J. Agric. Sci.* 144: 31–43.
2 Oerke, E.C. and Dehne, H.W. (2004). Safeguarding production losses in major crops and the role of crop protection. *Crop Prot.* 23: 275–285.

3 Scheyer, A., Briand, O., Morville, S., and Mirasel, P. (2007). Analysis of trace levels of pesticides in rainwater by SPME and GC-tandem mass spectrometry after derivatisation with PFFBr. *Anal. Bioanal. Chem.* 387: 359–368.

4 Wang, H., Wang, J., Timchalk, C., and Lin, Y. (2008). Magnetic electrochemical immunoassays with quantum dot labels for detection of phosphorylated acetylcholinesterase in plasma. *Anal. Chem.* 80: 8477–8484.

5 Health and the Environment Handbook for Health Professionals, Ministry of Public Work and Government Services Canada, Ontario, 1998.

6 Gutierrez, A. and Baran, N. (2009). Long-term transfer of diffuse pollution at catchment scale: respective roles of soil, and the unsaturated and saturated zones (Brévilles, France). *J. Hydrol.* 369: 381–391.

7 Saha, K., Agasti, S.S., Kim, C. et al. (2012). Gold nanoparticles in chemical and biological sensing. *Chem. Rev.* 112: 2739–2779.

8 Jain, P.K., Lee, K.S., El-Sayed, I.H., and El-Sayed, M.A. (2006). Calculated absorption and scattering properties of gold nanoparticles of different size, shape, and composition: applications in biological imaging and biomedicine. *J. Phys. Chem. B* 110: 7238–7248.

9 Jans, H. and Huo, Q. (2012). Gold nanoparticle-enabled biological and chemical detection and analysis. *Chem. Soc. Rev.* 41: 2849–2866.

10 Elghanian, R., Storhoff, J.J., Mucic, R.C. et al. (1997). Selective colorimetric detection of polynucleotides based on the distance-dependent optical properties of gold nanoparticles. *Science* 277: 1078–1081.

11 Lee, J.S., Jean, A.K.R.L., Hurst, S.J., and Mirkin, C.A. (2007). Silver nanoparticle-oligonucleotide conjugates based on DNA with triple cyclic disulfide moieties. *Nano Lett.* 7: 2112–2115.

12 Ling, J., Sang, Y., and Huang, H.C. (2008). Visual colorimetric detection of berberine hydrochloride with silver nanoparticles. *J. Pharm. Biomed. Anal.* 47: 860–864.

13 Rohit, J.V. and Kailasa, S.K. (2014). 5-Sulfo anthranilic acid dithiocarbamate functionalized silver nanoparticles as colorimetric probe for simple and selective detection of tricyclazole fungicide in rice samples. *Anal. Methods* 6: 5934–5941.

14 Rohit, J.V., Solanki, J.N., and Kailasa, S.K. (2014). Surface modification of silver nanoparticles with dopamine dithiocarbamate for selective colorimetric sensing of mancozeb in environmental samples. *Sens. Actuat. B* 200: 219–226.

15 Rohit, J.V. and Kailasa, S.K. (2014). Cyclen dithiocarbamate-functionalized silver nanoparticles as a probe for colorimetric sensing of thiram and paraquat pesticides via host–guest chemistry. *J. Nanopart. Res.* 16: 2585–2600.

16 Patel, G.M., Rohit, J.V., Singhal, R.K., and Kailasa, S.K. (2015). Recognition of carbendazim fungicide in environmental samples by using 4-aminobenzenethiol functionalized silver nanoparticles as a colorimetric sensor. *Sens. Actuators B* 206: 684–691.

17 Zheng, M., Wang, Y., Wang, C. et al. (2018). Silver nanoparticles-based colorimetric array for the detection of thiophanate-methyl. *Spectrochim. Acta A* 198: 315–321.

18 Jiang, X., Jin, H., Sun, Y., and Gui, R. (2019). Colorimetric and fluorometric dual-channel ratiometric determination of fungicide cymoxanil based on analyte-induced aggregation of silver nanoparticles and dually emitting carbon dots. *Microchim. Acta* 186: 580–592.

19 Siangproh, W., Somboonsuk, T., Chailapakul, O., and Songsrirote, K. (2017). Novel colorimetric assay for paraquat detection on-silica bead using negatively charged silver nanoparticles. *Talanta* 174: 448–453.

20 Rawat, K.A., Majithiya, R.P., Rohit, J.V. et al. (2016). Mg^{2+} ion as a tuner for colorimetric sensing of glyphosate with improved sensitivity via the aggregation of 2-mercapto-5-nitrobenzimidazole capped silver nanoparticles. *RSC Adv.* 6: 47741–47752.

21 Kumar, D.N., Rajeshwari, A., Alex, S.A. et al. (2015). Developing acetylcholinesterase-based inhibition assay by modulated synthesis of silver nanoparticles: applications for sensing of organophosphorus pesticides. *RSC Adv.* 5: 61998–62006.

22 Bala, R., Mittal, S., Sharma, R.K., and Wangoo, N. (2018). A supersensitive silver nanoprobe based aptasensor for low cost detection of malathion residues in water and food samples. *Spectrochim. Acta A* 196: 268–273.

23 Bhamore, J.R., Ganguly, P., and Kailasa, S.K. (2016). Molecular assembly of 3-mercaptopropinonic acid and guanidine acetic acid on silver nanoparticles for selective colorimetric detection of triazophos in water and food samples. *Sens. Actuators B* 233: 486–495.

24 Ma, S., He, J., Guo, M. et al. (2018). Ultrasensitive colorimetric detection of triazophos based on the aggregation of silver nanoparticles. *Colloids Surf. A* 538: 343–349.

25 Menon, S.K., Modi, N.R., Pandya, A., and Lodha, A.A. (2013). Ultrasensitive and specific detection of dimethoate using a p-sulphonato-calix[4] resorcinarene functionalized silver nanoprobe in aqueous solution. *RSC Adv.* 3: 10623–10627.

26 Li, Z., Wang, Y., Ni, Y., and Kokot, S. (2014). Unmodified silver nanoparticles for rapid analysis of the organophosphorus pesticide, dipterex, often found in different waters. *Sens. Actuators B* 193: 205–211.

27 Li, Y., Cui, Z., Li, D., and Li, H. (2011). Colorimetric determination of pyrethroids based on core–shell $Ag@SiO_2$ nanoparticles. *Sens. Actuators B* 155: 878–883.

28 Shrivas, K., Nirmalkar, N., Ghosale, A., and Thakur, S.S. (2016). Application of silver nanoparticles for a highly selective colorimetric assay of endrin in water and food samples based on stereoselective endorecognition. *RSC Adv.* 6: 29855–29862.

29 Ma, S., He, J., Guo, M. et al. (2016). Facile colorimetric detection of 6-benzylaminopurine based on p-aminobenzenethiol functionalized silver nanoparticles. *RSC Adv.* 6: 106608–106614.

30 Sun, Y. and Xia, Y. (2003). Gold and silver nanoparticles: a class of chromophores with colors tunable in the range from 400 to 750 nm. *Analyst* 128: 686–691.

31 Mehta, V.N., Basu, H., Singhal, R.K., and Kailasa, S.K. (2015). Simple and sensitive colorimetric sensing of Cd^{2+} ion using chitosan dithiocarbamate functionalized gold nanoparticles as a probe. *Sens. Actuators B* 220: 850–858.
32 Rastegarzadeh, S. and Abdali, S. (2013). Colorimetric determination of thiram based on formation of gold nanoparticles using ascorbic acid. *Talanta* 104: 22–26.
33 Hashemi, F., Rastegarzadeh, S., and Pourreza, N. (2018). A combination of dispersive liquid–liquid microextraction and surface plasmon resonance sensing of gold nanoparticles for the determination of ziram pesticide. *J. Sep. Sci.* 41: 1156–1163.
34 Kwon, Y.S., Nguyen, V.-T., Park, J.G., and Gu, M.B. (2015). Detection of iprobenfos and edifenphos using a new multi-aptasensor. *Anal. Chim. Acta* 868: 60–66.
35 Baek, S.H., Lee, S.W., Kim, E.J. et al. (2017). portable agrichemical detection system for enhancing the safety of agricultural products using aggregation of gold nanoparticles. *ACS Omega* 2: 988–993.
36 Zheng, J., Zhang, H., Qu, J. et al. (2013). Visual detection of glyphosate in environmental water samples using cysteamine-stabilized gold nanoparticles as colorimetric probe. *Anal. Methods* 5: 917–924.
37 Liu, G., Yang, X., Li, T. et al. (2015). Spectrophotometric and visual detection of the herbicide atrazine by exploiting hydrogen bond-induced aggregation of melamine-modified gold nanoparticles. *Microchim. Acta* 182: 1983–1989.
38 Liu, G., Wang, S., Yang, X. et al. (2016). Colorimetric sensing of atrazine in rice samples using cysteamine functionalized gold nanoparticles after solid phase extraction. *Anal. Methods* 8: 52–56.
39 Qu, Y., Qian, H., Mi, Y. et al. (2020). Rapid determination of the pesticide ametryn based on a colorimetric aptasensor of gold nanoparticles. *Anal. Methods* 12: 1919–1925.
40 Rohit, J.V., Basu, H., Singhal, R.K., and Kailasa, S.K. (2016). Dithiocarbamate-calix[4] arene functionalized gold nanoparticles as a selective and sensitive colorimetric probe for assay of metsulfuron-methyl herbicide via non-covalent interactions. *Sens. Actuators B* 237: 1044–1055.
41 Rohit, J.V. and Kailasa, S.K. (2017). Simple and selective detection of pendimethalin herbicide in water and food samples based on the aggregation of ractopamine-dithiocarbamate functionalized gold nanoparticles. *Sens. Actuators B* 245: 541–550.
42 Chen, N., Liu, H., Zhang, Y. et al. (2018). A colorimetric sensor based on citrate-stabilized Au NPs for rapid pesticide residue detection of terbuthylazine and dimethoate. *Sens. Actuators B* 255: 3093–3101.
43 Bala, R., Kumar, M., Bansal, K. et al. (2016). Ultrasensitive aptamer biosensor for malathion detection based on cationic polymer and gold nanoparticles. *Biosens. Bioelectron.* 85: 445–449.
44 Wang, X., Yang, Y., Dong, J. et al. (2014). Lanthanum-functionalized gold nanoparticles for coordination–bonding recognition and colorimetric detection of methyl parathion with high sensitivity. *Sens. Actuators B* 204: 119–124.

45 Wu, S., Li, D., Wang, J. et al. (2017). Gold nanoparticles dissolution based colorimetric method for highly sensitive detection of organophosphate pesticides. *Sens. Actuators B* 238: 427–433.
46 Bala, R., Sharma, R.K., and Wangoo, N. (2015). Highly sensitive colorimetric detection of ethyl parathion using gold nanoprobes. *Sens. Actuators B* 210: 425–430.
47 Bala, R., Sharma, R.K., and Wangoo, N. (2016). Development of gold nanoparticles-based aptasensor for the colorimetric detection of organophosphorus pesticide phorate. *Anal. Bioanal. Chem.* 408: 333–338.
48 Yang, L., Zhang, X., Li, H., and Jiang, L. (2017). Hyaluronan–tyrosine–gold nanoparticles as an enzyme-free colorimetric probe for the detection of phosphorothiolate pesticides. *Anal. Methods* 9: 6139–6147.
49 Li, H., Guo, J., Ping, H. et al. (2011). Visual detection of organophosphorus pesticides represented by mathamidophos using Au nanoparticles as colorimetric probe. *Talanta* 87: 93–99.
50 Fu, G., Chen, W., Yue, X., and Jiang, X. (2013). Highly sensitive colorimetric detection of organophosphate pesticides using copper catalyzed click chemistry. *Talanta* 103: 110–115.
51 D'souza, S.L., Pati, R.K., and Kailasa, S.K. (2014). Ascorbic acid functionalized gold nanoparticles as a probe for colorimetric and visual read-out determination of dichlorvos in environmental samples. *Anal. Methods* 6: 9007–9014.
52 Kim, M.S., Kim, G.W., and Park, T.J. (2015). A facile and sensitive detection of organophosphorus chemicals by rapid aggregation of gold nanoparticles using organic compounds. *Biosens. Bioelectron.* 67: 408–412.
53 Rohit, J.V., Basu, H., Singhal, R.K., and Kailasa, S.K. (2016). Development of *p*-nitroaniline dithiocarbamate capped gold nanoparticles-based microvolume UV–vis spectrometric method for facile and selective detection of quinalphos insecticide in environmental samples. *Sens. Actuators B* 237: 826–835.
54 Kailasa, S.K. and Rohit, J.V. (2017). Tuning of gold nanoparticles analytical applications with nitro and hydroxy benzylindole-dithiocarbamates for simple and selective detection of terbufos and thiacloprid insecticides in environmental samples. *Colloid. Surf. A* 515: 50–61.
55 Hu, Y., Wang, J., and Wu, Y. (2019). A simple and rapid chemosensor for colorimetric detection of dimethoate pesticide based on the peroxidase-mimicking catalytic activity of gold nanoparticles. *Anal. Methods* 11: 5337–5347.
56 Zhang, X., Sun, Z., Cui, Z., and Li, H. (2014). Ionic liquid functionalized gold nanoparticles: synthesis, rapidcolorimetric detection of imidacloprid. *Sens. Actuators B* 191: 313–319.
57 Tan, S., Zhao, H., Tian, D. et al. (2014). Piperidine–calix [4] arene modified gold nanoparticles: imidacloprid colorimetric sensing. *Sens. Actuators B* 204: 522–527.
58 Liu, J., Bai, W., Zhu, C. et al. (2015). Sensitive colorimetric detection of cyromazine in cucumber samples by using label-free gold nanoparticles and polythymine. *Analyst* 140: 3064–3069.

59 Bai, W., Zhu, C., Zhang, G. et al. (2016). Visual colorimetric detection of cyromazine in river water using citrate-stabilized gold nanoparticles. *Anal. Methods* 8: 5869–5873.

60 Liu, H., Gan, Z., Song, C. et al. (2020). Use of a smartphone for intelligent detection of cyromazine based on Tween 20 modified gold nanoparticles. *Anal. Methods* 12: 1469–1477.

61 Xu, Q., Du, S., Jin, G.-d. et al. (2011). Determination of acetamiprid by a colorimetric method based on the aggregation of gold nanoparticles. *Microchim. Acta* 173: 323–329.

62 Shi, H., Zhao, G., Liu, M. et al. (2013). Aptamer-based colorimetric sensing of acetamiprid in soil samples: sensitivity, selectivity and mechanism. *J. Hazard. Mater.* 260: 754–761.

63 Tian, Y., Wang, Y., Sheng, Z. et al. (2016). A colorimetric detection method of pesticide acetamiprid by fine-tuning aptamer length. *Anal. Biochem.* 513: 87–92.

64 Yang, W., Wu, Y., Tao, H. et al. (2017). Ultrasensitive and selective colorimetric detection of acetamiprid pesticide based on the enhanced peroxidase-like activity of gold nanoparticles. *Anal. Methods* 9: 5484–5493.

65 Qia, Y., Chen, Y., Xiu, F.-R., and Hou, J. An aptamer-based colorimetric sensing of acetamiprid in environmental samples: convenience, sensitivity and practicability. *Sens. Actuators B* https://doi.org/10.1016/j.snb.2019.127359.

66 Liu, W., Zhang, D., Tang, Y. et al. (2012). Highly sensitive and selective colorimetric detection of cartap residue in agricultural products. *Talanta* 101: 382–387.

67 Imene, B., Cui, Z.M., Zhang, X. et al. (2014). 4-Amino-3-mercaptobenzoic acid functionalized gold nanoparticles: synthesis, selective recognition and colorimetric detection of cyhalothrin. *Sens. Actuators B* 199: 161–167.

68 Kailasa, S.K. and Rohit, J.V. (2017). Multi-functional groups of dithiocarbamate derivative assembly ongold nanoparticles for competitive detection of diafenthiuron. *Sens. Actuators B* 244: 796–805.

69 Wang, Z., Huang, Y., Wang, D. et al. (2018). A rapid colorimetric method for the detection of deltamethrin based on gold nanoparticles modified with 2-mercapto-6-nitrobenzothiazole. *Anal. Methods* 10: 1774–1780.

70 Kang, J., Zhang, Y., Li, X. et al. (2018). Rapid and sensitive colorimetric sensing of the insecticide pymetrozine using melamine-modified gold nanoparticles. *Anal. Methods* 10: 417–421.

11

Miniaturized Liquid Extractions in MALDI–MS Analysis

Nazim Hasan and Shadma Tasneem

Jazan University, Department of Chemistry, Faculty of Science, Post Box No. 2097, Jazan, Saudi Arabia

11.1 Introduction

The sequence of nucleotides in the DNA with amino acids consists of various primary structures of biological molecules such as lipids, peptides, phosphopeptides, proteins, and phosphoproteins, which cover the central dogma of biological sciences. In many biological processes and key information pathway deals with peptides and proteins separation and extraction. The fundamental principle to study proteomics is based on genomic sequence information in terms of structure and specific active functional groups that control biological processes. In any vital cell, biological information is first generated through the DNA transcription into RNA, leading to the complex processing of RNA to mRNA and mRNA conversion to specific target functional and structured proteins. Proteins are biological molecules that dictate cellular structures, various cellular activities, and mechanisms of signaling between cells and tissues and catalyze chemical reactions that support metabolism. Protein structures determine their specific functions and dysfunctions. Dysfunction of proteins causes Alzheimer's and Huntington's diseases [1, 2]. Proteins are also major fighting biological molecules in the form of antibodies against viral and bacterial infections. Analysis of peptides, proteins, DNA, and RNA by various analytical techniques depends on sample preparation techniques such as extraction, separation, and preconcentration. The application of certain sample preparation techniques depends on the nature of biological molecules and the complex matrix from where they need to be extracted. Liquid–liquid extraction (LLE) is the most commonly used traditional and classical separation technique that depends on solubility of analytes in different liquid phases such as aqueous and organic phases. Nevertheless, LLE techniques have certain limitations: they require large amounts of organic solvents (toxic and expensive), are environmentally unfriendly, and are time-consuming. Multiple extraction procedures require some time, which makes them laborious, tedious, and difficult to automate. Miniaturized extraction techniques have been proposed for sample treatment in environmental and biological samples prior to MS analysis

Miniaturized Analytical Devices: Materials and Technology, First Edition.
Edited by Suresh Kumar Kailasa and Chaudhery Mustansar Hussain.

to improve the classical liquid extraction techniques. The solvent microextraction (SME) technique focuses on extraction and preconcentration, by using solvent volume of 0.5–100 ml, of a target species from sample matrices [3–5]. This is well suited for collecting, purifying, and refining environmental and biological samples in volatile, nonvolatile, polar, nonpolar, ionic, and metal analysis. Much effort has been made in recent years to establish green analytical methods for rapid preconcentration of analytes from environmental and biological samples using less amount of solvents and reagents and a dramatic reduction in laboratory waste. Since 1995, several liquid-phase microextraction (LPME) methods, including mass spectrometry, to assess organic, inorganic, and biological molecules in various samples have been coupled with specific analytical instruments [6–8]. Figure 11.1 displays the LPME classification based on their extraction modes. These miniaturized solvent extraction tools have been receiving increasing attention in various areas and are widely used for the isolation and concentration of trace-level targets with minimum solvent and sample volumes in modern analytical studies. LPME modality is based on the distribution of the analytes between a micro drop of an organic solvent at the tip of the needle of a microsyringe and an aqueous sample solution. The technique incorporates extraction, preconcentration, and sample application in a single stage, and mass spectrometry for analysis [9]. Nanoscience and nanotechnology are widely applied in almost all areas, with continuously evolving new topics. Nanomaterials exhibit unique optical, electrical, magnetic, and catalytic properties and have demonstrated a variety of fascinating applications in different analytical processes [10]. The expansion of the use of nanomaterials has also contributed to the development of suitable analytical methods for detection and classification. Nanomaterials have direct applications in liquid–solid sorption

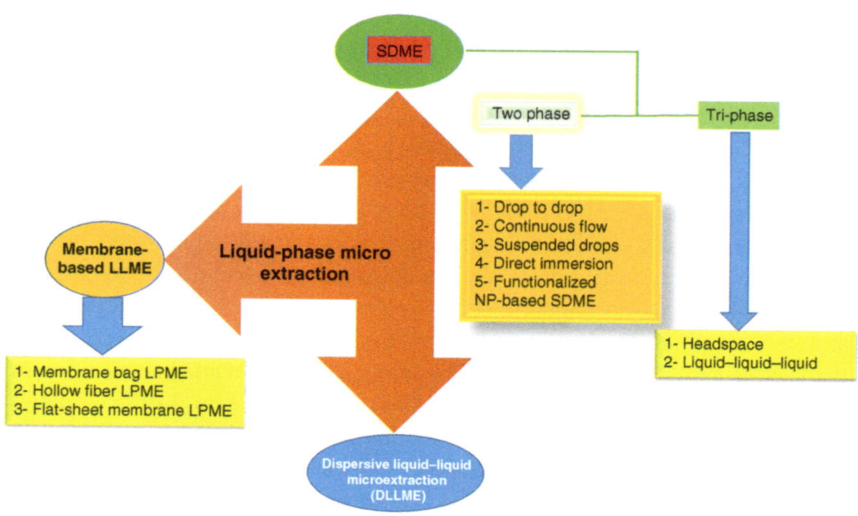

Figure 11.1 Different types of liquid microextraction.

processes due to their large surface area and their ability to incorporate various functional groups on their surface; they also have other desirable properties such as low diffusion resistance, high adsorption efficiency, and fast sorption kinetics. Various nanomaterials have been tested as new preconcentration sorbents in solid-phase microextraction (SPME) [11]. For liquid extraction, miniaturization based on noble metal nanoparticles (NPs) and metal-oxide NPs and their surface functionality may be exploited in combination with MALDI–mass spectrometry. Magnetic solid phases have additional benefits, as centrifuges are unnecessary for sorbents, and with an external magnetic field, the solid phase can be quickly separated from the sample solution and can be examined directly in MALDI–MS. This chapter deals with miniaturized liquid extraction techniques, which are effective for short-term preconcentration of analytes and can be analyzed using MALDI–MS. Miniaturization with single-drop microextraction [SDME] and liquid–liquid microextraction [LLME] methods were explored directly in conjunction with MALDI–MS for analysis of organic and biological molecules in different sample matrices. This chapter gives an overview and demonstrates miniaturized solvent extraction techniques and mechanisms integrated with nanoparticles in conjunction with MALDI–MS for the analysis of organic and biological molecules.

11.2 MALDI/SALDI–TOF–MS Instrumentation and Ionization Expected Mechanism Before Miniaturization of Liquid Extraction by Nanoparticles

11.2.1 MALDI–TOF–MS Techniques

Laser desorption and ionization (LDI) technique is used for analysis of organic or biological analytes using organic molecules as the matrix. In 1985, Hillenkamp and Karas coined the term matrix-assisted laser desorption ionization (MALDI) [12]. MALDI is based on a collective desorption cycle for analytes or dissolved matrix molecules. The ionizing mechanism of MALDI analysis is still to be investigated [13, 14]. The dynamic matrix molecules can absorb either electronic or vibrational light through ultraviolet (UV) and infrared (IR). A rapid expansion in the solid layer of the matrix, like the UV or IR light colloid, causes the matrix molecules to become desorbed along with analyte molecules. The proton transfer [15, 16], as shown in Figure 11.2a, is the point at which the load matrix analyte transfers and analytical ionization occurs. It creates a local high pressure due to the high matrix rates of the MALDI plumes that result in crash cooling. The analyte–matrix ratio should be low as well as adequate instances of contact and laser absorption between the analytes. The analytical matrix concentration should be between 1/1000 and 1/100 but depends on the analyte's size and complexity and is mostly determined by hit-and-test methods. Laser fluence also affects the amount of the

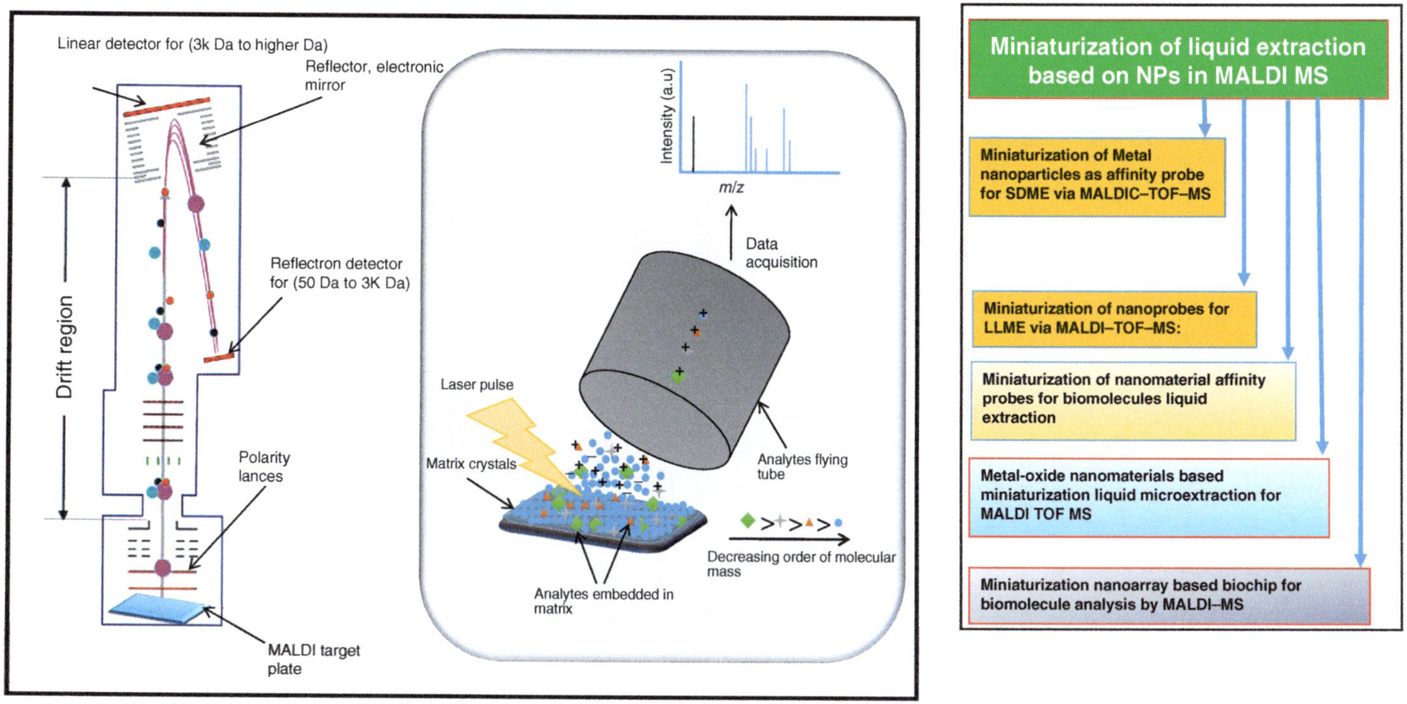

Figure 11.2 (a) MALDI–TOF instrumentation and laser desorption/ionization expected mechanism and (b) different NP-based miniaturization approaches for liquid extraction in MALDI–MS.

material desorbed; for proper desorption and ionization, this laser fluence is called the MALDI threshold [17]. The signal rises to a certain level above the threshold level, below which no signal appears. When high-energy lasers are used, the analytes are scattered in-source or in-post-source [18]. The factors of desorption and fragmentation of analytes depend not only on the laser fluence but also on the laser spot's matrix and size [19]. When the spot 's size reduces, the laser fluence must be ionized minimum, and fragmentation generally increases at the laser fluence. On the order of 100 μm, MALDI threshold plumes normally include a few femtomoles, which are clusters, molecules, and ions, for a laser spot scale. Almost 1 analyte can be ionized from 10 000 desorbed molecules. It will ionize only 10–100 analyte ions and reach the detector through a laser shot. The MALDI ionization process is not very effective, but the device allows for detection limitations at the attomole level even though mechanism-based limitation [20, 21]. Elevated-pressure and atmospheric pressure MALDI sources have been shown to stretch this limit to the sub-attomole range, which is attributed to increased collisional cooling of the ions, which reduces fragmentation during ionization. The full absorption of the laser energy should be available for this process. Usually, near-UV lasers are mostly used, including laser N_2 (337 nm) and laser Nd: YAG (355 nm). The N_2 laser has the benefit of a natural intensity fluctuation that, due to local hotspots, is expected to increase the signals [22]. This approach can be used for the calculation of all ions in a single test cycle fed into the mass spectrometer. The mass dividing analysis is based on the different flight times of the ions, starting from the same stage with equal energy. Moving ions in an electrostatic field obtain the kinetic energy E in which the power potential is U and the charge of the ion is Z as in Eq. (11.1):

$$\frac{m}{z} = zU \tag{11.1}$$

The electric potential energy is converted into the kinetic energy, and the kinetic energy is related to velocity (v) and mass (m) as $E = (1/2) \cdot mv^2$, the velocity of the ions is dependent on their mass (m) and charge (z):

$$\frac{m}{z} = 2U/v^2 \tag{11.2}$$

Mass spectrometers have a fixed field-free flight path with length (l) and the electrostatic extraction field is kept constant; this leads to a time relation. There is a quadratic relation between m/z and time (t) as shown in Eq. (11.4).

$$\frac{m}{z} = (2U/l^2) \cdot t^2 \tag{11.3}$$

$$\frac{m}{z} \infty\, t^2 \tag{11.4}$$

In short, the analyte's molecular weight is dominant, and the detector takes a longer duration for high-molecular-weight ions when the molecular weight of the lower ions is first applied. Usually, one or more microchannel plates (MCPs) detectors are used at the end of the "travel time" pipeline to detect various m/z ions. The MCPs follow the waterfall process whereby any ion transformed into an electron and the electron cascade is recognized by the detector as a cup collector or a phosphorus sensor. Such mutual detector signaling is transferred in connection with time-to-digital converter (TDC) via an analog-to-digital converter (ADC) or by constant fraction discriminator (CFD) with the combination of TDC. ADCs are most effective when several ions reach the detector within a limited time period (<10 ns, depending on the recovery speed of the detector and the electronics) [22].

11.2.2 Miniaturization-Based NPs in SALDI/MALDI–TOF–MS Application

Surface properties of inorganic nanomaterials dramatically change as the particle size decreases and often show excellent heat transfer efficiency after absorbing appropriate laser radiation. Many metal, metal-oxide, and silica nanoparticles and semiconductor quantum dots (QDs), as well as carbon-based graphene oxide nanoparticles, function as efficient absorbent and ionization substrates in surface-assisted laser desorption mass spectroscopy (SALDI–MS). In 1988, Koichi Tanaka et al. demonstrated the application of nitrogen laser and used ultrafine cobalt powder (diameter approximately 30 nm) in glycerol for analysis of high *m/z* proteins and polymers (25–100 kDa) by the LDI–TOF–MS technique [23]. Recently, some reviews have been published to explore the behavior of bare and functionalized nanoparticles as affinity, enrichment, and LDI probes for analyte analysis [24, 25]. Nanoparticle surfaces enrich the target analytes (drugs or small-molecular-weight analytes) and focused laser beam transfers the energy toward the surface of nanoparticles, which later moves to analytes [26–28]. As a result, desorption and ionization of analytes occur on the nanoparticle surface, which reflects the photothermal efficiency of nanoparticles to convert it into heat transfer to analytes. Photothermal energy conversion to heat highly controls the surface of nanoparticles, which leads to very low fragmentation of nanostructures and results in low background noises in the lower *m/z* region of SALDI–MS spectra for large analytes. Nanoparticle-based SALDI–MS miniaturization is a very promising approach for the analysis of medicinal drugs, metabolites, and environmental pollutants [26, 27]. Analysis of high-molecular-weight biological molecules such as peptides and proteins by SALDI–MS is compromised due to lack of adequate ionization via nanoparticles only. Nanoparticles combined with an organic matrix are an excellent platform to improve efficiency, specificity, sensitivity, and reproducibility of MALDI–TOF–MS laser absorption for ultratrace biological molecules. Nanoparticles display bifunctional surface properties: they function as affinity and enrichment probes for peptides and proteins; and they receive the UV laser energy and reserve it to control LDI with conventional organic matrices like α-cyano-4-hydroxycinnamic acid (CHCA) sinapic acid (SA) and 2,5-dihydroxybenzoic acid (DHB). Nevertheless, the application of various nanoparticles coupled with conventional organic matrices leads to effective and controls UV-energy transfer to LDI process without analyte fragmentation or denaturation after application of high-energy radiation. The NPs–conventional organic matrix combination provides a highly protonated and clean mass spectrum for biological molecule analytes by MALDI–TOF–MS. Application of nanoparticles for the analysis of biological molecules for miniaturized liquid extraction by MALDI–TOF–MS is shown in Figure 11.2b. In LPME, nanoparticles effectively preconcentrate the trace-level peptides, proteins, and bacterial surface proteins characterized by MALDI–MS within a short period.

11.3 Miniaturization of Metal Nanoparticles as Affinity Probe for SDME Via MALDI–TOF–MS

In real-world samples, analytes are usually found in trace level, which is a great challenge to preconcentrate, and various analytical techniques are required for analysis and characterization. Preconcentration of analytes depends on inter-chemical and -physical properties of the matrix. Direct complex sample analysis from real-world samples by MALDI–TOF–MS is usually deprived of the mass peak signals of the expected analyte. In this regard, required analyte preconcentration process strongly affects reproducibility of MALDI–TOF–MS. These analytical techniques to preconcentrate the analyte from a minimal volume sample by miniaturization of affinity probes such as nanomaterials have great potential. Murray et al. [29], in 1979, first developed a liquid-based microextraction system in which they used 200 μl of organic solvent as an extractant for GC. Leu and Dasgupta, in 1995, first used the liquid droplet for the analysis of air pollutants such as sulfur dioxide and ammonia [30]. The gas flow through the droplet was suggested in their system and contaminants diffused and dissolved into the droplet liquid. This piece of work has opened a new era of miniaturization extraction in analytical techniques. Further, an LLE system was developed to achieve solvent microextraction in the form of a microsyringe-based single drop (8 μl) by Cantwell and Jeannot [31]. In classical LLE, many organic solvents are used, which refer to the LPME technique. Single-drop microextraction (SDME) is the simplest technique of LPME in which a drop of an organic solvent is immersed in the aqueous phase sample through a sharp needle of a microsyringe. Small-molecular-weight organic or inorganic moieties from the aqueous solution can be easily infused into hydrophobic drops; nevertheless, higher molecular weight peptides or proteins need specific functional affinity for static drops. Bare or functionalized nanoparticle-based tiny static drops may show great affinity for preconcentration of analytes from complex samples. Miniaturized nanoparticle-based affinity probes are very effective to enrich peptides and proteins; on the other hand; simple SDME techniques of LPME have this limitation as a preconcentration probe. A certain period of extraction time, head drop of high-affinity nanoparticles is retracted into the syringe and the extractive phase can be easily applied on MALDI–TOF–MS with appropriate matrix to obtain an informative mass spectrum.

11.3.1 Affinity Probe of Functionalized Au and Ag Nanoparticles as SDME

Au nanoparticles show great surface affinity for biological molecules, especially peptides and proteins, due to their surface plasmonic response at different sizes and shapes. Bare or organic moiety-stabilized Au nanoparticles in a pure organic solvent can preconcentrate peptides or proteins through SDME. Sudhir et al. [32] used Au nanoparticles in toluene functionalized with tetraalkylammonium bromide (TAAB), and this affinity probe attributed negative and positive charges. This probe was applied as SDME for the enrichment of Methionine (Meth-enk) and Leucine encephalin (Leu-enk) peptide mixture with an LOD of 0.2 and 0.17 μl via MALDI–TOF–MS. Further, this group applied this Au nanoparticle-based miniaturization technique of SDME for a more complex solution of 1% Triton X-100 (nonionic surfactant) and 6.5 M urea (complex matrix interference) and successfully obtained MALDI–MS signal of selective peptides. Miniaturization techniques based on nanoparticles' affinity interchange interesting extracting probes as preconcentrating probes for analysis and characterization of biomolecule analytes by MALDI–TOF–MS. The SDME–MALDI–TOF–MS can be used as a sensitive detection system for peptides and proteins on the surface of nanoproducts tuned to different sizes, types, and functionalities. Shastri et al. used (4-mercapto phenyliminomethyl)-2-methoxyphenol (Schiff base) capped on Au nanoparticles [33]. These functionalized Au nanoparticles were dispersed in toluene and further used as multifunctional probes in SDME to preconcentrate proteins as well as to effectively work as a co-matrix with CHCA in MALDI–TOF–MS analysis. The Au nanoparticle samples have been functionalized to improve the biomolecular study and, by MALDI–TOF–MS, to be classified with a high spectrum such as gramicidin D (53.1 fmol), HW6 (450 fmol), ubiquitin (23.3 fmol), cytochromium c (40 fmol), and myoglobin (35.3 fmol). Surface plasmonic properties of Ag nanoparticles are also of great interest in biomolecular interactions and SDME miniaturization techniques coupled with MALDI–MS. Surface properties of Ag nanoparticles depend on surface functionalization with various organic moieties, different sizes, shapes, and electromagnetic interaction between Ag particles within hydrophobic proximity. Shrivas and Wu [34] modified and functionalized the surface of Ag nanoparticles with dodecanethiol and octadecanethiol (ODT) to achieve hydrophobic stability. Miniaturization of Ag nanoparticle–based affinity probes was effectively applied for the analysis of peptides and proteins from plasma and urine samples by SDME and MALDI–MS. Tandem nanoparticle-based SDME and MALDI–MS approach improved the extraction efficiency of minimum treatment of analytes within a short period. Shastri et al. synthesized bifunctional 1-ODT-stabilized Ag nanoparticles with two different capping reagents 4-aminothiopheneol (4-AMP) and 1-thioglycerol (1-TG) [35]. This study was based on two different functional groups on Ag nanoparticles, which led to strong interaction of nanoparticles with peptides and proteins. Miniaturized Ag@ODT/4-AMP and Ag@ODT/1-TG nanoparticles affinity–based interaction enhances the separation of cysteine (Cys), homocysteine (HCys), insulin, heart cytochrome c, ubiquitin, and lysozyme and further analysis by MALDI–TOF–MS. During this study, various parameters were optimized such as sample pH, stirring rate, salt concentration, extraction time, matrix type, and the amount of Ag NPs. These different functionalized Ag nanoparticles were applied successfully for real-world sample analysis such as insulin and cytochrome c in urine and milk and peptide-like Cys and HCys in urine samples by SDME and MALDI–TOF–MS.

11.3.2 Nanoparticles and Ionic Liquid (NP-IL) Hybrid Probes as SDME

Room-temperature ionic liquids (RTILs) are known as a designer solvent with tunable functionality, which can be controlled by the length of the alkyl carbon chain and groups incorporated by cations. ILs have exceptional physicochemical properties and are alternative to volatile organic solvents with relatively high density, very low vapor pressure, and non-flammability [36]. Guo et al. introduced IL-SDME as a rapid, simple, and environmentally friendly technique for the determination of sulfonamides from real-world water samples [37]. IL-SDME techniques proved to have a low detection limit with a linear range and high reproducibility. However, the nonvolatile nature of ILs makes IL-SDME technique incompatible with gas chromatography (GC) analysis [38, 39]. Nanoparticles have a high surface-area-to-volume ratio with small size, and various surface functionalities can attribute to preconcentrating probes for biomolecules coupled with SDME and further analysis by MALDI–TOF–MS. Nevertheless, nanoparticle-based SDME miniaturized techniques have some limitations toward the preconcentration probes such as small organic drop stability and temperature effects during the enrichment process of biological molecules. These limitations can be resolved by the application of nanoparticles and ILs composition as an affinity probe to enrich the biological molecules by the SDME technique. Bare or functionalized nanoparticles are easily dispersed in various ILs and can be exploited for preconcentration with a higher extraction efficiency of trace-level analytes from biological complex samples and further extracted mixture is applied directly to analysis by MALDI–MS. Ahmad and Wu used Pt nanoparticles with (1-butyl-3-methylimidazolium hexafluorophosphate, BMIHFP) ionic liquid for preconcentration and enrichment of bacterial protein ions from *E. coli* and *S. marcescens* by NPs-ILs–SDMI–MALDI–TOF–MS [40]. Enrichment of proteins from the bacteria was obtained based on the hydrophobic and electrostatic interactions between IL-dispersed Pt nanoparticles and proteins. Thus, this process shows nanoparticle- and IL-based SDME techniques have several intact bacterial protein ion peaks during MALDI–MS analysis without any influence on bacterial counts.

Carolina et al. used 1-hexyl-3-methylimidazolium hexafluorophosphate [C_6MIM] [PF_6] ionic liquid and CdSe/ZnS QDs dispersion for HS-SDME to enrich volatile trimethylamine (TMA) from the lake fish sample and to quantify by fluorescence spectrometry [41]. The limit of detection (LOD) of TMA from pond fish was 0.014 mg/l (0.35 μg TMA per gram of fish), and the relative standard deviation (RSD) observed for five samples was 3.5%.

Miniaturization of liquid extraction by nanoparticles acts as a high-affinity probe for biological molecules such as peptides and proteins even in a highly complex medium. Small size and various shapes of nanomaterial-dispersed suspension systems have large surface-to-volume ratios and can carry a high adsorption degree of target molecules even in a very small quantity of sample. The presence of ionic liquids with nanomaterials enhances affinity for small- and high-molecular-size biomolecules multiple times [42]. This hybrid matrix system efficiently works as an ionization source to ionize biological analytes and further analyze and characterize by MALDI–TOF–MS.

11.4 Miniaturization of Nanoprobes for LLME Via MALDI–TOF–MS

Green chemistry–based approaches to separate analytes through LLME were coupled with analytical techniques in the mid-1990s. LPME-based techniques have great selectivity and affinity for biomolecule analysis using gas chromatography–mass spectroscopy (GC–MS) [43, 44]. LLME approaches have multiple applications to separate analytes in the microliter quantity of extracting liquids. Due to various application parts and selectivity, LLME can be divided into dispersive liquid–liquid microextraction (DLLME), continuous-flow microextraction (CFME), and hollow-fiber liquid-phase microextraction (HF-LPME) [45–47]. The working mechanism of LLME affinity probes is based on high-density extractants such as organic nonpolar solvent (few μl quantities) and dispersed solvent (μl to ml) mixtures. In this process, high friction and fast shaking occur, forming very fine droplets in the entire aqueous sample solution. Such fine droplets generate high surface areas between the organic and hydrophilic phases and generate a balance and promote the extraction of analytes within a short period [48]. Nevertheless, LLME involves short sample preparation and easy separation techniques, but efficient preconcentration from a complex real-world sample for high-molecular-weight biological molecules (e.g. peptides and proteins) is very limited or not suitable. Miniaturization using nanoparticles may approach the LLME technique as a more efficient affinity probe for the analysis of biomolecules by MALDI–TOF–MS. Nanomaterials without functionality in different shapes and sizes or various surface-functionalized nanoparticles exhibit hydrophilic, hydrophobic, and electrostatic interactions to facilitate high-affinity and advantageous probes for analyte enrichment in complex samples. The advantage of nanomaterials is their physical and chemical properties, which depend on the size, surface, and shape of nanoparticles [11, 49].

11.4.1 Miniaturized Nanoparticles as LLME Enrichment Probes for Biomolecules

Surface-modified nanoparticles can be exploited and tuned for effective interaction toward the enrichment of biological moieties, which leads to a new era of miniaturized liquid extraction techniques coupled with MALDI–TOF–mass spectrometry [34, 50–54]. Surface modification of nanoparticle extraction processes is usually based on hydrophobic or electrostatic interactions of nanoparticles and biological molecules such as peptides and proteins. Shrivas et al. proposed surface modification of Ag nanoparticles using dodecanethiol (DT) and ODT as a capping agent. Hydrophobic capping agent–based nanoparticles were studied by MALDI–MS, which were dispersed in toluene and further applied for biological molecules with LLME and enriched by MALDI–MS. Since hydrophobicity increases as the alkyl chain length increases, Ag–ODT nanoparticles exhibit more enrichment efficiency compared to Ag–DT or bare Ag NPs (without capping agent) during LLME and MALDI–MS analysis [34]. In another study, two different Ag_2Se nanoparticles were stabilized by 11-mercaptoundecanoic acid (MUA) and ODT, respectively [51]. These two different functionalized Ag_2Se nanoparticles were applied as a preconcentration probe for hydrophobic peptides and proteins from soybean by LLME and nanoparticle-based enrichment biological molecules analysis by MALDI–TOF–MS. Ag_2Se bare nanoparticles, MUA–Ag_2Se, and ODT–Ag_2Se nanoparticles were dispersed in toluene after applied as an affinity probe for standard hydrophobic peptides such as valinomycin and gramicidin D. Since the interaction between the nanoparticles and peptides are based on hydrophobicity, ODT–Ag_2Se nanoparticles were considered the highest detection sensitivity probe for peptides using LLME and MALDI–MS. These functionalized nanoparticles and methods were applied successfully for real-world samples such as soybeans hydrophobic proteins and peptides, which explore soybean Bowman–Birk proteinase inhibitor and D-II inhibitor at 3879, 7855, and 8570 m/z mass peaks in MALDI–TOF–MS spectra. Bhat and Wu synthesized Pd nanoparticles stabilized by ODT in toluene and applied as preconcentration and extraction probes for proteins (insulin, lysozyme, and ubiquitin) by LLME [51]. Hydrophobic chains present in Pd nanoparticles enrich the proteins (pH $\sim$ pI), and these miniaturized LLME techniques were successfully applied for the analysis of real-time samples such as pancreas, mushroom, soybean, and milk proteins by MALDI–TOF–MS. Shrivas and Wu synthesized Co_3O_4 nanoparticles and stabilized by using the cetyltrimethylammonium (CTA) reagent [52]. Electrostatic interaction on Co_3O_4 nanoparticles, due to CTA, were exploited at pH > pI with respect of proteins sensitivity. Preconcentration of proteins such as insulin, chymotrypsinogen, and lysozyme were performed by Co_3O_4 nanoparticles within 10 minutes of incubation by LLME and analysis by MALDI–TOF–MS. In both studies, the presence of NaCl salt enhances MALDI–TOF–MS intensity [51, 52]. The salt ions interact with water molecules and change the binding behavior of proteins and the surface of functionalized nanoparticles. The higher salt concentration of NaCl decreases the water molecules on the surface and proteins of the nanoparticles and significantly enhances the stronger interaction with the hydrophobic carbon chain present on the nanoparticles [34].

11.4.2 Miniaturized Nanoparticle-Based LLME Affinity Probes for Bacterial Proteins

The bacterial aqueous sample is a very complex real-world sample for analysis by MALDI–TOF–MS. Nanoparticles have a great opportunity in the preconcentration of bacterial proteins on their surface and further analysis by MALDI–TOF–MS. Nanoparticle-based LLME techniques are suitable to enrich hydrophobic proteins, phosphoproteins, and peptides. Kailasa and Wu synthesized $Mg(OH)_2$ NPs stabilized by oleic acid in toluene and applied DLLME coupled with MALDI–TOF–MS [53]. This miniaturization technique demonstrated very effective and sensitive detection of a large number of analytes (0.1–2.0 mM for valinomycin and gramicidin D). It was also applied successfully for rapid enrichment and detection of hydrophobic bacterial (*E. coli* and *Bacillus subtilis*) membrane proteins prior to analysis by MALDI–TOF–MS. $Mg(OH)_2$ NPs were applied as affinity and enrichment probe for extraction of hydrophobic membrane proteins albeit at high concentrations of the salt (NaCl) and surfactant (Triton X-100). The broad surface area of functional nanostructures results in stronger interactions between analytical molecules, which facilitates efficient enrichment of low-abundance biomolecules from complex biological samples. The same group synthesized and functionalized $BaTiO_3$ NPs with a capping agent of 12-hydroxy octadecanoic acid (HOA). These nanoparticles were used as enrichment probes for phospholipids (PLs) and ionization matrix in MALDI–MS. LLME based on miniaturized $BaTiO_3$ NPs techniques was successfully applied to enrich hydrophobic proteins from *E. coli* bacteria before identification by MALDI–TOF–MS [54]. Figure 11.3 shows the synthesis and functionalization scheme for $BaTiO_3$ NPs. Interestingly, bare $BaTiO_3$ nanoparticles (hydrophilic nature) after functionalization with HOA move to the toluene solvent and converted into hydrophobic. Such nanoparticles were successfully used as a high affinity and enrichment probe in LLME miniaturized techniques for analysis of various phosphopeptides and *E. coli* membrane peptides and proteins in MALDI–MS. This method is linear in 1.0–5.0 μM and 1.0–10.0 μM concentrations of L-A phosphatidyl-L-serine (PS) and L-A-phosphatidic acid sodium

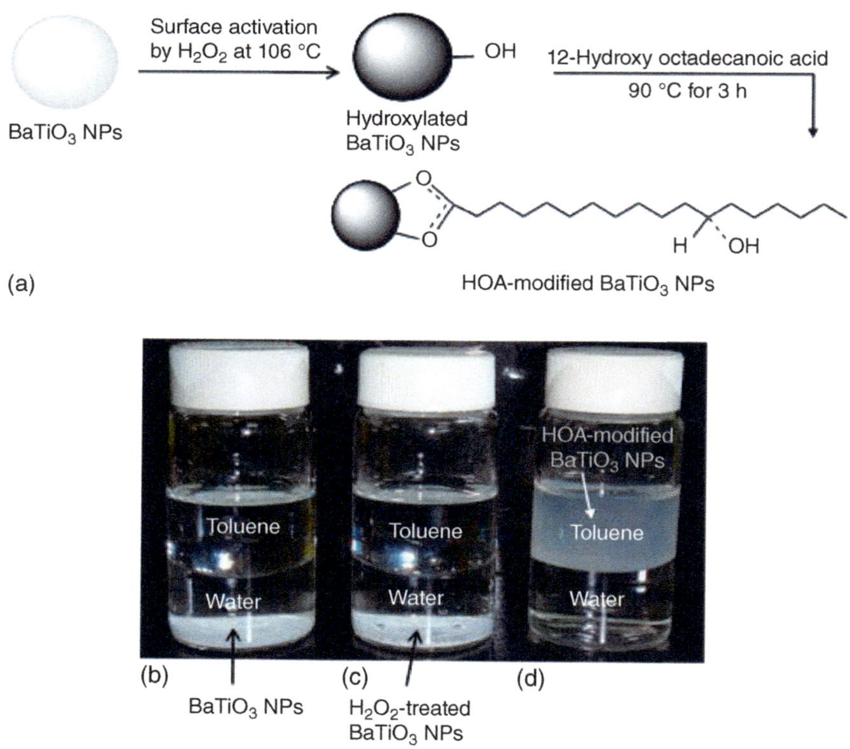

Figure 11.3 (a) Schematic representation of the synthesis of HOA-modified $BaTiO_3$ NPs, (b) photographs of (b) as-received $BaTiO_3$ NPs, (c) after H_2O_2-treated $BaTiO_3$ NPs, and (d) dispersed HOA-modified $BaTiO_3$ NPs in toluene. Source: Reprinted from Kailasa and Wu [54].

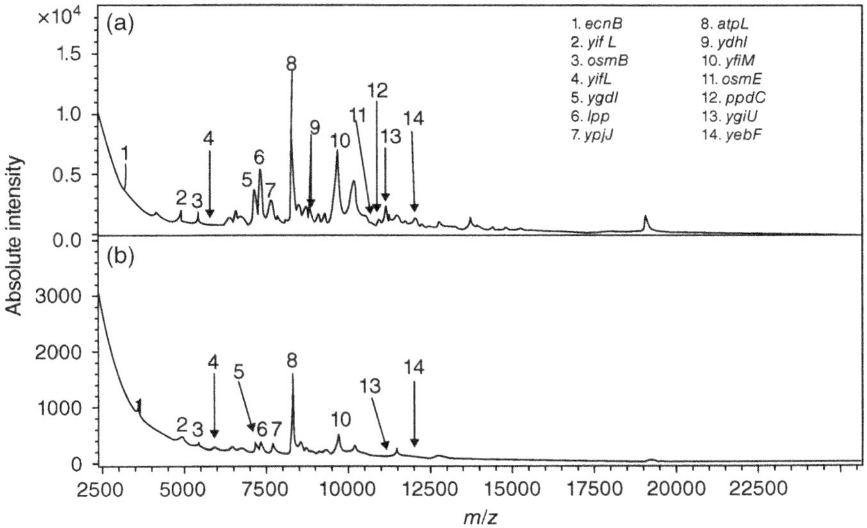

Figure 11.4 MALDI–mass spectra of identified hydrophobic proteins in *E. coli* by using (a) HOA-modified $BaTiO_3$ NP-assisted LLME, (b) without HOA-modified $BaTiO_3$ NPs. Source: Reprinted from Kailasa and Wu [54].

(PA). $BaTiO_3$ NP-assisted LLME combined with MALDI–MS offers a simple method for efficient extraction and determination of hydrophobic biological molecules, also in an aqueous sample of complex bacterial lysis. In the analysis with E. Coli bacteria, 14 large mass peaks with high signal intensity and mass peaks with an *m/z* of 3.4, 7.2, and 10.8 kDa, corresponding to ecnB (P56549) membrane proteins, lpp (P69776), and osmE (P23933), were discovered. As shown in Figure 11.4a, b, the mass peaks at 5.8, 7.0, and 8.892 kDa are categorized as hypothetical membrane proteins yifL (P39166), ygdI (P65292), and acetyl-acyl carrier proteins (acetyl-ACP, P0A6A8). *Haloferax mediterranei* microorganisms inhabit hypersaline environments and express a variety of red pigments on their cell membranes. Manikandan et al. used TiO_2 NPs and NiO NPs applied as oxidative stress on the extremophilic haloarchaea in *H. mediterranei* microorganisms for enhancing carotenoid production [55]. Thin-layer chromatography and further study by MALDI–TOF–MS have been used to isolate TiO_2 nanoparticles-based enhanced carotenoid pigments. Further, they used the Pt–ODT NP-assisted LLME for pigment extraction from the aqueous to nanoparticles suspended toluene medium as shown in Figure 11.5. During the study, it was observed that TiO_2 nanoparticles were more efficient to enhance the carotenoid production than NiO nanoparticles. Pt–ODT NPs are an effective enrichment in the hydrophobic chain of the Pt nanoparticle surface of halobacteria and then characterized by MALDI–TOF–MS (bacterioruberin, 2-isopentenyl-3, 4-dehydrorhodopin, and phosphatidylglycerol phosphate). Abdelhamid et al. synthesized CeO_2 cubic nanoparticles functionalized and stabilized by cetyltrimethylammonium bromide (CTAB) [56] as shown in Figure 11.6a. CeO_2@CTAB nanoparticles were applied in microextraction for highly sensitive

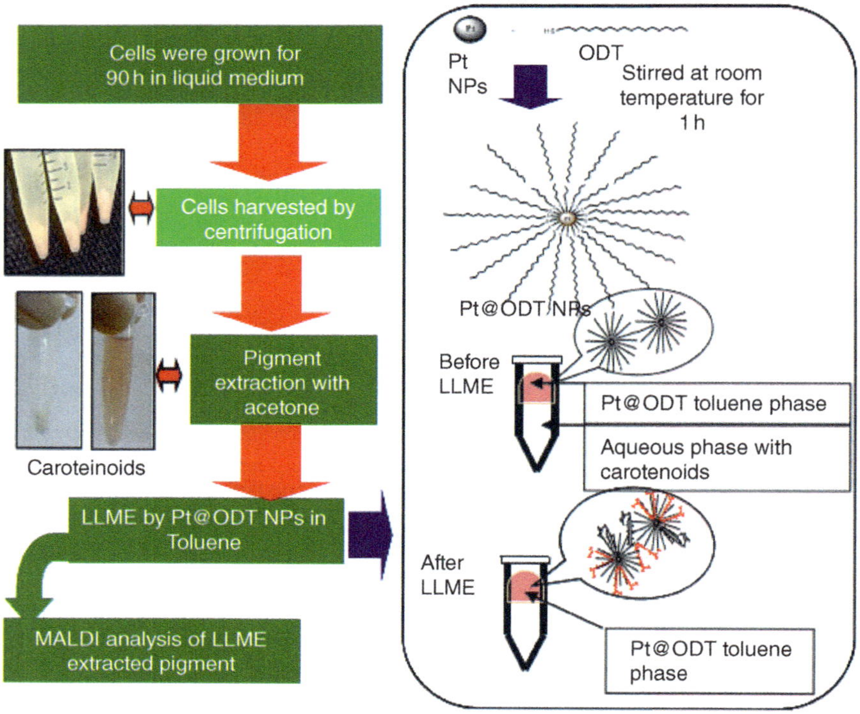

Figure 11.5 Schematic representation showing the steps involved in the concentration of pigments using Pt NPs functionalized with ODT in toluene by miniaturization of Pt-NPs-LLME technique. Source: Reprinted from Manikandan et al. [55].

detection of pathogenic bacteria such as *Pseudomonas aeruginosa* and *Staphylococcus aureus*. CeO_2@CTAB nanoparticles were used for ultrasound-enhanced surfactant-assisted dispersive liquid–liquid microextraction (UESA–DLLME) for an aqueous bacterial cell lysis solution and analyzed by MALDI–TOF–MS as shown in Figure 11.6b. It was explored that CeO_2@CTAB nanoparticles contained hydrophobic interaction and other factors that influence the bacterial protein enrichment on the nanoaffinity probe. CeO_2@CTAB nanoparticle interaction and affinity toward the complex lysate aqueous solution are based on zeta potential (electrostatic interactions), dipolar interaction and hydrogen bonding. This sensitive method was applied for the extraction/detection of pathogenic bacteria from real-world complex samples such as the blood of sheep, mouse and blood plasma samples using UESA–DLLME with CeO_2@CTAB coupled with MALDI–TOF–MS. Chen et al. applied DLLME for macrolide antibiotics such as erythromycin, spiramycin, tilmicosin, and tylosin analysis coupled with colloidal Ag nanoparticles and LDI–MS [57]. Various extraction solvent composition and pH were observed as factors in extraction of macrolide antibiotics. Dichloromethane solvent at pH 10.0 was found to be the optimum condition to extract macrolide antibiotics, and LOD was 2.0, 3.0, 3.0, and 2.0 nM for erythromycin, spiramycin, tilmicosin, and tylosin,

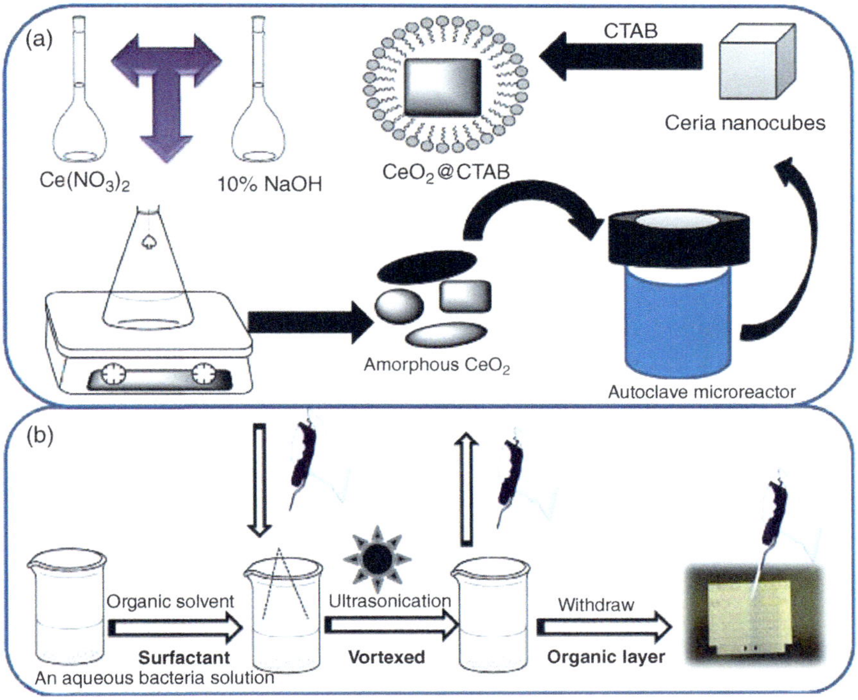

Figure 11.6 (a) Schematic of preparation of CeO_2 cubic nanoparticles, (b) schematic of ultrasound-enhanced surfactant-assisted dispersive liquid–liquid microextraction (UESA–DLLME). Source: Reprinted from Abdelhamid et al. [56].

respectively. Miniaturization of functionalized nanoparticles coupled with LLME provides great affinity and selective enrichment of biomolecule analytes even in the presence of complex matrix and hypersaline environment. After the enrichment on the surface of nanoparticles, a very small quantity of samples can be directly analyzed by the MALDI–TOF–MS technique within a short time period.

11.5 Miniaturization of Nanomaterial Affinity Probes for Biomolecules Liquid Extraction

Miniaturization of nanomaterials leads to high surface-area-to-volume ratios with high dispersion. High dispersion of nanoparticles can extract and enrich on their surface trace-level target biomolecules; hence attached analytes reach a high concentration in minimum volume. MALDI–TOF–MS is a versatile instrument for rapidly analyzing and characterizing biological molecules such as peptides and proteins. Traditional MALDI techniques in which organic compounds such as CHCA, SA, and DHB are used as a matrix for small-molecule (<700 Da) analysis generate several interference peaks. The MALDI–TOF–MS also has low immunity, weak ionization efficiency, and other ion interference. These limitations can be resolved by the enrichment or preconcentration of ultratrace peptides and proteins from highly complex mixtures by miniaturization of nanomaterials. Application of organic matrix (CHCA, SA, DHB etc.) in MALDI TOF MS have limitations during the co-crystallization process. Inhomogeneous mixture results in "hot spots" while nanomaterial facilitates in LDI excellent shot to shot analytes peaks reproducibility during the analytes analysis by MALDI TOF MS. In addition, a matrix that transfers energy from the laser to analytes, nanomaterials can also be used as adsorbents for clean-up and enhancement of analytes in complex samples that can be their detection limit in the LDI–MS analysis [58]. LDI-assisted nanomaterials typically have signal strength higher than that of traditional MALDI after enrichment, whereas in ionization, emulation between the analytes and the matrix is reduced [59]. Throughout this segment, nanoparticles from carbon, semiconductor, and metal oxides, as affinity and enrichment probes, are examined for direct analysis of peptides and proteins by MALDI–TOF–MS.

11.5.1 Metal Nanoparticle–Based Miniaturization Liquid Extraction Probes

During the past two decades, many research works have thoroughly investigated the application of metal nanomaterials as enhancement and affinity probes and matrices for the MALDI technique, and numerous studies have been published. A nanoparticle affinity probe functional to a mixed monolayer of cationic (11-trimethylammonium-undecanethiol) and anionic moieties (to-thiol carboxyl acids) was synthesized by Vanderpuije et al. [60]. Au nanoparticle affinity probes were used for the analysis of peptides such as glucagon, T-7 peptide, angiotensin I, ACTH (1–13), kinetensin, and bradykinin extraction and more complex trypsin-digested myoglobin protein analysis by MALDI–TOF–MS. Au nanoparticle affinity probes, based on the charge on nanoparticles, were explored based on peptide pI values and solution pH [61]. Au nanoparticles significantly coated with proteins or carbohydrates can be used in many research studies. Au nanoparticles were coated and bound with bovine serum albumin (BSA) proteins and digested by trypsin, and the resulting peptide fragments were analyzed by MALDI–TOF–MS after the addition of an isotope-labeled internal standard (IS) [61]. In one study, Au nanoparticles conjugated via a thiol-ending linker to ovalbumin peptide (OVA 323-339), the Lewis X-antigen trisaccharide, tetramannoside, or a mixture of both carbohydrates were analyzed by MALDI–TOF–MS [62]. Pseudo-molecular ions generated by collision-induced dissociation (CID) were studied in depth with Au ion adducts. Au nanoparticle adducts ionize small-molecular-weight peptides and proteins after digestion of particles and work effectively as a matrix, which may be called GALDI–MS [62, 63]. The Russell group used LDI–MS for peptide and protein analysis with bare Au NPs and 4-aminothiophenol (ATP)-capped Au NPs as the matrix and affinity probe [63, 64]. Some peptides and small proteins, such as bovine insulin, have successfully been identified without fragmentation, including posttranslationally modified peptides (e.g. phosphorylates). Besides, several phosphopeptides (i.e. LKRApYLG-NH 2, LRRApSLG, and LKRApTLG-NH2) were identified and differentiated from structural difficulties between phosphorylates. Moreover, the signal intensities of analytes were enhanced considerably using ATP–Au NPs [64]. The amount of [M+H]+ signals detected using 4-ATP capsule Au NPs in trifluoroacetic acid (TFA) compared with Au NPs capsule citrate is, evidently, significantly increased by 0.1 percent due to the 4-ATP monolayer in the Au NPs (surface pK_a ~6.9) at pH 2, which serves as a rich proton source and generates analytic ions with low inner energies (i.e. little or no fragmented). Liquid extraction of bacterial proteins by gold nanoparticles as enrichment and affinity probes is a very effective and direct applicable technique using MALDI–TOF–MS. Kun-Chan Ho functionalized Au nanoparticle surface by human immunoglobulin (IgG) and explored the selective interaction of IgG and the bacterial cell wall [65]. This miniaturization of the Au–IgG nanoaffinity probe was successfully applied for *Staphylococcus saprophyticus* and *S. aureus* pathogens, and cell wall proteins and pathogens were effectively characterized by MALDI–TOF–MS analysis. Au–IgG approach effectively identified the pathogen *S. saprophyticus* from a real-world urine sample with LOD at 3×10^7 cfu/ml [65]. Chen et al. applied carbohydrate-encapsulated Au NPs for target protein enrichment as an effective affinity and separation sample to achieve MALDI–TOF–MS epitope mapping [66]. Carbohydrate-functionalized Au NPs were used for extracting a mixture of proteins from *P. aeruginosa* lectin I, enolase, alcohol dehydrogenase, and myoglobin. Au NP miniaturized technique is also capable of isolating and enriching large-molecular-weight proteins by bioconjugation and have the potential to identify unknown target proteins and discover novel receptors by MALDI–TOF–MS in biological complex samples [66]. Nile Red–adsorbed Au nanoparticles were used as selective probes and matrices for the analysis of aminothiols such as glutathione (GSH), Cys, and HCys through SALDI–MS [67]. Nile Red Au nanoparticle matrices provide several advantages such as ease of preparation, sensitivity, and selectivity for small-molecular-weight aminothiols. Gold nanoparticles coated with thin films of titania [68] were used to selectively bind traces of phosphopeptides from complex samples with lower LOD <50 nM. This miniaturization approach enhanced the reproducibility of phosphopeptide analysis by LDI–MS because Au nanoparticles shift the wavelength absorption band in the SPR spectrum. Functional groups on Au nanoparticle surface have high affinity and enrichment properties to trap the trace-level analyte moieties from the complex sample and enhances the signal intensity of MALDI–MS spectra. Recently, Kailasa and Wu functionalized Au nanoparticle surfaces by dopamine dithiocarbamate (DDTC) and used as the matrix and affinity probe for rapid enrichment of amino acid (GSH), drugs (desipramine and enrofloxacin), and peptides (valinomycin and gramicidin D) [69]. Miniaturized Au–DDTC nanoparticle probe was also applied for the study of phosphopeptides enrichment from casein proteins (α, β-casein and nonfat) using tryptic digestion in microwave before MALDI–MS analysis. Au–DDTC-functionalized NPs behave as heat-absorbing materials to enhance tryptic digestion as well as a phosphopeptide enrichment affinity probe. CdS:Mn/ZnS QDs were also functionalized with DDTC and were used as a novel matrix for the analysis of drugs such as efavirenz, tobramycin, and aspartame using MALDI–TOF–MS. Functionalized CdS:Mn/ZnS

QDs with dopamine DDTC exhibit rich proton sources, which are highly attractive and promising materials for efficient MALDI–TOF–MS analysis [70]. This miniaturized QD-based approach exhibited great potential for quantification of efavirenz from blood samples of patients with HIV with good linearity. As heat-absorbing materials [69, 70], QDs–DDTC was demonstrated to be a facile, fast, and efficient microwave-assisted proteolysis approach for rapid identification of digested proteins with MALDI–TOF–MS using as affinity probes. Chiang et al. determined Cys, GSH, and HCys using liquid microextraction with unmodified Au nanoparticles (14 nm) and *N*-2-mercaptopropionylglycine (MPG)-modified Au nanoparticles as probes in SALDI–MS [71]. The deciding factors for precise quantification of aminothiols are strong linearity, superior reproducibility and high precise particle concentration, time of reaction, pH, and salt. The simple methodology schematic representation of liquid extraction miniaturization techniques is shown in Figure 11.7. This miniaturized method was used to analyze GSH in cell lysates as a real-world sample and to assess the drug resistance of MCF-7 breast cancer cells by quantifying GSH in cell lysates. Teng et al. synthesized nanoparticles bound by S-Au bonding on the magnetic particle surface [59]. This liquid extraction probe carries negative charges on the magnetic core base with Au nanoparticles and behaves as preconcentrating probes for positively charged proteins (cytochrome c and myoglobin) by electrostatic interactions prior to analysis by MALDI–MS.

Ag nanoparticles has great affinity for the bacterial cell wall, nevertheless, at high Ag nanoparticles concentration nanoparticles dominant on bacterial surface and start to work as an antibacterial probe to rupture the pathogen cell wall. Gopal

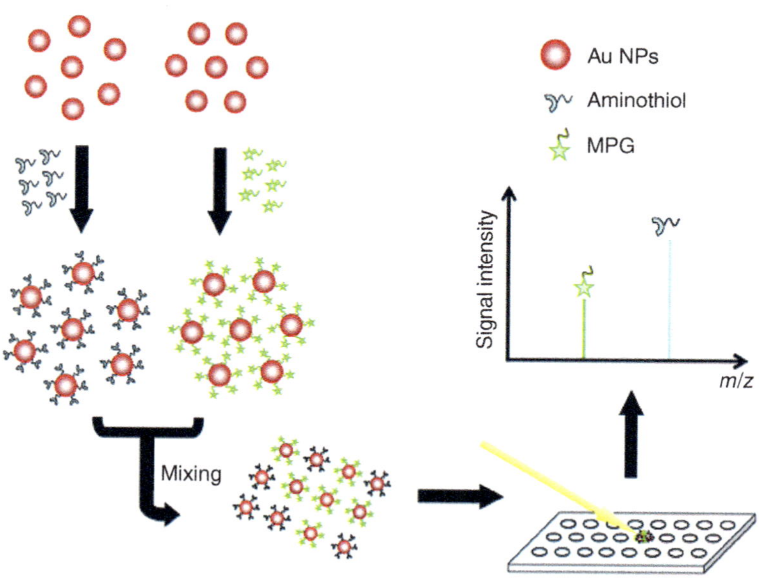

Figure 11.7 Schematic representation of the determination of aminothiols through surface-assisted laser desorption/ionization mass spectrometry using N-2-mercaptopropionylglycine (MPG)–gold nanoparticles (Au NPs) as the internal standard and Au NPs as the matrix. Source: Reprinted from Chiang et al. [71].

et al. applied bifunctional bare Ag nanoparticles on two pathogens *E. coli* and *S. marcescens* and identified the critical concentration of affinity probes (CCAP) at different nanoparticle concentrations by MALDI–TOF–MS. Qualitative mass data was used to get quantitative results. Data was compared to observe CCAP and antibacterial activity points after using MYSTAT software-based Tukey–Kramer multiple comparison tests [72]. Ag nanoparticles' affinity and bactericidal effects involve various mechanisms against microorganisms. Ag nanoparticles and their surface functionalization directly impact bacterial activities after adhering to the surface of the cell wall. Ag nanoparticles damage the cell wall and inhibit its formation to seize transport activities. One of the mechanisms is oxidative stress caused by the generation of reactive oxygen species (ROS) and free radicals. Free radicals increase porosity of the bacterial cell membrane and lead to cell death. Another proposed mechanism for Ag nanoparticles is based on silver ions generated from the nanoparticle surface, and these ions react with thiol groups of enzymes, proteins, phosphoproteins, and DNA and interrupt bacterial growth activities and lead to cell death [73–75]. Fernanda Monedeiro et al. applied various concentrations of Ag nanoparticles and incubation periods against *E. coli* and salivary *Klebsiella oxytoca* and *Staphylococcus saccharolyticus* bacterial suspensions and obtained biomarker proteins by MALDI–TOF–MS [76]. This approach can be useful in monitoring response to the treatment based on Ag NPs at different nanoparticle concentrations and incubation periods.

The miniaturization techniques used in the study of insulin, ubiquitin, lysozymes, and microwave-digested lysozymes and BSA are based on the use of Pt nanoparticle affinity and enhancement in MALDI–MS signals [77]. In this miniaturized approach, Pt nanoparticles behaved as a heat-absorbing material and the affinity and enrichment probe in the microwave and enhanced digestion of lysozyme and BSA proteins before analysis by MALDI–TOF–MS. Gopal et al. used five different metal and metal-oxide nanoparticles in preincubation-assisted bacterial analysis by MALDI–TOF–MS [78]. Interaction of different nanoparticles with *P. aeruginosa* (Gram +ve) and *S. aureus* (Gram −ve) was studied and used as the affinity and enrichment probe as well as bactericidal effects.

These activities and proteins were analyzed by MALDI–MS. Pt and NiO nanoparticles worked as the affinity and enrichment probe on the bacterial cell wall and enhanced the protein signal intensity in MALDI–MS. TiO_2, ZnO, and Ag nanoparticles were used as bactericidal and centrifuged nanoparticles for proteins and peptides in analysis by MALDI–MS. Bare (surface-unmodified) NPs also exhibit promising surface behavior in bacterial proteins analysis, which is simple, fast, economical, and highly effective in bacterial proteomics using MALDI–MS. Kawasaki et al. synthesized FePtCu nanoparticles and modified with a sulfonated group, which gives a strongly negative surface to trap trace-level positively charged species from a complex mixture, and further analyzed by LDI–MS [79]. This miniaturized sulfonated FePtCu nanoparticle technique was successfully applied to analyze the proteins (5 kDa) with high sensitivity (5 fmol) and used as the surface-assisted laser desorption ionization (SALDI)–MS matrix. After enrichment as an affinity probe, FePtCu nanoparticles were easily separated from the complex

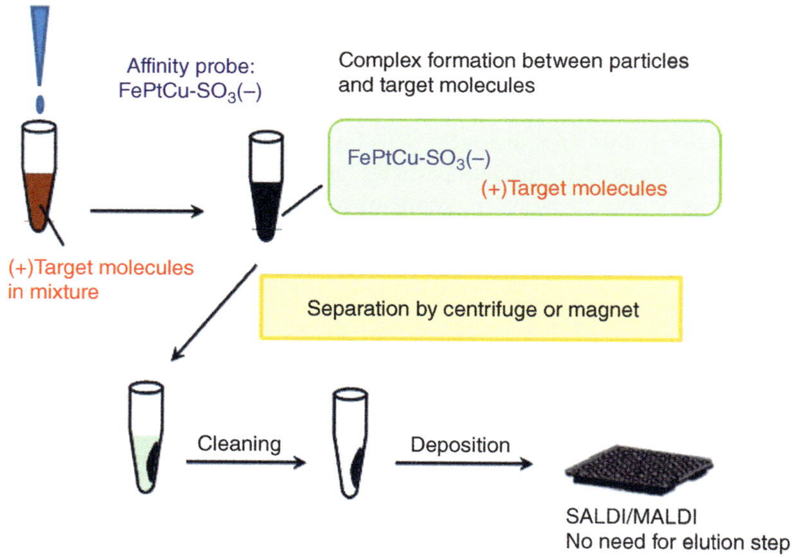

Figure 11.8 Sulfonated FePtCu-SO_3(–) NPs as affinity probe and liquid extraction analysis by LDI–MS. Source: Reprinted from Kawasaki et al. [79].

mixture by an external magnet or centrifugation. The simple extraction technique is shown in Figure 11.8. The surface of FePtCu nanoparticles carries negative charge, and works as an electrostatic enrichment probe for oligopeptide mixtures and lysozyme analysis by SALDI–MS from the complex human blood serum sample. The lysozyme protein mass peaks were observed at *m/z* 14304, 7153, and 4769, which corresponded to FePtCu nanoparticles that acted as a matrix in LDI–MS to produce single-proton adducts and multiple-proton adducts.

The most appealing aspects of the enrichment and interaction affinity probe in MALDI–MS are the multipurpose nanoparticles for bioassays. Ahmed et al. functionalized Pt NPs with human serum immunoglobulin G (h-IgG) and used as a biosensor to detect *Bacillus thuringiensis* and *B. subtilis* [80]. Both bacteria were isolated from the rhizospheric soil and root, which belong to the microorganisms associated with plants. Interestingly, Pt–IgG functionalized nanoparticle-based MALDI–TOF–MS identified many protein peaks from the soil samples and a couple of peaks from the plant root samples. *B. thuringiensis* mass spectrum peaks from the plant sample coupled with Pt–IgG nanoprobes matched well with the standard bacteria mass peaks at m/z 3923, 5019, 5581, 6675, 7185, 9747, and 11 384, which also corresponded to plant-associated bacteria isolates. Pt nanodots were also applied for the analysis of the cell populations of eukaryotic cells the alga *Chlamydomonas reinhardtii* and the yeast *Saccharomyces cerevisiae* [81]. Pt nanodot-based MALDI–MS effectively analyzed and characterized the proteins of the cell population in the range of 3200–640 cells/ml for *C. reinhardtii,* while 40 000–3200 cells/ml for *Saccharomyces cerevisiae.* Enhancement of protein mass signals in the presence of Pt nanodots is dependent on their dynamic nature to carry the absorbed energy to cells before ionization from an applied laser in MALDI–MS. Kawasaki et al. synthesized and characterized Pt nanoflowers and used as a matrix for analysis of different small-molecular-weight peptides and phosphopeptides in SALDI–MS [82]. Pt nanoflower as high sensitivity and affinity probe limit of detection was achieved at lower laser energy and analyte concentration (few femtomoles) without any organic matrix in LDI–MS analysis.

11.5.2 Semiconductor Quantum Dots (QDs)-Based Miniaturization Liquid Extraction Probes in MALDI–TOF Analysis

Semiconductor QDs are nanocrystals or nanomaterials of a new class of fluorescent tiny particles exhibiting unique photophysical and photochemical properties due to surface phenomena. QDs are in a size ranging from 1 to 10 nm and typically their diameter is about 2–10 nm in which quantum effects reflect high affinity and interaction toward peptides and proteins. QDs are composed of a group of atoms in three dimensions termed "artificial atoms" [83] and usually synthesized in the form of binary (CdS, CdSe, CdTe, InAs, and InP QDs) and ternary (CdSe/ZnS, CuInSe) metal alloys [84–86]. QDs are quite different from the bulk particle materials. Different chemical and physical properties are attributed to the size effects and reciprocal value of their radii and bandgap difference. QDs exhibiting discrete energy levels between the electron-filled valence band and the empty conduction band can be precisely modulated and functionalized by different organic and biological moieties [87, 88]. The dimensions and composition of QDs dictate their wavelength and color of emission. The spectrum of emission can be very narrow ranging from 20 to 40 nm at a half maximum intensity [89, 90]. Semiconductor QDs and their functionality provide high affinity and enrichment probes in proteomics, which is explored by mass spectrometry techniques. The presence of suitable organic ligands on the QD surface particles controls and enhances effective trapping and behaves as affinity probes to show unique physicochemical properties in proteomics. Kailasa et al. synthesized semiconductor ZnS NPs and functionalized by different ligand groups such as sodium citrate, 3-mercaptopropionic acid (3-MPA), cysteamine, and 2-mercaptoethane sulfonate [91]. All these functionalized ZnS NPs were used as MALDI–TOF–MS matrix, affinity and enrichment probes. Among the functionalized ZnS NPs, ZnS-3-MPA showed optimum efficiency against CDs, ubiquitin, and insulin in SALDI–TOF–MS for high sensitivity detection. The detection limits were found to be 20–55, 91, and 85 nM for CDs, ubiquitin, and insulin, respectively. The practicability of this method was demonstrated by the detection of ubiquitin-like proteins in oyster mushroom and also in the analysis of analytes in biological samples such as human urine and plasma. In one report, ZnS-N_3 QDs were functionalized and used as high-affinity probes for milk proteins and ubiquitin-like proteins from mushroom in real sample analysis directly by MALDI–TOF–MS [92]. Similarly, CdSe QDs were functionalized by mercaptoundecanoic acid (MUA) to serve as a matrix for selective ionization of proteins using LDI–MS (QD-LDI–MS) in high-resolution, fast analysis of amino acids and peptides [93]. The CdSe QDs were used for preconcentration studies and enrichment of peptides from lysozyme obtained from chicken-egg white. The CdSe QDs demonstrated the potential to absorb radiation from microwave and capture peptides from lysozyme protein digestion in microwave. The sensitivity of the system was substantially improved and the LOQ is <100 p.m. with an RSD of 10%. Abdelhamid et al. successfully modified the surface of CdTe QDs by Cys and applied for small-molecule analysis such as α-, β-, and γ-cyclodextrin gramicidin D, perylene, pyrene, and triphenylphosphine [94]. Cys@CdTe QDs showed synergistic properties because they absorbed the wavelength that matched that of N_2 laser and works as MALDI–TOF–MS matrix. Before ESI–MS detection, CdS QDs were used as acceleration and enrichment probes for microwave-assisted tryptic protein digestion [95]. Mainly, these QD-based approaches with MALDI–MS are a good bioanalytical tool for ultrafast analysis of different biomolecules in proteome research. Cd^{2+} ion-doped multiwalled CNTs modified with CdS NPs were synthesized in a chemical reduction process, used in AP–MALDI and MALDI–MS as samples for the analysis of peptides and proteins [96].

Similarly, in SALDI–MS, MPA-capped ZnSe QDs were used as the matrix and affinity probes for direct peptide and protein analysis [97]. The signals were improved due to ionizing efficiency of target analytes with increased signal intensities by adsorbing positively charged peptidase or proteins to the surfaces of negatively charged ZnSe QDs by electrostatic interactions. Specific biomolecules including Leu Enk, Met-enk, HW6, peptides such as angiotensin II, cytochrome c, myoglobin, and lysozyme were successfully examined. Fluorescent semiconductor QDs (Cys-capped ZnSe) were applied as affinity probes to enrich target peptides and proteins from complex samples before the MALDI–MS study [98]. This method is simple and straightforward in identifying peptides, proteins, and digested proteins without complex sample separation and washing processes. This approach also examined IgG and protein G interactions. Signals of adducts $[IgG+2H]^{2+}$, $[IgG+3H]^{3+}$, [protein G+H]+, $[protein\ G+2H]^{2+}$, $[IgG+protein\ G+2H]^{2+}$, and $[IgG+protein\ G+3H]^{3+}$ were found at *m/z* 74 885, 49 984, 26 023, 13 019, 86 585, and 58 297, respectively, in MALDI–MS. Rao et al. synthesized and miniaturized nitrogen-doped graphene QDs (N-GQDs), which is a facile and fast method to dope other atoms into GQDs [99]. The prepared N-GQDs were used successfully as a novel matrix in MALDI–TOF–MS for the analysis of perfluoroalkyl sulfonate (PFOS) and amino acids in the negative mode. N-GQDs exhibit high sensitivity, with very low signal background interference, and high efficiency, which is due to π-conjugation in graphene to enhance desorption/ionization in MALDI–TOF–MS. A similar study was done in which N and S atoms were doped with carbon dots (N, S-CDs) and successfully applied as a matrix for the analysis of various small-molecular-weight amino acids [100]. Miniaturization of carbon as CDs were further used to quantitatively determine endogenous glucose and classify breast cancer cell lines using CDs as a direct matrix for MALDI–TOF–MS. Nevertheless, QDs behave as efficient energy mediators for the UV-laser energy absorption and generate an appropriate amount of heat in their respective environment. As this condition reaches the QD surface, rapid desorption/ionization of peptides and proteins with narrow ion velocity distribution happens in MALDI– TOF–MS. High affinity of QDs or nanoparticle surfaces depends on covalent, electrostatic, and hydrophobic or hydrophilic probes, which are largely exploited to analyze biological molecules from real-world complex samples [101, 102].

11.5.3 Metal-Oxide Nanomaterial–Based Miniaturization Liquid Microextraction for MALDI–TOF–MS

The miniaturization of metal–oxide nanoparticle surface offers unique chemical and physical properties. As the size of particles decreases, the surface-to-volume ratio increases. During this miniaturization process, interface atoms undergo strain or stress and structural perturbations happen, which alter the particles' chemical, conducting, and electronic properties. In proteomics, liquid extraction based on nanoparticles and MALDI–TOF–MS have great possibilities for isolation and detection of peptides, phosphopeptides, proteins, and bacterial proteins from crude biological complex samples. Metal-oxide nanoparticles open a new era for the scientific community to exploit them as affinity and concentrating probes to isolate trace levels of biomolecules before characterization using MALDI–TOF–MS techniques. Bare or surface-modified metal-oxide nanoparticles by appropriate organic or biological moieties not only alter the surface behavior but also enhance efficiency of the approaches to isolate target biological peptides or proteins.

11.5.3.1 Phosphopeptides Enrichment by Liquid Microextraction Analysis by MALDI–TOF–MS

Protein phosphorylation and dephosphorization is the most important post-translational modification and ubiquitous activity in eukaryotic and prokaryotic cells. Maximum cellular processes are highly regulated and controlled by two classes of enzymes kinases and phosphatases. Kinases modify other proteins by adding phosphate groups that carry signal transduction to induce specific cellular activities known as phosphorylation. Phosphorylation plays a vital role in controlling many cellular processes such as cell division, cell cycle control, and signal transduction to act as a switch to turn on and off a protein activity [103]. In phosphoproteomics, four different types of phosphoamino acids are present: O-phosphomonoesters (most abundant), N-phosphoamidates, acylphosphates, and S-phosphates (S-phosphothioesters). Phospho groups are mostly bonded with serine, threonine, and tyrosine amino acids [103]. Direct analysis of phosphopeptides by MALDI–TOF–MS is not straightforward due to highly dynamic, poor ionization, and low abundance nature in a real-world complex sample. Bare and functionalized metal-oxide nanoparticles are well known for enrichment of phosphopeptides due to electrostatic interactions and high affinity of phospho groups for nanoparticles. Enrichment of metal-oxide nanoparticles can be directly analyzed by MALDI–TOF–MS after the application of an appropriate matrix for phosphoproteome characterization. Many oxide nanoparticles or functionalized oxide nanoparticles were used for phosphopeptide enrichment and analysis by various mass spectrometry techniques [104, 105]. Here, we discuss the most prominent and important miniaturization techniques for phosphoproteins digestion on nanoparticles and liquid-based phosphopeptides extraction using nanoparticles as a high enrichment affinity probe before analysis by MALDI–MS techniques. Zhou et al. synthesized zirconium oxide (ZrO_2) nanoparticles and applied in liquid extraction of specific phosphopeptide enrichment probes from a complex peptide mixture [106]. In this miniaturization process, ZrO_2 NPs showed superior selectivity for phosphorylated peptides enrichment even at the femtomole level compared to conventional immobilized metal affinity chromatography. This liquid extraction affinity probe approach detected 248 phosphorylation sites in the mouse liver containing 140 phosphopeptides before MALDI–TOF–MS. ZrO_2 aerogel was synthesized by green sol-gel process and behaved as high-affinity probes for selective enrichment of phosphopeptides [107]. This miniaturized technique was applied for liquid extraction of phosphopeptides from the digestion of β-casein (4 pmol) in the presence of digested BSA with high amounts of nonphosphopeptides. This approach was also applied for phosphopeptide enrichment from a real-world milk sample (100 ng), and 17 phosphopeptides were identified and characterized by MALDI–TOF–MS. Nelson and coworkers synthesized mesoporous ZrO_2, TiO_2, and HfO_2 nanoparticles and applied for phosphopeptide enrichment [108]. In this liquid microextraction of phosphopeptides with these three kinds of nanoparticles, mesoporous ZrO_2 and HfO_2 NPs were found to be superior over TiO_2 NPs for phosphopeptide enrichment due to large surface area. This approach was highly specified, and in single enrichment analysis, the phosphopeptide digested complex solution with 7% of α-casein identified 21 out of 22 phosphorylation sites for α-casein. NiO nanoparticles were used in a simple and highly sensitive liquid microextraction for phosphopeptide enrichment from trypsin-predigested phosphoproteins in a microwave oven (210 W) [109]. After enrichment in the microwave for 30 seconds, these techniques were combined with centrifugation on-particles ionization/enrichment (CF-OPIE) of multiple phosphopeptides by MALDI–TOF–MS. NiO nanoparticles possess weak magnetic properties, and in the microwave environment, nanoparticles carry very active and positively charged surfaces and enrich multiple phosphopeptides at low pH. Phosphopeptide enrichment by TiO_2 NPs in a microwave oven as well as on-particles/ionization during MALDI TOF MS are the key application of this methodology. The LOD for this miniaturization technique for α-casein and β-casein was found to be 2.0 nM. This approach was applied successfully for real-world complex samples such as human blood serum, nonfat milk, and egg white to explore their phosphopeptides. TiO_2 nanoparticles were used as a multifunctional nanoprobe for tryptic digestion of phosphoproteins (α-casein, β-casein, and milk) in the microwave (within 45 s) [110]. TiO_2 NPs as multifunctional nanoprobes such as desalting, accelerating, and affinity probes can effectively enrich and accelerate the digestion reactions of phosphoproteins and real-world complex samples in the microwave and phosphopeptide characterization by MALDI–TOF–MS.

In another study, $CoFe_2O_4$–ZnO magnetic nanoparticles were used for rapid microwave-assisted tryptic digestion of phosphoproteins, and this miniaturization approach efficiently enriched phosphopeptides from the complex sample [111]. $CoFe_2O_4$–ZnO magnetic nanoparticles absorb microwave radiation, which enhanced plethora of heat in milieu of nanoparticle surface. Tryptic digestion of phosphoproteins such as β-casein, ovalbumin, and real-world samples like human serum and egg white with $CoFe_2O_4$–ZnO nanoparticles was enhanced dramatically, rapidly, and efficiently. Phosphopeptides enriched on the surface of $CoFe_2O_4$–ZnO nanoparticles were further analyzed by MALDI–TOF–MS. The limits of detection for β-casein and ovalbumin by $CoFe_2O_4$–ZnO nanoparticles were 0.2 and 20 fmol, respectively, within 30 seconds. Kailasa and Wu used $BaTiO_3$ nanoparticles as a miniaturized novel substrate for phosphopeptide enrichment in the microwave tryptic digestion of α-casein and nonfat milk [112]. After using $BaTiO_3$ nanoparticles, microwave-based tryptic phosphoprotein digestion was achieved within 50 seconds and phosphopeptides were selectively enriched on $BaTiO_3$ particles and characterized by MALDI–TOF–MS. This method successfully identified 24 phosphopeptides from α-casein and 21 phosphopeptides from nonfat milk. Human saliva contains more information about endogenous peptides including phosphopeptides. Saliva is a complex mixture of proteins and peptides at trace level those released by diseased tissues. These proteins and peptides have gained considerable interest as a disease at the initial stage by biomarker discovery. Chen and Chen synthesized Fe_3O_4@ZnO MNPs and used as high-affinity probes to selectively enrich phosphoproteins from complex saliva samples in the microwave [113]. Later on, Fe_3O_4@ZnO-trapped protein mixture was applied as tryptic digestion in the microwave and digested phosphopeptide enrichment was successfully elucidated by MALDI–TOF–MS. Kupcik et al. synthesized 1D TiO_2 nanotubes, and TiO_2 nanotubes decorated by magnetic Fe_3O_4 nanoparticles were successfully applied for phosphopeptide enrichment from the complex digested mixture of BSA and α-casein [114]. This miniaturized approach was applied for phosphopeptide enrichment from the tryptic-digested Jurkat T-cell lysate complex mixture. TiO_2 nanotubes and TiO_2 NTs@Fe_3O_4 NPs were 28.7 and 25.3 % more efficient for enrichment of phosphopeptides compared to standard TiO_2 microspheres. Several miniaturization techniques were applied to study thoroughly phosphopeptide enrichment affinities and characterization via MALDI–TOF–MS [104, 115].

11.5.3.2 Miniaturization of Metal-Oxide Nanoparticles for Bacterial Proteins Liquid Microextraction Analysis by MALDI–TOF–MS

Miniaturization of nanoparticle surface milieu toward bacterial cell wall proteins to achieve liquid microextraction and antibacterial activity. Interaction of nanoparticles with bacterial cell wall is based on electrostatic attraction, van der Waals forces, hydrophobic interactions, and ligand–receptor activities. Using these interactions, nanoparticles may present on the bacterial cell wall surface, but at a critical concentration, nanoparticles cross the bacterial cell wall membrane and cluster along the metabolic pathway formed, which influences bacterial activities. Nanoparticles behave as an affinity probe toward a basic component of a bacterial cell such as enzymes, lysosomes, ribosomes, and DNA. The presence of nanoparticles inside the bacterial cell leads to oxidative stress, deactivation of proteins, inhibition of enzyme activity, drastic changes in cell membrane permeability and gene expression, thereby bacterial cell halt [116–118]. Using this nanoparticle-based miniaturization, current research has proposed mechanisms against microorganisms such as oxidative stress [117], non-oxidative [119], and metal ion release [120].

ROS carry strong positive redox potential, which depends on the type of nanoparticles and their critical concentration. Nanoparticles application may annihilate bacterial cells by four different ROS mechanisms such as generation of superoxide radicals (O_2^-), singlet oxygen (O_2), hydroxyl radicals (.OH), and hydrogen peroxide (H_2O_2). ROS production during the application of oxide nanoparticles is due to nanoparticle crystal defect sites, restructuring the nanoparticles and oxygen vacancies inside the crystal. Under normal conditions, bacterial cells balance the production and neutralization of ROS. However, in the presence of oxide nanoparticles, excessive ROS production happens, and redox balance by the oxidation of bacterial cells, which generates excessive oxidative stress and damages bacterial individual components, leads to cell death [117, 118, 121]. Some metal-oxide nanoparticles release metal ions after absorbing through the bacterial cell membrane, and these metal ions make the chelate bonds between the functional group of proteins (–SH, –NH, and –COOH). As metal ions chelate with proteins, bacterial enzyme activities come to halt and the normal physiological process affects the cell structure. Nevertheless, metal ions have less impact on the pH inside the lipid vesicles during antibacterial activities of metal-oxide nanoparticles and exhibit weak antibacterial effects. Therefore, metal-oxide nanoparticles dissolve metal ions and do not play any significant role as the main antibacterial mechanism [122]. Bare metal-oxide nanoparticles or surface-functionalized miniaturized nanoparticles enrich the proteins, which can be directly analyzed and characterized by MALDI–TOF–MS. Metal-oxide nanoparticles can be segregated by using a strong external magnet or centrifugation after enrichment of bacterial proteins and subsequent nanoparticles washing process. Ho et al. synthesized Fe_3O_4 magnetic nanoparticles and covalently immobilized human immunoglobulin (IgG) on nanoparticle surface [123]. IgG-functionalized Fe_3O_4 nanoparticles work as affinity and extracting probes to selectively concentrate bacterial cells such as *Staphylococcus aureus* (*S. aureus*) and *Staphylococcus saprophyticus* (*S. saprophyticus*) from the complex sample solutions. Nanoparticles were separated by a strong external magnet and directly analyzed by MALDI–MS. The lowest detectable concentration of bacteria by this miniaturization approach was 3×10^5 cfu/ml. The lowest detection limit of *S. saprophyticus* bacteria in the spiked urine sample was calculated to be 3×10^7 cfu/ml by MALDI–MS. Lin et al. [124] used vancomycin-modified magnetic Fe_3O_4 nanoparticles as affinity probes for the bacterial cell wall as shown in Figure 11.9. Vanc-Fe_3O_4 magnetic nanoparticles were observed to selectively bind toward the surface walls of Gram-positive bacteria (*S. aureus and S. saprophyticus*), and the mixture was separated by a strong external magnet and directly characterized by MALDI–MS. The antibiotic vancomycin, the most potent drug, has a specificity and interaction for Gram-positive bacteria by the D-Alanine (Ala) (D-Ala) linkage. This approach was applied for a real-world complex urine sample, which was spiked with two different Gram-positive bacteria *S. aureus* and *S. saprophyticus*, and the limits of detection that could be detected by MALDI–MS were 7.4×10^4 cfu/ml (3.0 ml) and 7.8×10^4 cfu/ml (3.0 ml), respectively. In another

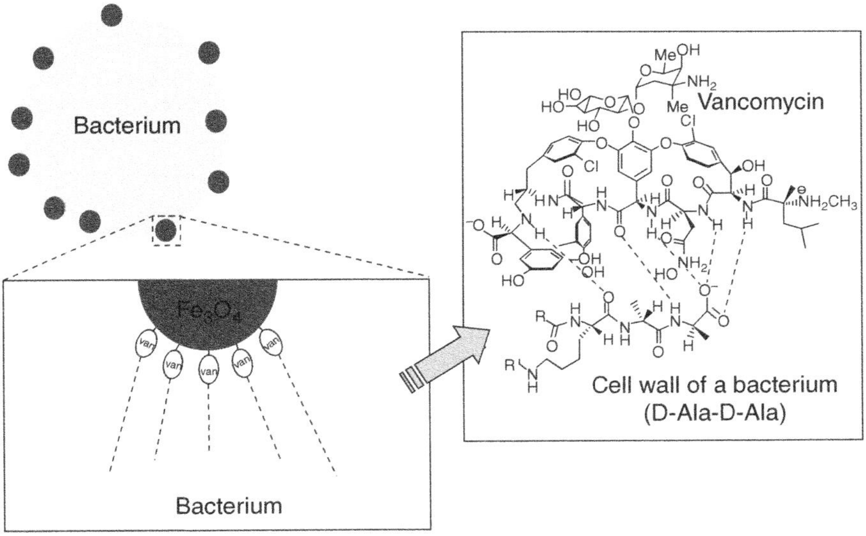

Figure 11.9 Cartoon illustrations of the proposed method for anchoring vancomycin-immobilized magnetic nanoparticles onto the surface of a Gram-positive bacterial cell and the binding of vancomycin to the terminal of D-Ala-D-Ala units of the peptides on the cell wall of a Gram-positive bacterium. Source: Reprinted from Lin et al. [124].

study, vancomycin-modified magnetic nanoparticles (15–20 nm) combined with membrane filtration were used to concentrate Gram-positive bacteria from tap and reservoir water samples [125]. Concentrated bacteria after applying vancomycin-modified magnetic nanoparticles were rapidly characterized by whole-cell MALDI–MS. The capture efficiency for *Bacillus cereus* (*B. cereus*), *Enterococcus faecium* (*E. faecium*), and *S. aureus* in water was 26.7–33.3% within approximately 2 hours. Recently, we synthesized Fe_3O_4 magnetic nanoparticles and further modified with 3-aminopropyltriethoxysilane (APTES). Further, using conventional carbodiimide chemistry, APTES-mag-NPs were functionalized with β-lactam antibiotic drugs such as amoxicillin. This approach was applied for *S. aureus* (Gram +) and *E. coli* (Gram −) bacteria to analyze higher molecular weight penicillin-binding proteins (PNBs) (20–55 kDa) by MALDI–MS [126]. This miniaturization techniques and liquid microextraction PNB method are shown systematically in Figure 11.10. β-Lactam-functionalized APTES-mag-NPs were also applied for bacteria enrichment, and a strong external magnet was used to separate the bacteria. High-molecular-weight femAB (49570.4 Da) operon-based proteins were successfully enriched on the amoxicillin-functionalized-mag-NPs and identified by MALDI–MS. Many high-molecular-weight PNBs of *S. aureus* and *E. coli* containing 10^4 and 10^3 cfu/ml were detected by MALDI–MS, respectively. Heat stress on bacteria is also a new miniaturization technique to characterize microorganisms in the presence of oxide nanoparticles. NiO nanoparticles and *E. coli* suspension were applied at different temperatures to generate heat stress, and effects were analyzed by MALDI–MS [127]. Heat stress on *E. coli* (10^7 cfu/ml) in

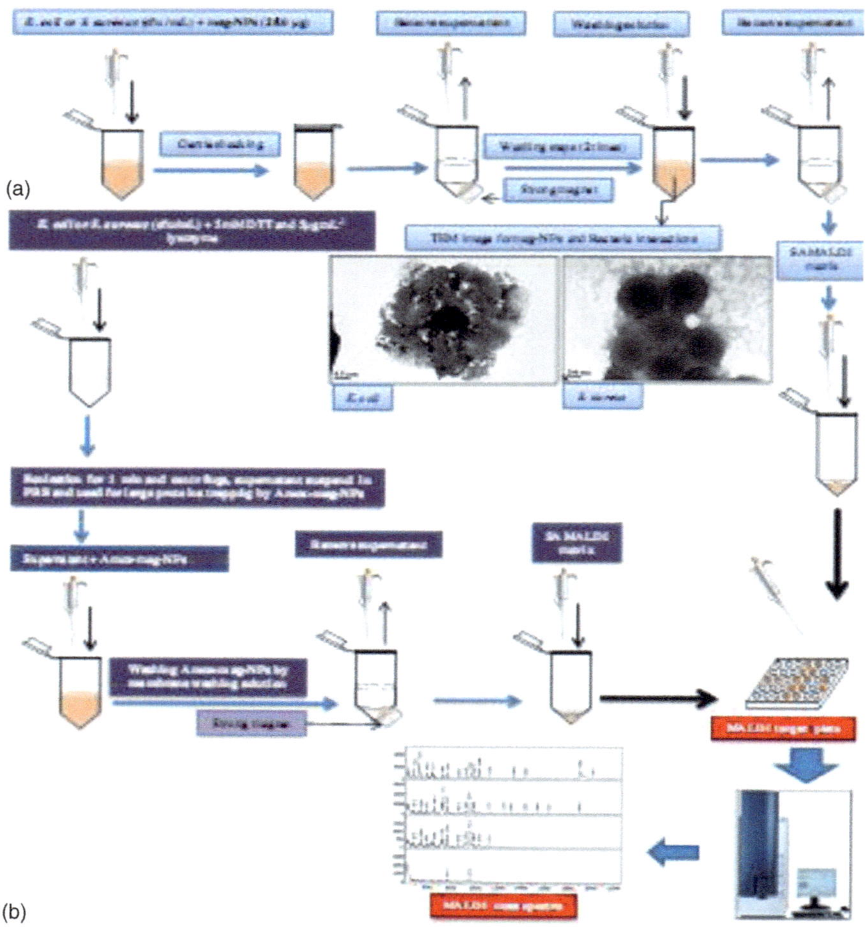

Figure 11.10 Schematic showing workflow followed in this chapter: (a) mag-NPs and bacterial interaction, (b) penicillin-binding proteins enrichment by Amox-mag-NPs. Source: Hasan et al. [126].

the presence of NiO nanoparticles generate high-molecular-weight chaperonin proteins at temperatures from 40 to 80 °C. Effects of nanoparticles on *E. coli* at various temperatures were checked by bacterial viability and MALDI–MS. For bacterial detection under harsh conditions, this technique is sensitive and efficient with MALDI–MS. Gedda et al. green synthesized CaO nanoplates from shrimp shells and applied them as they possess effective antibacterial activities against *E. coli* and *S. aureus* bacteria and further analyzed by MALDI–MS [128]. It was found that as the concentration of CaO nanoplates from 5, 10, 25, and 50 μg/ml increased, bacterial viability drastically decreased for both bacteria, and the signal intensity of bacterial proteins suppressed or disappeared in MALDI–MS. These fierce activities of CaO nanoplates toward bacterial cells reflect high toxicity effects after generating a high amount of ROS. This research data shows that at high concentrations CaO nanoplates show bifunctional activities such as antibacterial effects and depletion in MALDI–TOF proteins signal intensity. Gopal et al. synthesized ZnO nanoparticles and used as affinity probes for signal enhancement in MALDI–MS for *E. coli* bacteria analysis at a lower concentration of nanoparticles (10, 5, and 1 g/l ZnO NPs) [129]. At a high concentration of ZnO (more than 50 g/l), nanoparticles caused MALDI–MS signal depletion and severe bactericidal effects against *E. coli*. A comparative study between metal (Pt and Ag NPs) and metal-oxide (ZnO, NiO, and TiO_2) nanoparticles was done for different bacteria [78]. Pt and NiO nanoparticles were used as affinity and enrichment probes for bacterial proteins signal intensity enhancement in MALDI–MS. TiO_2, ZnO, and Ag nanoparticles exhibited bactericidal effects and depleted signal intensity during MALDI–MS analysis. Anatase form of TiO_2 nanoparticles exhibits bactericidal activities and affinity for the bacteria and proteins. Gopal et al. synthesized anatase TiO_2 nanoparticles and applied for real-world sample analysis [130] and near-UV and laser effects on *S. aureus* bacteria [131]. This miniaturized approach at a low concentration of TiO_2 performed very effectively to obtain a more informative MALDI–MS spectrum for *S. aureus*. In the second study, anatase TiO_2 showed anti-staphylococcal activity under the laser and near UV-irradiation. Under laser irradiation (one order reduction in bacterial counts) and near-UV irradiation (five order inhibition), anatase TiO_2 nanoparticles reflect the radiation effects on their surface to generate ROS for biocidal activities. TiO_2 nanoparticles enhance the signal in MALDI–MS compared to the control *S. aureus* (10^7 cfu/ml) to make a simple and easy bacterial characterization. Abdelhamid and Wu proposed graphene-magnetic@chitosan (GMCS) nanosheets synthesis and demonstrated multifunctional activities to detect pathogenic bacteria (*P. aeruginosa* and *S. aureus*) by MALDI–MS [132]. This miniaturization technique was used for direct detection of low concentrations of *P. aeruginosa* (5.0×10^2 cfu/ml) and *S. aureus* (4.0×10^2 cfu/ml) in mouse blood samples after spiking with the bacteria. GMCS nanosheets aggregated in the sample vial after attachment to the surface of the bacteria wall and GMCS were easily separated by a strong external magnet. Separated samples were mixed with sinapinic acid matrix solution (50.0 mM in aqueous acetonitrile (50 : 50)) and 2.0 μl sample mixture directly applied on MALDI–MS target plate for bacterial characterization.

11.5.3.3 Miniaturization Nanoarray-Based Biochips for Biomolecule Analysis by MALDI–MS

Functionalized nanoarrays or nanoparticles possess unique surface properties that can be used on a chip and miniaturized for further applications of liquid extraction of biological analytes by MALDI–MS. Biological analytes bind to a nanosized pattern of materials, which are referred to as biochips, and improve the analysis and sensitive detection by mass spectrometry. Under the on-line protocol, biochips selectively enrich analytes from complex biological samples, and after the washing process, biochips can be used directly in MALDI–TOF–MS to elucidate the mass spectrum of the attached biological analytes. The selectivity, affinity, and enrichment for biological molecules depend on different metal or nonmetal nanoarrays and patterns. In 1994, for the first time, Zuluzec introduced the concept of lab-on-a-chip for MALDI–MS analysis. Zuluzec used positively charged modified nylon-66 membranes to preconcentrate the adsorbed peptides and proteins [133]. Shen et al. used *N*-nitrilotriacetic acid (NTA) as an SAM on Au surface for target enrichment and affinity array to biomolecules such as phosphopeptides, peptides, and oligohistidine-tagged proteins by MALDI–TOF–MS [134]. Blacken et al. fabricated a steel target plate by three different hydrophobic metal oxides such as zirconium, titanium, and hafnium by landing of gas-phase ions [135]. Zirconium oxide steel plate showed high affinity for phosphopeptides than titanium and hafnium oxides. Enriched phosphopeptides on the oxide chip surface was characterized by MALDI–MS. More recently, Urban et al. have published a review on the current lab-on-a-chip domain associated with functionalized MALDI–MS targets [136]. Judy et al. recently prepared heat-treated titanium chips at 600, 700, 800, and 900 °C, and examined their different surface affinity for pathogenic bacteria such as *S. aureus* and *P. aeruginosa* [137]. These bacterial titanium chips were used to capture the bacteria after immersion into a bacterial suspension. After sensing and enrichment of bacteria on the chips, these chips were directly applied as the target plate in MALDI for sensitive and rapid detection of bacteria. In their study, titanium bacterial chips heat-treated at 800 and 900 °C were found to have high affinity and effectively sensing bacteria. The limits of detection of 800 and 900 °C biochips were 10^3 cfu/ml for both *P. aeruginosa* and *S. aureus* checked by MALDI–MS signals and plate counting methods. In another study, two different titanium metal pickling protocols were applied on titanium chips as standard protocol (40 g/l HF and 400 g/l HNO_3) and new protocol (4% HF and 40% HNO_3) and treated at high temperatures of 600–900 °C [138]. The high-temperature-treated titanium bacterial chip of one set was ablated by Nd-YAG laser to form the self-assembled monolayer (SAM) underneath by removing the oxide film from the chip surface. Hydrophilicity and hydrophobicity were examined for each titanium chip and were used as affinity probes for *P. aeruginosa* and *S. aureus* and as the MALDI–MS target plate for direct analysis and an informative mass spectrum was obtained by the MALDI–MS technique. *S. aureus* pathogenic bacteria showed affinity for the hydrophilic surface, while *P. aeruginosa* showed hydrophobic affinity for titanium bacterial chips and characterized by the obtained MALDI MS spectrum. Manikandam et al. explored more titanium chips treated at 1000 °C to form SAM beneath the titanium oxide layer and applied as titania cancer chip for sensitive and direct analysis of cancer by MALDI–MS techniques [139]. This miniaturization on the titanium chip surface reflects sensitivity and adsorption of highly complex cancer cells and rapid analysis via direct application of titanium cancer chips as MALDI target plates. Figure 11.11 shows the MALDI–MS spectrum cancer cells after treatment by the control titanium (Figure 11.11a), titanium oxide (TO) (Figure 11.11b), and SAM of titanium cancer chip (Figure 11.11c). SAM layers on the titanium cancer chip were found to be super hydrophilic with high surface roughness, which plays a major role in adsorption of the Nero2A CCL-131 cancer cell and rapidly analysis by MALDI–MS. LOD was 10 cell/ml of the mouse neuroblastoma cell line Nero2A CCL-131 by the SAM titanium cancer chip. Gedda and Wu modified Zn foil surface by a chemical

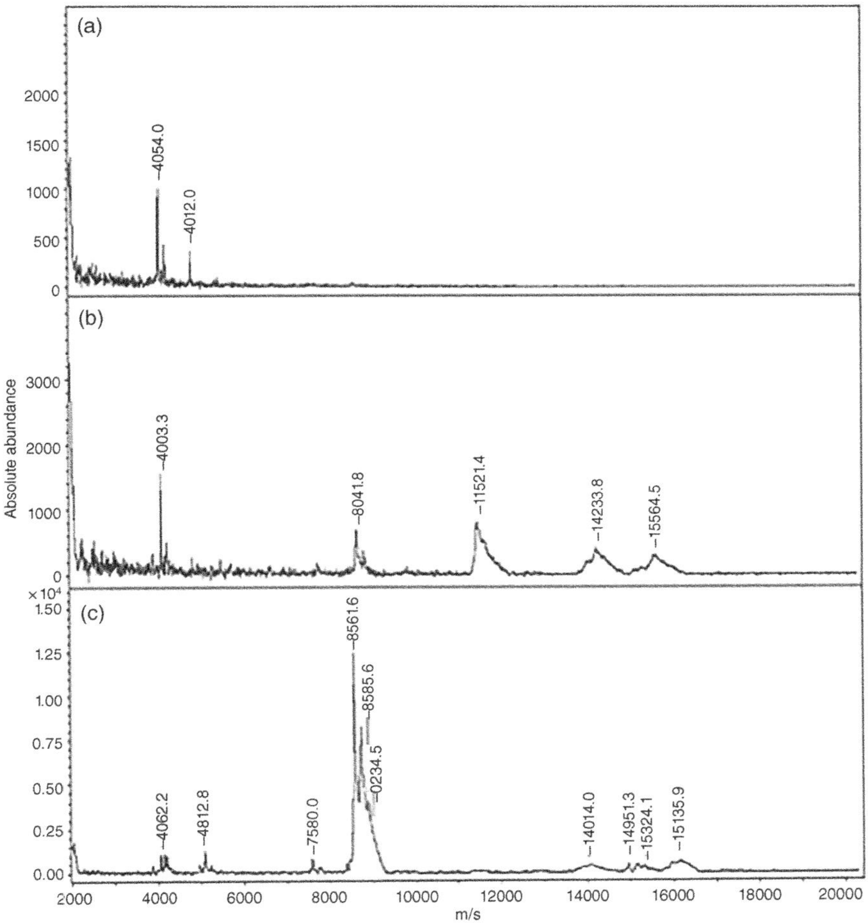

Figure 11.11 MALDI–MS spectra showing the comparative cancer cell capture efficiency of the (a) control, (b) TO, and (c) SAM cancer chip surface after a minimal exposure time of 3 hours. Source: Reprinted from Manikandan et al. [139].

process with ZnO nanorods (ZnO NR) [140]. The miniaturized Zn foil with ZnO NR array chip was used as a sensitive affinity probe for preconcentration and separation of proteins, peptides and bacterial analysis by MALDI–MS. ZnO NR array chips enhance the MALDI–MS signals compared to conventional MALDI–MS techniques. The hydrophobic surface of ZnO NR array chips was obtained by the functionalization of steric acid and it was proposed for *E. coli* bacteria enrichment for analysis in nanoliter droplets using MALDI–MS. This approach was applied as an efficient affinity and sensor probe to separate bacteria from real-world samples such as seawater and urine. Nevertheless, current research reflects the miniaturization of biochips as MALDI–MS targets to have great affinity and sensing capability toward enrichment of biological molecules even in a highly complex environment.

11.6 Conclusion

Nanoparticle-based miniaturization for preconcentration of peptides, proteins, prokaryotic and eukaryotic cells from small volumes of biological samples and their analysis through MALDI–MS is a promising approach in the current research scenario. Unique surface properties of nanomaterials not only decrease the laborious pretreatment of biological complex samples but also improve sensitivity, enrichment, sensor capabilities, and analysis by mass spectrometry within a short period. Miniaturization of analytical methodologies coupled with the dual application of nanoscience and MALDI–MS for proteomics analysis is ideally suited for separation, extraction, and enrichment processes at the nanoscale level. Nanoparticle-based preconcentration of biological analytes depends upon electrostatic, hydrophobic/hydrophilic, covalent, and ionic interactions between functionalized nanoparticles and biological molecules. This chapter dealt with many miniaturization analytical techniques based on metal nanoparticles and QDs for liquid extraction of biomolecules with direct analysis by MALDI–TOF–MS. Nanomaterial surfaces not only increase the selectivity by enrichment of biological analytes but also reduce analyte fragmentation and mass spectrum background noises during MALDI–MS analysis. Nanomaterial-based liquid extraction techniques such as SDME and LLE, on-particle phosphoprotein digestion, phosphopeptide enrichment in the microwave environment, and metal-oxide nanoarray biochips for biomolecules and bacterial analysis by MALDI–MS were explored on the basis of the current literature and published reports. This has been also explored for enrichment of many types of microorganisms after using various nanoparticles, surface-modified nanoparticles, and metal-based biochips prior to analysis by MALDI–TOF–MS. Nevertheless, each miniaturization approach with particular nanomaterials deals with different goals to work as affinity probes for preconcentration of various biological molecules.

References

1 Sigurdsson, E.M. (2018). Tau immunotherapies for Alzheimer's disease and related tauopathies: progress and potential pitfalls. *J. Alzheimers Dis.* 64: S555–S565.

2 Cuiling, Z., Giulietta, M., Hu, S., and Junwei, L. (2019). Advances in mRNA vaccines for infectious diseases. *Front. Immunol.* 10: 594–606.

3 Kailasa, S.K. and Wu, H.F. (2012). *Inorganic Contaminants: Sample Preparation Approaches in Comprehensive Sampling and Sample Preparation*, vol. 3 (eds. J. Pawliszyn, X.C. Le, X.F. Li and H.K. Lee). UK: Elsevier.

4 Kataoka, H. (2010). Recent developments and applications of microextraction techniques in drug analysis. *Anal. Bioanal. Chem.* 396: 339–364.

5 Han, D. and Row, K.H. (2012). Trends in liquid-phase microextraction, and its application to environmental and biological samples. *Microchim. Acta* 176: 1–22.

6 Jeannot, M.A., Przyjazny, A., and Kokosa, J.M. (2010). Single drop microextraction-development, applications and future trends. *J. Chromatogr. A* 1217 (16): 2326–2336.

7 Kokosa, J.M. (2013). Advances in solvent-microextraction techniques. *Trends Anal. Chem.* 43: 1–13.

8 Kocurov, L., Balogh, I.S., Sandrejova, J., and Andruch, V. (2012). Recent advances in dispersive liquid–liquid microextraction using organic solvents lighter than water. A review. *Microchem. J.* 102: 11–17.

9 Bagheri, H. and Khalilian, F. (2005). Immersed solvent microextraction and gas chromatography-mass spectrometric detection of s-triazine herbicides in aquatic media. *Anal. Chim. Acta* 537: 81–87.

10 Lucena, R., Simonet, B.M., Cárdenas, S., and Valcárcel, M. (2011). Potential of nanoparticles in sample preparation. *J. Chromatogr. A* 1218: 620–637.

11 Pyrzynska, K. (2013). Use of nanomaterials in sample preparation. *TrAC-Trends. Anal Chem.* 43: 100–108.

12 Karas, M., Bachmann, D., and Hillenkamp, F. (1985). Influence of the wavelength in high-irradiance ultraviolet laser desorption mass spectrometry of organic molecules. *Anal. Chem.* 57: 2935–2939.

13 Dreisewerd, K. (2003). The desorption process in MALDI. *Chem. Rev.* 103: 395–426.

14 Kruger, R. and Karas, M. (2002). Formation and fate of ion pairs during MALDI analysis: anion adduct generation as an indicative tool to determine ionization processes. *J. Am. Soc. Mass Spectrom.* 13: 1218–1226.

15 Combie, G.M. and Knochenmuss, R. (2006). Enhanced MALDI ionization efficiency at the metal-matrix interface: practical and mechanistic consequences of sample thickness and preparation method. *J. Am. Soc. Mass Spectrom.* 17 (5): 737–745.

16 Knochenmuss, R. (2009). A bipolar rate equation model of MALDI primary and secondary ionization processes, with application to positive/negative analyte ion ratios and suppression effects. *Int. J. Mass Spectrom.* 285: 105–113.

17 Ens, W., Mao, Y., Mayer, F., and Standing, K.G. (1991). Properties of matrix-assisted laser desorption. Measurements with a time-to-digital converter. *Rapid Commun. Mass Spectrom.* 5: 117–123.

18 Luo, G., Marginean, I., Ye, L., and Vertes, A. (2008). Competing ion decomposition channels in matrix-assisted laser desorption ionization. *J. Phys. Chem. B* 112 (23): 6952–6956.

19 Wiegelmann, M., Dreisewerd, K., and Soltwisch, J. (2016). Influence of the laser spot size, focal beam profile, and tissue type on the lipid signals obtained by MALDI-MS imaging in oversampling mode. *J. Am. Soc. Mass Spectrom.* 27: 1952–1964.

20 Nordhoff, E., Lehrach, H., and Gobom, J. (2007). Exploring the limits and losses in MALDI sample preparation of attomole amounts of peptide mixtures. *Int. J. Mass Spectrom.* 268: 139–146.

21 Garaguso, I. and Borlak, J. (2008). Matrix layer sample preparation: an improved MALDI-MS peptide analysis method for proteomic studies. *Proteomics* 8 (13): 2583–2595.

22 Watson, J.T. and Sparkman, O.D. (2008). *Introduction to Mass Spectrometry*, 4e. Wiley.

23 Tanaka, K., Waki, H., Ido, Y. et al. (1988). Protein and polymer analyses up to m/z 100 000 by laser ionization time-of-flight mass spectrometry. *Rapid Commun. Mass Spectrom.* 2: 151–153.

24 Picca, R.A., Calvano, C.D., Cioffi, N., and Palmisano, F. (2017). Mechanisms of nanophase-induced desorption in LDI-MS. A short review. *Nanomaterials* 7: 75.

25 Chu, H.W., Unnikrishnan, B., Anand, A. et al. (2018). Nanoparticle-based laser desorption/ionization mass spectrometric analysis of drugs and metabolites. *J. Food Drug Anal.* 26: 1215–1228.

26 Ha, M.S., Seo, H., Bae, D.H., and Yeo, W.S. (2014). Detection of enrofloxacin and its metabolite ciprofloxacin using gold nanoparticles and laser desorption/ionization time-of-flight mass spectrometry. *Anal. Sci.* 30: 451–455.

27 Guinan, T., Ronci, M., Vasani, R. et al. (2015). Comparison of the performance of different silicon-based SALDI substrates for illicit drug detection. *Talanta* 132: 494–502.

28 Iwaki, Y., Kawasaki, H., and Ryuichi, A. (2012). Human serum albumin-modified Fe3O4 magnetic nanoparticles for affinity-SALDI-MS of small-molecule drugs in biological liquids. *Anal. Sci.* 28: 893–900.

29 Murray, D.A.J. (1979). Rapid microextraction procedure for analyses of trace amounts of organic comparisons with macroextraction methods. *J. Chromatogr.* 177: 135–140.

30 Liu, S. and Dasgupta, P.K. (1995). Liquid droplet. A renewable gas sampling interface. *Anal. Chem.* 67: 2042–2049.

31 Cantwell, F.F. and Jeannot, M.A. (1996). Solvent microextraction into a single drop. *Anal. Chem.* 68: 2236–2240.

32 Sudhir, P.R., Wu, H.F., and Zhou, Z.C. (2005). Identification of peptides using gold nanoparticle-assisted single-drop microextraction coupled with AP-MALDI mass spectrometry. *Anal. Chem.* 77 (22): 7380–7385.

33 Shastri, L., Kailasa, S.K., and Wu, H.F. (2010). Nanoparticle-single drop microextraction as multifunctional and sensitive nanoprobes: binary matrix approach for gold nanoparticles modified with (4-mercaptophenyliminomethyl)-2-methoxyphenol for peptide and protein analysis in MALDI-TOF MS. *Talanta* 81: 1176–1182.

34 Shrivas, K. and Wu, H.F. (2008). Modified silver nanoparticle as a hydrophobic affinity probe for analysis of peptides and proteins in biological samples by using liquid-liquid microextraction coupled to AP- MALDI-ion trap and MALDI-TOF mass spectrometry. *Anal. Chem.* 80 (7): 2583–2589.

35 Shastri, L., Abdelhamid, H.N., Nawaz, M., and Wu, H.F. (2015). Synthesis, characterization and bifunctional applications of bidentate silver nanoparticle assisted single drop microextraction as a highly sensitive preconcentrating probe for protein analysis. *RSC Adv.* 5 (52): 41595–41603.

36 Dietz, M.L. (2006). Ionic liquids as extraction solvents: where do we stand? *Sep. Sci. Technol.* 41: 2047–2063.

37 Guo, X.P., Yin, D.Q., Peng, J.F., and Hu, X.L. (2012). Ionic liquid-based single-drop liquid-phase microextraction combined with high-performance liquid chromatography for the determination of sulfonamides in environmental water. *J. Sep. Sci.* 35: 452–458.

38 Anderson, J.L. and Armstrong, D.W. (2003). High-stability ionic liquids. A new class of stationary phases for gas chromatography. *Anal. Chem.* 75: 4851–4858.

39 Anderson, J.L. and Armstrong, D.W. (2005). Immobilized ionic liquids as high-selectivity/high-temperature/high-stability gas chromatography stationary phases. *Anal. Chem.* 77: 6453–6462.

40 Ahmad, F. and Wu, H.F. (2011). Characterization of pathogenic bacteria using ionic liquid via single drop microextraction combined with MALDI-TOF MS. *Analyst* 136: 4020–4027.

41 Carrillo-Carrión, C., Simonet, B.M., and Valcárcel, M. (2012). (CdSe/ZnS QDs)-ionic liquid-based headspace single drop microextraction for the fluorimet- ric determination of trimethylamine in fish. *Analyst* 137: 1152–1159.

42 Abdelhamid, H.N. (2018). Ionic liquid-assisted laser desorption/ionization – mass spectrometry: matrices, microextraction, and separation. *Methods Protoc.* 1: 23.

43 Rutkowska, M., Dubalska, K., Konieczka, P., and Namiesnik, J. (2014). Microextraction techniques used in the procedures for determining organomercury and organotin compounds in environmental samples. *Molecules* 19: 7581–7609.

44 Nuhu, A.A., Basheer, C., and Saad, B. (2011). Liquid-phase and dispersive liquid–liquid microextraction techniques with derivatization: recent applications in bioanalysis. *J. Chromatogr. B* 879: 1180–1188.

45 Rezaee, M., Assadi, Y., Hosseini, M.R.M. et al. (2006). Determination of organic compounds in water using dispersive liquid–liquid microextraction. *J. Chromatogr. A.* 1116: 1–9.

46 Spietelun, A., Marcinkowski, Ł., Guardia, M.D.L., and Namieśnik, J. (2014). Green aspects, developments and perspectives of liquid phase microextraction techniques. *Talanta* 119: 34–45.

47 Alsharif, A.M.A., Tan, G.H., Choo, Y.M., and Lawal, A. (2017). Efficiency of hollow fiber liquid-phase microextraction chromatography methods in the separation of organic compounds: a review. *J. Chromatogr. Sci.* 55 (3): 378–391.

48 Kailasa, S.K., Mehta, V.N., and Wu, H.F. (2014). Recent developments of liquid-phase microextraction techniques directly combined with ESI- and MALDI-mass spectrometric techniques for organic and biomolecule assays. *RSC Adv.* 4: 16188–16205.

49 Rios, A., Zougagh, M., and Bouri, M. (2013). Magnetic (nano) materials as a useful tool for sample preparation in analytical methods: a review. *Anal. Methods* 5: 4558–4573.

50 Kailasa, S.K. and Wu, H.F. (2010). Surface modified silver selinide nanoparticles as extracting probes to improve peptide/protein detection via

nanoparticles-based liquid phase microextraction coupled with MALDI mass spectrometry. *Talanta* 83: 527–534.

51 Bhat, A.R. and Wu, H.F. (2010). Synthesis, characterization and application of modified Pd nanoparticles as preconcentration probes for selective enrichment/analysis of proteins via hydrophobic interactions from real-world samples using nanoparticle-liquid-liquid microextraction coupled to matrix-assisted laser desorption/ionization time-of-flight mass spectrometry. *Rapid Commun. Mass Spectrom.* 24: 3547–3552.

52 Shrivas, K. and Wu, H.F. (2012). Rapid and highly sensitive protein extraction via cobalt oxide nanoparticle-based liquid–liquid microextraction coupled with MALDI mass spectrometry. *Analyst* 137: 890–895.

53 Kailasa, S.K. and Wu, H.F. (2012). Dispersive liquid–liquid microextraction using functionalized $Mg(OH)_2$ NPs with oleic acid as hydrophobic affinity probes for the analysis of hydrophobic proteins in bacteria by MALDI MS. *Analyst* 137: 4490–4496.

54 Kailasa, S.K. and Wu, H.F. (2013). Surface modified $BaTiO_3$ nanoparticles as the matrix for phospholipids and as extracting probes for LLME of hydrophobic proteins in *Escherichia coli* by MALDI–MS. *Talanta* 114: 283–290.

55 Manikandan, M., Hasan, N., and Wu, H.F. (2013). Rapid detection of haloarchaeal carotenoids via liquid–liquid microextraction enabled direct TLC MALDI-MS. *Talanta* 107: 167–175.

56 Abdelhamid, H.N., Bhaisare, M.L., and Wu, H.F. (2014). Ceria nanocubic-ultrasonication assisted dispersive liquid–liquid microextraction coupled with matrix assisted laser desorption/ ionization mass spectrometry for pathogenic bacteria analysis. *Talanta* 120: 208–217.

57 Chen, K.Y., Yang, T.C., and Chang, S.Y. (2012). Determination of macrolide antibiotics using dispersive liquid–liquid microextraction followed by surface-assisted laser desorption/ionizationmass spectrometry. *J. Am. Soc.Mass Spectrom.* 23: 1157–1160.

58 Ma, Y.R., Zhang, X.L., Zeng, T. et al. (2013). Polydopamine-coated magnetic nanoparticles for enrichment and direct detection of small molecule pollutants coupled with MALDI-TOF-MS. *ACS Appl. Mater. Interfaces* 5: 1024–1030.

59 Teng, C.H., Ho, K.C., Lin, Y.S., and Chen, Y.C. (2004). Gold nanoparticles as selective and concentrating probes for samples in MALDI MS analysis. *Anal. Chem.* 76: 4337–4342.

60 Vanderpuije, B.N.Y., Han, G., Rotello, V.M., and Vachet, R.W. (2006). Mixed monolayer protected gold nanoclusters as selective peptide extraction agents for MALDI-MS analysis. *Anal. Chem.* 78: 5491–5496.

61 Ju, S. and Yeo, W.S. (2012). Quantification of proteins on gold nanoparticles by combining MALDI-TOF MS and proteolysis. *Nanotechnology* 23: 135701–135707.

62 Nicolardi, S., Van Der Burgt, Y.E.M. et al. (2017). Structural characterization of biofunctionalized gold nanoparticles by ultrahigh-resolution mass spectrometry. *ACS Nano* 11: 8257–8264.

63 McLean, J.A., Stumpo, K.A., and Russell, D.H. (2005). Size-selected (2–10 nm) gold nanoparticles for matrix assisted laser desorption ionization of peptides. *J. Am. Chem. Soc.* 127: 5304–5305.

64 Castellana, E.T. and Russell, D.H. (2007). Tailoring nanoparticle surface chemistry to enhance laser desorption ionization of peptides and proteins. *Nano Lett.* 7: 3023–3025.

65 Ho, K.C., Tsai, P.J., Lin, Y.S., and Chen, Y.C. (2004). Using biofunctionalized nanoparticles to probe pathogenic bacteria. *Anal. Chem.* 76: 7162–7168.

66 Chen, Y.J., Chen, S.H., Chien, Y.Y. et al. (2005). Carbohydrate-encapsulated gold nanoparticles for rapid target-protein identification and binding-epitope mapping. *ChemBioChem* 6: 1169–1173.

67 Huang, Y.F. and Chang, H.T. (2006). Nile red-adsorbed gold nanoparticle matrixes for determining aminothiols through surface-assisted laser desorption/ionization mass spectrometry. *Anal. Chem.* 78: 1485–1493.

68 Lin, H.Y., Chen, C.T., and Chen, Y.C. (2006). Detection of phosphopeptides by localized surface plasma resonance of titania-coated gold nanoparticles immobilized on glass substrates. *Anal. Chem.* 78: 6873–6878.

69 Kailasa, S.K. and Wu, H.F. (2012). One-pot synthesis of dopamine dithiocarbamate functionalized gold nanoparticles for quantitative analysis of small molecules and phosphopeptides in SALDI- and MALDI-MS. *Analyst* 137: 1629–1638.

70 Kailasa, S.K. and Wu, H.F. (2012). Functionalized quantum dots with dopamine dithiocarbamate as the matrix for the quantification of efavirenz in human plasma and as affinity probes for rapid identification of microwave tryptic digested proteins in MALDI-TOF-MS. *J. Proteome* 75: 2924–2933.

71 Chiang, C.K., Lin, Y.W., Chen, W.T., and Chang, H.T. (2010). Accurate quantitation of glutathione in cell lysates through surface-assisted laser desorption/ionization mass spectrometry using gold nanoparticles. *Nanomedicine* 6: 530–537.

72 Gopal, J., Wu, H.F., and Lee, C.H. (2011). The bifunctional role of Ag nanoparticles on bacteria a MALDI-MS perspective. *Analyst* 136: 5077–5083.

73 Dakal, T.C., Kumar, A., Majumdar, R.S., and Yadav, V. (2016). Mechanistic basis of antimicrobial actions of silver nanoparticles. *Front. Microbiol.* 7: 1831–1848.

74 Prabhu, S. and Poulose, E.K. (2012). Silver nanoparticles: mechanism of antimicrobial action, synthesis, medical applications, and toxicity effects. *Int. Nano Lett.* 2: 32–42.

75 Sondi, I. and Salopek-Sondi, B. (2004). Silver nanoparticles as antimicrobial agent: a case study on *E. coli* as a model for Gram-negative bacteria. *J. Colloid Interface Sci.* 275: 177–182.

76 Monedeiro, F., Pomastowski, P., Milanowski, M. et al. (2019). Monitoring of bactericidal effects of silver nanoparticles based on protein signatures and VOC emissions from *Escherichia coli* and selected salivary bacteria. *J. Clin. Med.* 8 (11): 2024–2047.

77 Shrivas, K., Agrawal, K., and Wu, H.F. (2011). Application of platinum nanoparticles as affinity probe and matrix for direct analysis of small biomolecules and microwave digested proteins using matrix-assisted laser desorption/ionization mass spectrometry. *Analyst* 136: 2852–2857.
78 Gopal, G., Manikandan, M., Hasan, N. et al. (2013). A comparative study on the mode of interaction of different nanoparticles during MALDI-MS of bacterial cells. *J. Mass Spectrom.* 48: 119–127.
79 Kawasaki, H., Akira, T., Watanabe, T. et al. (2009). Sulfonate group-modified FePtCu nanoparticles as a selective probe for LDI-MS analysis of oligopeptides from a peptide mixture and human serum proteins. *Anal. Bioanal. Chem.* 395: 1423–1431.
80 Ahmad, F., Siddiqui, M.A., Babalola, O.O., and Wu, H.F. (2012). Biofunctionalization of nanoparticle assisted mass spectrometry as biosensors for rapid detection of plant associated bacteria. *Biosens. Bioelectron.* 35: 235–242.
81 Manikandan, M., Wu, H.F., and Hasan, N. (2012). Cell population based mass spectrometry using platinum nanodots for algal and fungal studies. *Biosens. Bioelectron.* 35: 493–497.
82 Kawasaki, H., Yonezawa, T., Watanabe, T., and Arakawa, R. (2007). Platinum nanoflowers for surface-assisted laser desorption/ionization mass spectrometry of biomolecules. *J. Phys. Chem. C* 111: 16278–16283.
83 Klimov, V.I. (2007). Spectral and dynamical properties of multiexcitons in semiconductor nanocrystals. *Annu. Rev. Phys. Chem.* 58: 635–673.
84 Mohanta, D., Bora, H., Dutta, N., and Choudhur, A. (2008). Development principles and production of paired PbS quantumdot. *Eur. Phys. J. Appl. Phys.* 41: 129–132.
85 Jun-Ichi, I., Atsushi, O., and Kenji, S. (2010). Large-scale first principles electronic structure calculations for nano-meter size Si quantumdots. *J. Surf. Sci. Nanotech.* 8: 48–51.
86 Allen, P.M. and Bawendi, M.G. (2008). Ternary I-III-VI quantum dots luminescent in the red to near-infrared. *J. Am. Chem. Soc.* 130: 9240–9241.
87 Sun, M., Yang, L., Jose, P. et al. (2013). Functionalization of quantum dots with multidentate zwitterionic ligands: impact on cellular interactions and cytotoxicity. *J. Mater. Chem. B* 1: 6137–6146.
88 Bilan, R., Fleury, F., Nabiev, I., and Sukhanova, A. (2015). Quantum dot surface chemistry and functionalization for cell targeting and imaging. *Bioconjug. Chem.* 26: 609–624.
89 Bruchez, M.P. (2005). Turning all the lights on: quantum dots in cellular assays. *Curr. Opin. Chem. Biol.* 9: 533–537.
90 Jaiswal, J.K. and Simon, S.M. (2004). Potentials and pitfalls of fluorescent quantum dots for biological imaging. *Trends Cell Biol.* 14: 497–504.
91 Kailasa, S.K., Kiran, K., and Wu, H.F. (2008). Comparison of ZnS semiconductor nanoparticles capped with various functional groups as the matrix and affinity probes for rapid analysis of cyclodextrins and proteins in surface-assisted laser desorption/ionization time-of-flight mass spectrometry. *Anal. Chem.* 80: 9681–9688.

92 Wu, H.F., Kailasa, S.K., and Shastri, L. (2010). Electrostatically self-assembled azides on zinc sulfide nanoparticles as multifunctional nanoprobes for peptide and protein analysis in MALDI-TOF MS. *Talanta* 82: 540–547.

93 Shrivas, K., Kailasa, S.K., and Wu, H.F. (2009). Quantum dots laser desorption/ionization MS: multifunctional CdSe quantum dots as the matrix, concentrating probes and acceleration for microwave enzymatic digestion for peptide analysis and high resolution detection of proteins in a linear MALDI-TOFMS. *Proteomics* 9: 2656–2667.

94 Abdelhamid, H.N., Chen, Z., and Wu, H.F. (2017). Surface tuning laser desorption/ionization mass spectrometry (STLDI-MS) for the analysis of small molecules using quantum dots. *Anal. Bioanal. Chem.* 409: 4943–4950.

95 Shrivas, K., Kailasa, S.K., and Wu, H.F. (2009). Quantum dots – electrospray ionization mass spectrometry: 3-mercaptopropanic acid capped CdS quantum dots as accelerating and enrichment probes for microwave tryptic digestion of proteins. *Rapid Commun. Mass Spectrom.* 23: 3603–3607.

96 Shrivas, K. and Wu, H.F. (2010). Multifunctional nanoparticles composite for MALDI-MS: Cd^{2+}-doped carbon nanotubes with CdS nanoparticles as the matrix, preconcentrating and accelerating probes of microwave enzymatic digestion of peptides and proteins for direct MALDI-MS analysis. *J. Mass Spectrom.* 45: 1452–1460.

97 Wu, H.F. and Chung, F.T. (2011). 3-Mercaptopropionic acid modified ZnSe quantum dots as the matrix for direct surface-assisted laser desorption/ionization mass spectrometric analysis of peptides/proteins from sodium salt solution. *Rapid Commun. Mass Spectrom.* 25: 1779–1786.

98 Shastri, L., Kailasa, S.K., and Wu, H.F. (2009). Cysteine-capped ZnSe quantum dots as affinity and accelerating probes for microwave enzymatic digestion of proteins via direct matrix-assisted laser desorption/ionization time-of-flight mass spectrometric analysis. *Rapid Commun. Mass Spectrom.* 23: 2247–2252.

99 Rao, Z., Geng, F., Zhou, Y. et al. (2017). N-doped graphene quantum dots as a novel highly-efficient matrix for the analysis of perfluoroalkyl sulfonates and other small molecules by MALDI-TOF MS. *Anal. Methods* 9: 2014–2020.

100 Wang, Y., Gao, D., Chen, Y. et al. (2016). Development of N, S-doped carbon dots as a novel matrix for the analysis of small molecules by negative ion MALDI-TOF MS. *RSC Adv.* 6: 79043–79049.

101 Bailes, J., Vidal, L., Ivanov, D.A. et al. (2009). Quantum dots improve peptide detection in MALDI MS in a size dependent manner. *J. Nanobiotechnol.* 7 (10): 1–12.

102 Wu, H.F., Gopal, J., Abdelhamid, H.N. et al. (2012). Quantum dot applications endowing novelty to analytical proteomics. *Proteomics* 12: 2949–2961.

103 Ardito, F., Giuliani, M., and Perrone, D. (2017). The crucial role of protein phosphorylation in cell signaling and its use as targeted therapy (review). *Int. J. Mol. Med.* 40: 270–280.

104 Kailasa, S.K. and Wu, H.F. (2014). Recent developments in nanoparticle-based MALDI mass spectrometric analysis of phosphoproteomes. *Microchim. Acta* 181: 853–864.

105 Li, X.S., Yuan, B.F., and Feng, Y.Q. (2016). Recent advances in phosphopeptide enrichment: Strategies and techniques. *TrAC Trends Anal. Chem.* 78: 70–83.

106 Zhou, H., Tian, R., Ye, M. et al. (2007). Highly specific enrichment of phosphopeptides by zirconium dioxide nanoparticles for phosphoproteome analysis. *Electrophoresis* 28: 2201–2215.

107 Zhang, L., Xu, J., Sun, L. et al. (2011). Zirconium oxide aerogel for effective enrichment of phosphopeptides with high binding capacity. *Anal. Bioanal. Chem.* 399: 3399–3405.

108 Nelson, C.A., Szczech, J.R., Dooley, C.J. et al. (2010). Effective enrichment and mass spectrometry analysis of phosphopeptides using mesoporous metal oxide nanomaterials. *Anal. Chem.* 82 (17): 7193–7201.

109 Hasan, N. and Wu, H.F. (2011). Highly selective and sensitive enrichment of phosphopeptides via NiO nanoparticles using a microwave-assisted centrifugation on-particle ionization/enrichment approach in MALDI-MS. *Anal. Bioanal. Chem.* 400: 3451–3462.

110 Hasan, N., Wu, H.F., Li, Y.H. et al. (2010). Two-step on-particle ionization/enrichment via a washing- and separation-free approach: multifunctional TiO_2 nanoparticles as desalting, accelerating, and affinity probes for microwave-assisted tryptic digestion of phosphoproteins in ESI-MS and MALDI-MS: comparison with microscale TiO_2. *Anal. Bioanal. Chem.* 396: 2909–2919.

111 Nawaz, M.I., Hasan, N., and Wu, H.F. (2016). $CoFe_2O_4$–ZnO nanoparticles for rapid microwave-assisted tryptic digestion of phosphoprotein and phosphopeptide analysis by matrix-assisted laser desorption/ionization mass spectrometry. *Rapid Commun. Mass Spectrom.* 30: 1443–1453.

112 Kailasa, S.K. and Wu, H.F. (2012). Rapid enrichment of phosphopeptides by $BaTiO_3$ nanoparticles after microwave-assisted tryptic digest of phosphoproteins, and their identification by MALDI-MS. *Microchim. Acta* 179: 83–90.

113 Chen, W.Y. and Chen, Y.C. (2010). Functional Fe_3O_4@ZnO magnetic nanoparticle-assisted enrichment and enzymatic digestion of phosphoproteins from saliva. *Anal. Bioanal. Chem.* 398: 2049–2057.

114 Kupcik, R., Rehulka, P., Bilkova, Z. et al. (2017). New interface for purification of proteins: 1D TiO_2 nanotubes decorated by Fe_3O_4 nanoparticles. *ACS Appl. Mater. Interfaces* 9 (34): 28233–28242.

115 Dunn, J.D., Reid, G.E., and Bruening, M.L. (2010). Techniques for phosphopeptide enrichment prior to analysis by mass spectrometry. *Mass Spectrom. Rev.* 29: 29–54.

116 Xu, Y., Wei, M.T., Ou-Yang, H.D. et al. (2016). Exposure to TiO_2 nanoparticles increases *Staphylococcus aureus* infection of HeLa cells. *J. Nanobiotechnol.* 14 (34): 1–16.

117 Li, Y., Zhang, W., Niu, J., and Chen, Y. (2012). Mechanism of photo-generated reactive oxygen species and correlation with the antibacterial properties of engineered metal-oxide nanoparticles. *ACS Nano* 6 (6): 5164–5173.

118 Malka, E., Perelshtein, I., Lipovsky, A. et al. (2013). Eradication of multi-drug resistant bacteria by a novel Zn-doped CuO nanocomposite. *Small* 9 (23): 4069–4076.

119 Leung, Y.H., Ng, A.M., Xu, X. et al. (2014). Mechanisms of antibacterial activity of MgO: non-ROS mediated toxicity of MgO nanoparticles towards *Escherichia coli*. *Small* 10 (6): 1171–1183.

120 Zakharova, O.V., Godymchuk, A.Y., Gusev, A.A. et al. (2015). Considerable variation of antibacterial activity of Cu nanoparticles suspensions depending on the storage time, dispersive medium, and particle sizes. *Biomed. Res. Int.* 2015 (412530): 1–16.

121 Peng, Z., Ni, J., Zheng, K. et al. (2013). Dual effects and mechanism of TiO_2 nanotube arrays in reducing bacterial colonization and enhancing $C_3H_{10}T_{1/2}$ cell adhesion. *Int. J. Nanomedicine* 8: 3093–3105.

122 Yu, J., Zhang, W., Li, Y. et al. (2014). Synthesis, characterization, antimicrobial activity and mechanism of a novel hydroxyapatite whisker/nano zinc oxide biomaterial. *Biomed. Mater.* 10 (1): 015001.

123 Ho, K.C., Tsai, P.J., Lin, Y.S. et al. (2004). Using bio functionalized nanoparticles to probe pathogenic bacteria. *Anal. Chem.* 76: 7162–7168.

124 Lin, Y.S., Tsai, P.J., Weng, M.F. et al. (2005). Affinity capture using vancomycin-bound magnetic nanoparticles for the MALDI-MS analysis of bacteria. *Anal. Chem.* 77: 1753–1760.

125 Li, S., Guo, Z., Wu, H.F. et al. (2010). Rapid analysis of Gram-positive bacteria in water via membrane filtration coupled with nanoprobe-based MALDI-MS. *Anal. Bioanal. Chem.* 397: 2465–2476.

126 Hasan, N., Guo, Z., and Wu, H.F. (2016). Large protein analysis of *Staphylococcus aureus* and *Escherichia coli* by MALDI TOF mass spectrometry using amoxicillin functionalized magnetic nanoparticles. *Anal. Bioanal. Chem.* 408 (23): 6269–6281.

127 Hasan, N., Ahmad, F., and Wu, H.F. (2013). Monitoring the heat stress response of *Escherichia coli* via NiO nanoparticle assisted MALDI–TOF mass spectrometry. *Talanta* 103: 38–46.

128 Gedda, G., Pandey, S., Lin, Y.C., and Wu, H.F. (2015). Antibacterial effect of calcium oxide nano-plates fabricated from shrimp shell. *Green Chem.* 17 (6): 3276–3280.

129 Gopal, J., Wu, H.F., and Lee, Y.H. (2010). Matrix-assisted laser desorption ionization-time-of-flight mass spectrometry as a rapid and reliable technique for directly evaluating bactericidal activity: probing the critical concentration of ZnO nanoparticles as affinity probes. *Anal. Chem.* 82: 9617–9621.

130 Gopal, J., Narayana, J.L., and Wu, F.H. (2011). TiO_2 nanoparticle assisted mass spectrometry as biosensors of *Staphylococcus aureus*, key pathogen in nosocomial infections from air, skin surface and human nasal passage. *Biosens. Bioelectron.* 27: 201–206.

131 Gopal, J., Narayana, J.L., Hasan, N. et al. (2015). MALDI mass spectrometry for probing the anti-staphylococcal capability of TiO_2 nanoparticles via near-UV and laser irradiation. *New J. Chem.* 39 (6): 4909–4918.

132 Abdelhamid, H.N. and Wu, H.F. (2013). Multifunctional graphene magnetic nanosheet decorated with chitosan for highly sensitive detection of pathogenic bacteria. *J. Mater. Chem. B* 1: 3950–3961.

133 Zaluzec, E.J., Gage, D.A., Allison, J., and Watson, J.T. (1994). Direct matrix-assisted laser desorption ionization mass spectrometric analysis of proteins immobilized on nylon-based membranes. *J. Am. Soc. Mass Spectrom.* 5: 230–237.

134 Shen, J., Ahmed, T., Vogt, A. et al. (2005). Preparation and characterization of nitrilotriacetic-acid-terminated self-assembled monolayers on gold surfaces for matrix-assisted laser desorption ionization-time of flight-mass spectrometry analysis of proteins and peptides. *Anal. Biochem.* 345: 258–269.

135 Blacken, G.R., Volny, M., Diener, M. et al. (2009). Reactive landing of gas-phase ions as a tool for the fabrication of metal oxide surfaces for in situ phosphopeptide enrichment. *J. Am. Soc. Mass Spectrom.* 20: 915–926.

136 Urban, P.L., Amantonico, A., and Zenobi, R. (2011). Lab-on a-plate: extending the functionality of MALDI MS and LDI-MS targets. *Mass Spectrom. Rev.* 30: 435–478.

137 Gopal, J., Hasan, N., and Wu, H.F. (2013). Fabrication of titanium based MALDI bacterial chips for rapid, sensitive and direct analysis of pathogenic bacteria. *Biosens. Bioelectron.* 39: 57–63.

138 Hasan, N., Gopal, J., and Wu, H.F. (2014). Surface pretreatment effects on titanium chips for the adhesion of pathogenic bacteria in the MALDI-TOF MS. *Appl. Surf. Sci.* 314: 52–63.

139 Manikandan, M., Gopal, J., Hasan, N., and Wu, H.F. (2014). A nano-patterned self-assembled monolayer (SAM) rutile titania cancer chip for rapid, low cost, highly sensitive, direct cancer analysis in MALDI-MS. *Talanta* 130: 78–89.

140 Gedda, G. and Wu, H.F. (2019). Fabrication of surface modified ZnO nanorod array for MALDI-MS analysis of bacteria in a nanoliter droplet: a multiple function biochip. *Sensors Actuators B Chem.* 288: 667–677.

12

Mechanisms and Applications of Nanopriming: New Vista for Seed Germination

Karen P. Pachchigar[1], Darshan T. Dharajiya[2], Sumeet N. Jani[1], Jaykishan N. Songara[1], and Gaurav S. Dave[3]

[1] *Department of Biotechnology, College of Basic Science and Humanities, Sardarkrushinagar Dantiwada Agricultural University (SDAU), Sardarkrushinagar-385506, Gujarat, India*
[2] *Bio Science Research Centre, SDAU, Sardarkrushinagar-385506, Gujarat, India*
[3] *Pulses Research Station, SDAU, Sardarkrushinagar-385506, Gujarat, India*

12.1 Introduction to Agriculture and Green Nanotechnology

Agriculture is the most important as well as vulnerable sector that produces and provides raw materials for food and feed industries. Continuing global climate change gives rise to various challenges for sustainable agriculture [1]. Agricultural practices in the world are changing dramatically, i.e. usage of chemical fertilizers, usage of pesticides for pest control, usage of fortified crops, and usage of genetically modified crops. All these practices are used for better productivity and higher yield, which can fulfill the current agriculture-based requirements of society. All the aforementioned techniques are having their own pros and cons. Nanotechnology has the potential to revolutionize agriculture and plays an important role in food and crop production. Plant-based materials are considered the best candidates for synthesizing biocompatible nanoparticles (NPs) due to their biochemical diversity of plant extract, nontoxic phytochemical constitutes, nonpathogenicity, low cost, and flexibility in reaction parameters compared to chemical synthesis methods. Hence, nowadays nanotechnology has become widely popular among researchers [2]. In commercial agriculture, rapid and uniform seed germination and seedling emergence are important determinants of successful stand establishment. Nanopriming is a modified seed priming technique with application of NPs to seeds for enhancing their germination rate as well as faster and better growth [3].

Nanotechnology has become an interdisciplinary field of research in most of the new research fields, and many "nano"-related words such as nanoscale, nanotechnology, nanostructures, and nanotubes are found in dictionaries [4]. Nanotechnology has a considerable impact on industry, economy, and human life all over the world. As per the definition given by the US Environmental Protection Agency (EPA), "nanotechnology is the science of understanding and control of

Miniaturized Analytical Devices: Materials and Technology, First Edition.
Edited by Suresh Kumar Kailasa and Chaudhery Mustansar Hussain.

matter at dimensions of roughly 1–100 nm, where unique physical properties make novel applications possible" [5]. The given definition is slightly rigid with regard to size dimensions. Greater emphasis could have been placed on the problem-solving capability of the materials. Other attempts to define nanoparticles from the point of view of agriculture include "particulate between 10 and 1000 nm in size dimensions that are simultaneously colloidal particulate" [6].

Green nanotechnology is a new field of research involving the design of nanomaterials, nanoscale, and processes through green chemistry and engineering leading to the development of new functions without harmful subsequences on biosphere and humanity [7]. NPs are abundant in nature and are created in many natural operations such as volcanic eruptions, photochemical reactions, forest fires, erosion, and shedding of plants and animal hair [4]. Most organisms, especially those that interact strongly with their surroundings, are affected by exposure to NPs. NP formulations by green routes are proven to be eco-friendly and beneficial to farmers, stakeholders, and the agro-industry [8]. In the modern agricultural production system, rapid and uniform seed germination is required for successful seedling establishment and for yield achievement. The increasing applications of nanoparticles in diverse agricultural sectors have made it a crucial subject of study. Various applications of nanotechnology in agriculture (Figure 12.1) include nanopesticides (to enhance the efficacy to kill pests and to reduce the environmental load of pesticides),

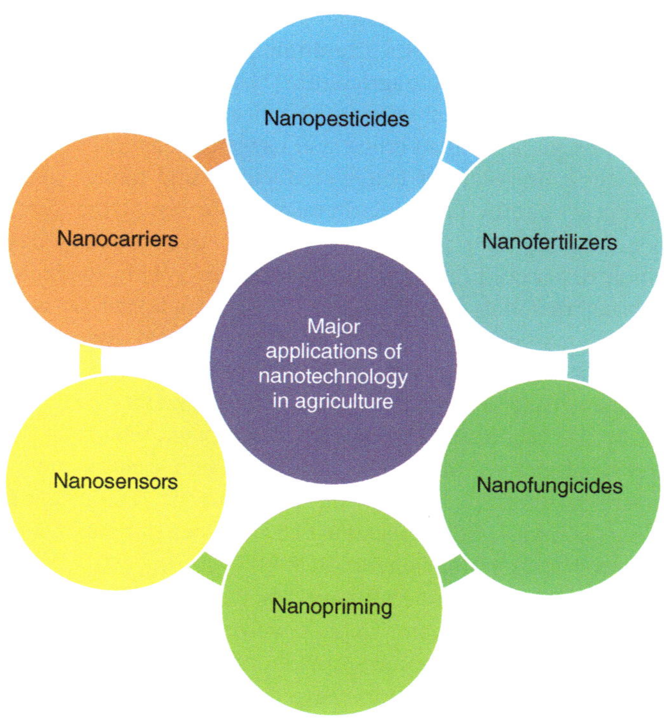

Figure 12.1 Major applications of nanotechnology in agriculture.

nanofertilizers (to improve nutrient effectiveness), nanofungicides (an eco-friendly alternative to conventional synthetic fungicides and having more efficiency), and nanosensors (to detect residues and plant diseases) [9]. Apart from this, nanoparticles are being used in nanopriming (to improve germination and seedling growth of low vigor seeds) [3], nanocarriers (to deliver agrochemicals and plant hormones) [10] and as an antimicrobial agents [9] in recent times.

Although NP-based studies are fruitful in numerous fields over a decade, such as nanomedicine, nanoindustrial application, nanopharmacy, and nanopesticide, a higher advancement of NP-based studies in releasing seed dormancy and enhancing seed germination as well as seedling development have recently come to the limelight in the form of seed nanopriming technology [11]. Nanopriming, a technique based on the combination of seed priming and nanoparticle treatment, has been a useful tool for enhancing seed quality, seedling establishment, and crop yield as well as for increasing tolerance to environmental stresses, compared to unprimed or other agent-primed seeds.

12.2 Nanopriming for Better Crop Germination

Seed priming is a technique that partially hydrates seeds in natural or synthetic compounds under specific environments to a point where germination-related metabolic processes begin, but radicle emergence does not occur. Seed priming has been found to be useful for enhancing seed quality, seedling establishment, and crop yields as well as for increasing tolerance to environmental stresses. Priming technique relies on the uptake of water, which triggers metabolic activity in controlled conditions. Major types of priming techniques are: hydro-, halo-, osmo-, nutria-, chemo-priming, and others (with plant growth regulator and matrix support) [12]. Liquid system, solid particulate system, and controlled hydration are chief mediators of these priming methods. Liquid system involves the regulation of water potential using high-molecular-weight compounds, i.e. polyethylene glycol (PEG 8000) or inorganic salts. Solid particulate system uses solid particulate materials such as fine-grade vermiculite and water. The advantage of seed treatment in reducing the germination time and improving emergence uniformity is well established under laboratory conditions [13]. However, different priming treatments have different properties, effectiveness, and optimized priming agents for each crop species. Therefore, there is a growing need to develop new priming agents to enhance seed germination of various crop plants.

Nanopriming is a modified seed priming technique with application of NPs or nanoformulations to seeds for enhancing their germination for faster and improved growth. Origin of this concept lies with several problems associated with agriculture, i.e. lower germination rate, infection of different plant pathogens, and prolonged dormancy in various plants. Some options are available, with some severe environmental hazards. So, green synthesis of nanoparticles may provide a safe and efficient solution for these problems. In several crops, nanoformulations provide a potent protection against some severe pathogens. Apart from their pesticidal activity, nanoparticles have proved to be an efficient seed priming tool for some crops with enhanced germination rate, i.e. silver nanoparticles (Ag NPs) for rice seeds.

Normally, the seed germination ratio of each crop is different because of several factors. Nanopriming as a "win–win approach" not only induces seed germination but also helps the farmers who usually spray insecticides and pesticides during plant growth. NPs naturally possess antimicrobial activities against plant pathogens that comprehensively cause increased plant growth and yield. Many studies have demonstrated that nanopriming can significantly increase germination percentage and seedling vigor of crop plants compared to unprimed and hydro-primed seeds. Moreover, nanopriming can enhance seed water uptake and starch metabolism of the treated seeds [3].

12.3 Anticipated Mechanisms Underlying Nanopriming: Plant Physiology and Molecular-Level Interactions

Different mechanisms underlying NP-induced seed germination have been proposed, including creation of nanopores for enhanced water uptake, rebooting ROS/antioxidant systems in seeds, generation of hydroxyl radicals for cell wall loosening, and nanocatalysts for fastening starch hydrolysis.

Seed priming is a pre-sowing treatment applied to seeds in a specific concentration of defined priming agent (e.g. halide solution, osmoregulators, hormones or GRs, and other herb extracts) for a specific period. Priming creates a physiological state in the seed that strengthens its growth capacity against biotic or abiotic stresses [14]. Priming can accelerate germination rate, vigor index of seedling, root and shoot elongation, wet and dry weight of seedling, photosynthetic rate, and other growth-related traits of plants [15]. The purpose of seed priming is to treat seeds with diverse agents, such as water, that activate physiological processes of seeds. If time vs. seed water content is plotted, then primed seeds and non-primed seeds demonstrate three phases. The first phase (Phase I: imbibition) represents the entry of water in the seed by adsorption (Figure 12.2). The second phase (Phase II: activation) is initiation of biochemical processes and demonstrates hydration in non-primed seeds. In primed seeds, hydration treatment permits controlled imbibition and induction of the pre-germinative metabolism, but radicle emergence is prohibited (Figure 12.2). The third phase (Phase III: growth) displays the germination and post germination phases, where sugar consumption by embryo under plant growth regulation leads to radicle emergence (Figure 12.2) [16, 17]. Moreover, priming actuates metabolism before germination that contains a vast scope of physiological operations such as repair pathways of DNA and reactive oxygen species (ROS) [18]. Therefore, seed priming is a low-cost, simple, and suitable technique for enhancing seed germination and seedling establishment. It eventually increases plant yield, particularly under stress conditions [19]. The success of priming depends on a combination of

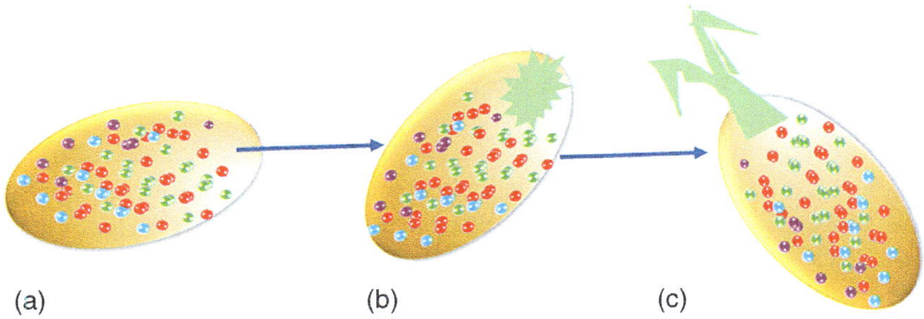

Figure 12.2 General mechanism of nanopriming. (a) Seed covered or layered with nanoparticles, (b) increased internal ABA/GA level and improvised GA activity, and (c) fostered radicle protrusion.

interacting factors such as plant, water, priming treatment, priming period, temperature, vigorness of the seed, and seed storage conditions.

Several studies have shown that NPs have the ability to infiltrate the coat of seeds and increase water uptake, which augments germination and growth [20]. Priming with micronutrients has been recognized as a possible technique for enhancing germination rate and the development of seedling vigor. The type and the magnitude of the effect of NPs on plant germination depend on the plant species, concentration of NPs in the environment, contact time, intrinsic properties of nanomaterials, and interaction between living environment and plant [21, 22]. The use of nanotechnology to optimize the formulation of chemical fertilizers can attain remarkable achievements, including reduced energy consumption and production costs and minimal environmental damage.

As discussed earlier, nanopriming induces seed germination by various mechanisms. Physiologically, it imparts an enhanced water uptake, rebooting ROS/antioxidant systems in seeds, and generation of hydroxyl radicals for cell wall loosening. This comprehensively contributes to better plant growth and crop physiology. Overall, various mechanisms directly and indirectly involved in nanopriming have been summarized as follows:

1. Imbibition and vigorous seedling growth
2. Osmotic adjustment and membrane dynamics
3. Antioxidant pool and ROS signaling
4. Hormonal crosstalk and metabolic flux

12.3.1 Imbibition and Vigorous Seedling Growth

Seed imbibition, water uptake of seeds is the primary event hallmark for the initiation of the germination process. Nanopriming provokes seed germination by modulating abscisic acid/gibberellic acid (ABA/GA) homeostasis. It is well known that higher ABA/GA concentration is a crucial factor in maintaining seed dormancy. So, seed dormancy is usually challenged and broken by metal nanoparticle treatment by modulating level of ABA/GA level [23]. Seed imbibition (Phase-1) is characterized by rapid and swift uptake of water, followed by provoking seed cellular machinery (Phase-2) for radicle protrusion phase (Phase-3). This fascinating dynamic of imbibition to protrusion and radicle emergence is a highly balanced event under regulation of ABA/GA; nevertheless, plant growth is ceased by their cell wall. With electron spin resonance (ESR), Ag NPs by acting as "Fenton nanoparticles" contribute to development of free hydroxyl radical ·OH, contributed chiefly to cell wall loosening, testa and endosperm weakening, which is essential for radicle protrusion in seed germination [3]. NP-primed seeds also showed higher activities of many enzymes for metabolism of proteins (proteases), carbohydrates (α- and β-amylases), and lipids (isocitrate lyase) involved in mobilization of stored reserves in seeds [24, 25]. ROS involvements in program cell death (PCD) during seed germination and seedling establishment are well reported in the aleurone layer of cereal grains and are related to H_2O_2 interactions with GA and ABA hormones [26].

12.3.2 Osmotic Adjustment and Membrane Dynamics

Osmotic adjustment (OA) of plasma membranes is well known for their homeostasis attainment. OA is chiefly mediated by various late embryo abundant (LEA) proteins, proline, and other osmoregulator like glycine betaine [27]. Under nanopriming, NP-mediated osmotic environment changes were reported with uptake and downregulation of sugar. Nanopriming with Ag NPs and M NPs caused an increase in the sugar level inside the seed aleurone layer and that flux for changing amylase activity. Increases in soluble sugar in the internal compartment of the seed drive for more water uptake and improves water uptake by cell [3]. This water uptake lowers osmotic potential and increases water potential in the exterior part that ultimately facilitates water movement into the seeds by osmosis. Plasma membrane "flip-flop" movement probably could allow binding and entry of nanosized metal particles into the cell through various mechanisms. As shown in Figure 12.3, NPs enter the cell via (i) diffusion through its complexification with specific or nonspecific proteins, or by a wide variety of compounds present in the radical exudates; (ii) NPs interaction with cell membrane can be altered by hydrophilic and hydrophobic components; (iii) NPs present in the cytoplasm can be embedded by carrier proteins or specific biomolecules; or (iv) NPs contained in vesicles can serve as regulators for intracellular traffic in specific areas between organelles [28]. So, this harmony for membrane dynamics for NPs uptake and their translocation possibly via carrier protein or through membrane channel could be responsible for their provoking Phase-2 to Phase-3 radicle protrusion for seed germination. Various studies suggested that in the upper surface of the seed coat, engineered NPs take the parenchymatous intercellular spaces to penetrate; these intercellular spaces support the dispersion of liquid suspension to the cotyledon [21, 29, 30]. It has been reported that silver (Ag) and aluminum oxide (Al_2O_3)-based NPs do not disturb the germination process and growth of *Arabidopsis thaliana* although ZnO NPs hamper their germination process [30]. On the seed surface, other types of entrance and aquaporins are associated, which regulates the admission of the NPs [31, 32].

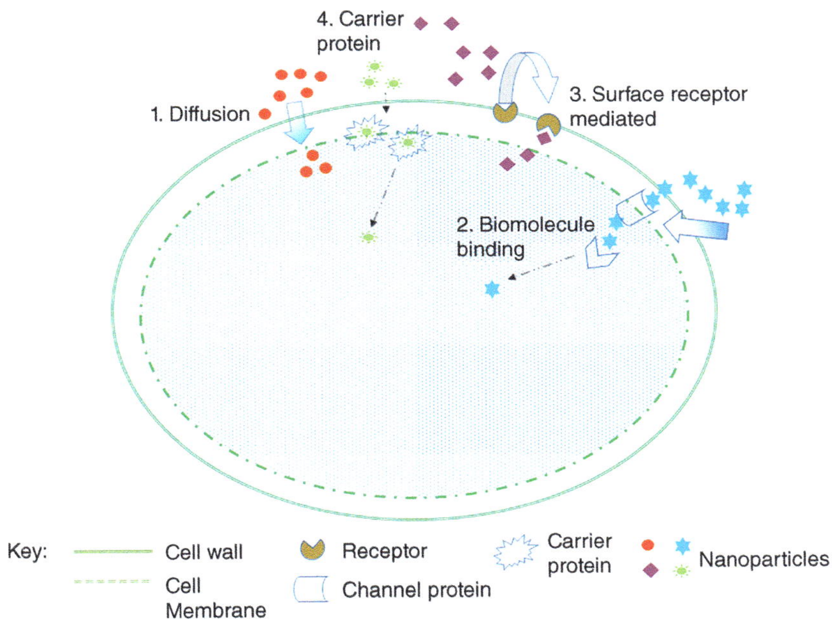

Figure 12.3 Mechanism of NP transport and uptake by plant cell. (1) Simple diffusion phenomenon by transport, (2) uptake of NPs by channel protein like aquaporin and their further binding with cytosolic biomolecule, (3) binding of NPs with cell wall/membrane surface receptor, and (4) carrier protein–mediated NP uptake.

12.3.3 Antioxidant and ROS Signaling

Antioxidant pool in primed seedlings harbors different antioxidants such as peroxidase (POD), ascorbate peroxidase (APX), superoxide dismutase (SOD), and catalase (CAT). Enhanced activities of CAT, SOD, and POD in seedlings emerged from primed seeds have been reported under normal and stress conditions [33–35]. Antioxidant pools often safeguard cellular membranes from detrimental effects of ROS such as H_2O_2, hydroxyl radicals, superoxide radicals, and singlet oxygen [36]. Studies have shown accumulation of high levels of ROS within the intracellular compartment of unprimed seeds compared to lesser ROS in primed seeds at the cost of lower antioxidant potential of cell. Higher levels of CAT were noted in primed seeds with Ag NPs for counteracting ROS level from cellular environment [37]. Additionally, H_2O_2 is one of the key ROS molecules acting as a signaling molecule for the regulation of seed germination, and the precise regulation of H_2O_2 content by antioxidant machinery is essential to achieve a balance between oxidative signaling that stimulates germination and oxidative damage that prevents or delays germination. Thus, higher levels of H_2O_2 observed in nanoprimed seeds can act as signal molecules. Previously, ROS have been found to involve in endosperm weakening, cell wall loosening in growing tissues, breaking down polysaccharides, DNA, RNA, and fatty acids [38], and in protein carbonylation (which increases protein susceptibility to proteolytic cleavage) – thus aiding the mobilization of those molecules during germination has been reported [39].

Cellular pool maintains ROS and antioxidants level harmony for quick switch from dormant to speedy germination of seeds. Although ROS causes detrimental effects on cell membrane and stability of cellular machinery, optimum ROS level for "oxidative window for germination" is essential for seed germination to begin [40]. High amount of antioxidant enzymes was observed in silver nanoprimed seeds under a tightly controlled mechanism to balance ROS in the oxidative window range for stimulating seed germination. Photosynthesized Ag NPs in rice have well described the plausible mechanism of ROS and antioxidant relationship whereby nanoprimed seeds showed elevated levels of SOD, CAT activity for scavenging ROS [3]. So, overall under control regulation of antioxidant pool, elevated ROS exhibits tightly controlled regulation to produce ROS in the optimum range of oxidative window as signaling molecules for triggering essential metabolic activity of seeds for promoting seed germination and seedling development.

12.3.4 Hormonal Crosstalk and Metabolic Flux

Shifting from quiescent to germination and progressing from the embryonic to maturation stage is a crucial event during the seed cycle. This phenomenon of seed germination is well restrained under GA/ABA balanced coordination. GA causes weakening of endosperm, reserve hydrolyzed sugar mobilization, [41] and elongation of embryo cells [41]. GA/ABA balance predominantly modulates water uptake of seeds, thereby channelizing seed dormancy and shift water potential thresholds for radicle emergence due to elevated water uptake for seeds [23]. Probably, phytohormone couples, antioxidant pools downregulate ROS for better water uptake, mobilizing for storage protein, endoplasm weakening, and starch granule degradation with α-amylase. It is well evident that α-amylase plays a vital role in regulation of GA for seed germination activity. α-Amylase notably hydrolyzes the starch granules, implying a higher induction of α-amylase activity in nanoprimed seeds [42].

Ag-NP-primed rice seeds exhibited elevated upregulated aquaporin-based PIP1 and PIP2 genes level and facilitated better water uptake. So, nanoprimed seeds, compared to unprimed seeds, exhibit elevated ROS levels, thereby facilitating NP uptake and transport via aquaporin under ABA-controlled upregulation of PIP2, NIP1, TIP3, and TIP4 genes [37]. Further, GA-directed increased amylase activity induced cellular soluble sugar level and lowered osmotic potential. Zero-valent Fe-NPs-mediated OH radical could induce cell wall loosening to enhance the root elongation in *Arabidopsis* plants by 150–200% compared to that of the control [43]. Thus, seed priming with NPs at low doses is likely to produce moderate ROS for activating cell wall loosening and trigger early metabolic events inside seeds. Nanopriming mechanism whereby phytohormone-coupled antioxidant pool regulated ROS activity to elicit sugar leve for osmotic potential, starch hydrolysis, and cell wall loosening for seed germination is represented in Figure 12.4.

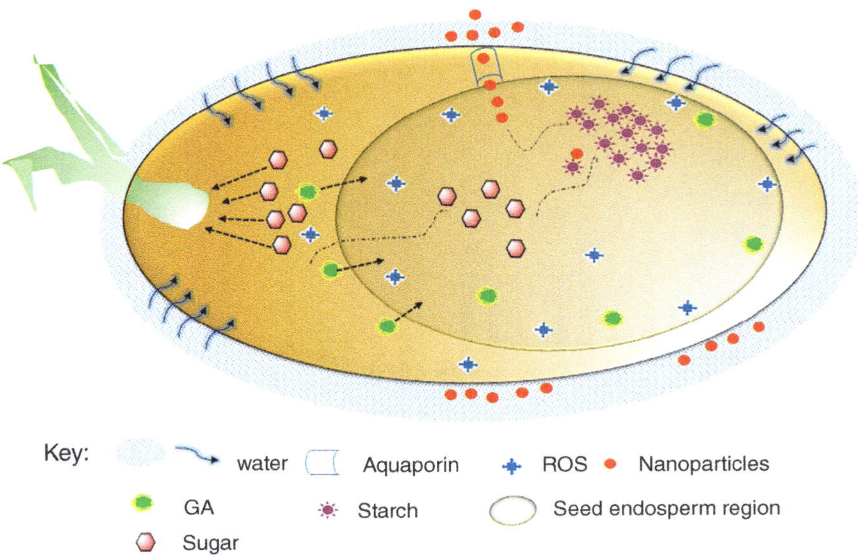

Figure 12.4 Proposed mechanism of NP-induced seed germination. Uptake of NPs through aquaporin- or other receptor-induced phytohormone GA/ABA activation with ROS and water uptake induced cell wall loosening. Activated GA acts on the aleurone layer of the seed endosperm and NPs by binding with starch, via hydrolysis-release sugar that is used by growing seed embryos.

12.4 Current Status of Nanopriming

Recent studies on nanopriming in various plant species have been summarized in Table 12.1. The studies indicated that different types of NPs have been used in nanopriming which includes Ag NPs [44–47], Al_2O_3 NPs[48], chitosan NPs [49], Fe_2O_3 NPs [50–52], FeO NPs [53], MgO NPs [54], Mn NPs [55], S NPs [56], Si NPs [57, 58], TiO_2 NPs [59, 60], and ZnO NPs [61–63, 67]. Commonly studied plant for the study of nanopriming was wheat (*Triticum aestivum* L.) [61, 62, 65, 67]. Effect of nanopriming on some other plants, namely, watermelon (*Citrullus lanatus* (Thunb.) Matsum. & Nakai) [44], pearl millet (*Pennisetum glaucum* L.)[45], sheep fescue (*Festuca ovina*) [46], Egyptian Roselle (*Hibiscus sabdariffa* L.) [48], French bean (*Phaseolus vulgaris* L.) [49], rice (*Oryza sativa* L.) [53], mung bean (*Vigna radiata* L.) [54], sweet pepper (*Capsicum annuum* L.) [55], sunflower (*Helianthus annuus* L.) [56], maize (*Zea mays* L.) [59], parsley (*Petroselinum crispum* L.) [60], and lupine (*Lupinus termis*) [66] have also been studied. Nanopriming showed some good positive results, among which most common were enhanced seed germination, photosynthesis, growth, yield, and antioxidant system [44, 47, 56, 60]. Some studies included the effects of nanopriming on plants during stressed conditions, which increased tolerance against particular abiotic stress. Increased tolerance against drought stress in sheep fescue [46], salinity stress in wheat [61], pearl millet [45],

Table 12.1 Summary of nanopriming work with list of crops, NPs used, their physiological and biochemical effects on seed germination.

NPs	Concentration (optimum/range)	Plant species	Positive effects/results obtained	References
Ag	31.3 ppm	Watermelon (*C. lanatus*)	Enhanced seed germination, soluble sugar contents during germination, growth and yield (up to 31.6% and 35.6%)	[44]
	20 mM	Pearl millet (*P. glaucum*)	Enhanced salinity tolerance by increased defense system, RWC and proline contents; improved plant height, growth-related attributes, and ultimately fresh and dry weight	[45]
	25–75%	Sheep fescue (*F. ovina*)	Increased tolerance against drought stress, enhanced root and shoot length, fresh and dry weight, vigor index, and reduced germination time	[46]
	1 ppm	Wheat (*T. aestivum*)	Enhanced seed germination, growth parameters and photosynthetic efficiency of seedlings, increased IBA, NAA, BAP contents and reduced ABA content	[47]
Al_2O_3	0.01%	Egyptian Roselle (*H. sabdariffa*)	Enhanced fresh weight, dry weight, shoot length, root length, and leaf area, photosynthetic pigments, soluble sugars, protein, amino acid, proline, and the activities of defense enzymes (SOD, CAT, POD, and APX)	[48]
Chitosan (Cs)	200 ppm	French bean (*P. vulgaris*)	Improved in growth, yield, antioxidant system, and biochemical content of seeds	[49]
Fe_2O_3	20–160 ppm	Watermelon (*C. lanatus*)	Increased nonenzymatic antioxidant potential and induced jasmonates-linked defense responses	[50]
	25–600 ppm	Wheat (*T. aestivum*)	Noticeable increase in seed germination and shoot length; higher (up to 45.7%) Fe accumulation in grain	[51]
	500 ppm	Sorghum (*S. bicolor*)	Increased growth through increased photosystem II efficiency, chlorophyll index, photosynthetic rate and RWC with decreased lipid peroxidation; increased salinity tolerance	[52]

FeO	20 ppm	Rice (*O. sativa*)	Enhanced seed germination and seedling vigor, increased root length (50%), dry weight, sugar and amylase content, antioxidant enzyme activity, and Fe acquisition	[53]
MgO	100 ppm	Mung bean (*V. radiata*)	Enhanced seed germination and seedling vigor	[54]
Mn	1 ppm	Sweet pepper (*C. annuum*)	Increased salt stress tolerance, enhanced seedling growth, and antioxidant enzymes concentration	[55]
S	12.5–200 μM	Sunflower (*H. annuus*)	Enhanced Mn detoxification through counterbalancing oxidative damage and increasing antioxidant defense system; stimulated CAT and SOD activities and enhanced the antioxidant compounds (ASA, TFC, and TPC)	[56]
Si	1200 ppm	Wheat (*T. aestivum*)	Enhanced growth and chlorophyll contents, biomass and yield, antioxidant enzyme activity; reduced oxidative stress and Cd uptake	[57]
	0.4 mM	Sunflower (*H. annuus*)	Reduced germination time, improved root length, seedling vigor index, germination percentage	[58]
TiO_2	60 ppm	Maize (*Z. mays*)	Enhanced germination percentage, germination energy, seedling vigor index, lengths of root and shoot, fresh and dry weights of seedling, K^+ concentration, RWC, TPC, proline content, SOD, CAT, and PAL activities	[59]
	30 mg/ml	Parsley (*P. crispum*)	Increased seed germination, germination rate index, root and shoot length, fresh weight, vigor index, and chlorophyll content of seedlings	[60]
ZnO	50–500 ppm	Wheat (*T. aestivum*)	Enhanced growth biomarkers of wheat plants (shoot height, FM, and DM), photosynthesis, and growth under saline conditions	[61]
	10 ppm	Wheat (*T. aestivum*)	Improved seed germination and photosynthetic performance	[62]

(Continued)

Table 12.1 (Continued)

NPs	Concentration (optimum/range)	Plant species	Positive effects/results obtained	References
	80 ppm	Corn (*Z. mays*)	Increased germination, root length, and dry biomass production	[63]
	100 ppm	Maize (*Z. mays*)	Improved seed germination rate and seedling parameters (shoot length, shoot width, root length, root width, leaf length, width, vigor index, and dry matter production)	[64]
	25–100 ppm	Wheat (*T. aestivum*)	Increased growth characteristics, photosynthesis, and biomass; reduced Zn deficiency in plants	[65]
	60 ppm	Lupine (*L. termis*)	Enhanced salt tolerance, stimulated growth, photosynthetic pigments, organic solutes, TPC, ASA, Zn, and activities of SOD, CAT, POD, and APX enzymes	[66]
ZnO, Fe	0–100 ppm for ZnO and 0–20 ppm for Fe	Wheat (*T. aestivum*)	Increased plant height, spike length, and dry weights of shoots, roots, spikes, and grains; increased photosynthesis, decreased Cd toxicity	[67]

ABA, Abscisic acid; APX, ascorbate peroxidase; ASA, ascorbic acid; BAP, 6-benzylaminopurine; CAT, catalase; DM, dry mass; FM, fresh mass; IBA, indole-3-butyric acid; NAA, 1-naphthalene acetic acid; NPs, nanoparticles; PAL, phenylalanine ammonia lyase; POD, peroxidase; RWC, relative water content; SOD, superoxide dismutase; TFC, total flavonoids content; and TPC, total phenolic content.

and sorghum [52] and salt stress in sweet pepper [55] and lupine [66] have also been proven by the application of nanopriming.

12.5 Conclusion

With the phenomenon of "plenty of room at bottom," NPs offer tremendous opportunities for exploiting their role in seed priming. Various metal nanoparticles and other nanoformulations have exhibited their rife role to accelerate germination rate, vigor index of seedling, root and shoot elongation, wet and dry weight of seedling, photosynthetic rate, and other growth related traits of plants. Underlying mechanism of nanopriming suggested various means for uptake to translocation of NPs and their further involvement in seed germination involving binding with starch, their hydrolysis followed by endosperm weakening under paramount control of GA/ABA coupled ROS-mediated downstream signaling. Nevertheless, the picture of nanopriming is currently hazy and needs attention to be exploited for proper understanding. Previous reports provided new opportunities and hope to burgeoning seed industry to improvised lacking domain of seed germination and more vigorous seeds for future. Further, exploration in research opportunities needs to be satisfied for application of nanoparticles to seeds and upgradation in agricultural practices.

References

1 Food and Agriculture Organization of the United Nations (2018). Transforming food and agriculture to achieve the SDGs-20 interconnected actions to guide decision-makers.

2 Mittal, A.K., Chisti, Y., and Banerjee, U.C. (2013). Synthesis of metallic nanoparticles using plant extracts. *Biotechnol. Adv.* 31 (2): 346–356.

3 Mahakham, W., Sarmah, A.K., Maensiri, S., and Theerakulpisut, P. (2017). Nano-priming technology for enhancing germination and starch metabolism of aged rice seeds using phytosynthesized silver nanoparticles. *Sci. Rep.* 7 (1): 1–21.

4 Buzea, C., Pacheco, I.I., and Robbie, K. (2007). Nanomaterials and nanoparticles: sources and toxicity. *Biointerphases* 2 (4): MR17–MR71.

5 United States Environmental Protection Agency (2021). Nanotechnology & Nanomaterials Research. https://www.epa.gov/sites/production/files/2013-12/documents/nanotechnology-fact-sheet.pdf (accessed at EPA-Research on Nanomaterial on 22 March 2021).

6 Mukhopadhyay, S.S. (2014). Nanotechnology in agriculture: prospects and constraints. *Nanotechnol. Sci. Appl.* 7: 63.

7 McKenzie, L.C. and Hutchison, J.E. (2004). Green nanoscience. *Chim. Oggi* 22 (9): 30–33.

8 Sastry, R.K., Rashmi, H.B., Rao, N.H., and Ilyas, S.M. (2010). Integrating nanotechnology into agri-food systems research in India: a conceptual framework. *Technol. Forecast. Soc. Chang.* 77 (4): 639–648.

9 Ghidan, A.Y. and Al Antary, T.M. (2020). Applications of nanotechnology in agriculture. In: *Applications of Nanobiotechnology* (eds. M. Stoytcheva and R. Zlatev). London: IntechOpen https://doi.org/10.5772/intechopen.88390.

10 Shakiba, S., Astete, C.E., Paudel, S. et al. (2020). Emerging investigator series: polymeric nanocarriers for agricultural applications: synthesis, characterization, and environmental and biological interactions. *Environ. Sci. Nano* 7 (1): 37–67.

11 Ibrahim, E.A. (2016). Seed priming to alleviate salinity stress in germinating seeds. *J. Plant Physiol.* 192: 38–46.

12 Sher, A., Sarwar, T., Nawaz, A. et al. (2019). Methods of seed priming. In: *Priming and Pretreatment of Seeds and Seedlings*, 1–10. Singapore: Springer.

13 Chavan, N.G., Bhujbal, G.B., and Manjare, M.R. (2014). Effect of seed priming on field performance and seed yield of soybean [*Glycine max* (L.) Merill] varieties. *Bioscan* 9 (1): 111–114.

14 Conrath, U. (2011). Molecular aspects of defence priming. *Trends Plant Sci.* 16 (10): 524–531.

15 Tian, Y., Guan, B., Zhou, D. et al. (2014). Responses of seed germination, seedling growth, and seed yield traits to seed pretreatment in maize (*Zea mays* L.). *Sci. World J.* 2014 https://doi.org/10.1155/2014/834630.

16 Nascimento, W.M. and Aragão, F.A.S.D. (2004). Muskmelon seed priming in relation to seed vigor. *Sci. Agric.* 61 (1): 114–117.

17 Rajjou, L., Duval, M., Gallardo, K. et al. (2012). Seed germination and vigor. *Annu. Rev. Plant Biol.* 63: 507–533.

18 Paparella, S., Araújo, S.S., Rossi, G. et al. (2015). Seed priming: state of the art and new perspectives. *Plant Cell Rep.* 34 (8): 1281–1293.

19 Singh, H., Jassal, R.K., Kang, J.S. et al. (2015). Seed priming techniques in field crops-A review. *Agric. Rev.* 36 (4).

20 Maroufpoor, N., Mousavi, M., Hatami, M. et al. (2019). Chapter 5 - Mechanisms involved in stimulatory and toxicity effects of nanomaterials on seed germination and early seedling growth. In: *Advances in Phytonanotechnology* (eds. M. Ghorbanpour and S.H. Wani), 153–181. Elsevier.

21 Ma, X., Geiser-Lee, J., Deng, Y., and Kolmakov, A. (2010). Interactions between engineered nanoparticles (ENPs) and plants: phytotoxicity, uptake and accumulation. *Sci. Total Environ.* 408 (16): 3053–3061.

22 Miralles, P., Johnson, E., Church, T.L., and Harris, A.T. (2012). Multiwalled carbon nanotubes in alfalfa and wheat: toxicology and uptake. *J. R. Soc. Interface* 9 (77): 3514–3527.

23 Rodríguez-Gacio, M.D.C., Matilla-Vázquez, M.A., and Matilla, A.J. (2009). Seed dormancy and ABA signaling: the breakthrough goes on. *Plant Signal. Behav.* 4 (11): 1035–1048.

24 Varier, A., Vari, A.K., and Dadlani, M. (2010). The subcellular basis of seed priming. *Curr. Sci.*: 450–456.

25 Sisodia, A., Padhi, M., Pal, A.K. et al. (2018). Seed priming on germination, growth and flowering in flowers and ornamental trees. In: *Advances in Seed Priming* (eds. A. Rakshit and H. Singh), 263–288. Singapore: Springer.

26 Gomes, M. and Garcia, Q. (2013). Reactive oxygen species and seed germination. *Biologia* 68 (3): 351–357.

27 Saha, P., Chatterjee, P., and Biswas, A.K. (2010). NaCl pretreatment alleviates salt stress by enhancement of antioxidant defense system and osmolyte accumulation in mungbean (*Vigna radiata* L. Wilczek). *Indian J. Exp. Biol.* 48: 593–600.

28 Fincheira, P., Tortella, G., Duran, N. et al. (2020). Current applications of nanotechnology to develop plant growth inducer agents as an innovation strategy. *Crit. Rev. Biotechnol.* 40 (1): 15–30.

29 Van Dongen, J.T., Ammerlaan, A.M., Wouterlood, M. et al. (2003). Structure of the developing pea seed coat and the post-phloem transport pathway of nutrients. *Ann. Bot.* 91 (6): 729–737.

30 Lee, C.W., Mahendra, S., Zodrow, K. et al. (2010). Developmental phytotoxicity of metal oxide nanoparticles to *Arabidopsis thaliana*. *Environ. Toxicol. Chem.: Int. J.* 29 (3): 669–675.

31 Abu-Hamdah, R., Cho, W.J., Cho, S.J. et al. (2004). Regulation of the water channel aquaporin-1: isolation and reconstitution of the regulatory complex. *Cell Biol. Int.* 28 (1): 7–17.

32 Khodakovskaya, M., Dervishi, E., Mahmood, M. et al. (2009). Carbon nanotubes are able to penetrate plant seed coat and dramatically affect seed germination and plant growth. *ACS Nano* 3 (10): 3221–3227.

33 Goswami, A., Banerjee, R., and Raha, S. (2013). Drought resistance in rice seedlings conferred by seed priming. *Protoplasma* 250 (5): 1115–1129.

34 Zheng, M., Tao, Y., Hussain, S. et al. (2016). Seed priming in dry direct-seeded rice: consequences for emergence, seedling growth and associated metabolic events under drought stress. *Plant Growth Regul.* 78 (2): 167–178.

35 Hussain, S., Khan, F., Hussain, H.A., and Nie, L. (2016). Physiological and biochemical mechanisms of seed priming-induced chilling tolerance in rice cultivars. *Front. Plant Sci.* 7: 116.

36 Posmyk, M.M., Corbineau, F., Vinel, D. et al. (2001). Osmoconditioning reduces physiological and biochemical damage induced by chilling in soybean seeds. *Physiol. Plant.* 111 (4): 473–482.

37 Chandrasekaran, U., Luo, X., Wang, Q., and Shu, K. (2020). Are there unidentified factors involved in the germination of nanoprimed seeds? *Front. Plant Sci.* 11: 832.

38 Buetler, T.M., Krauskopf, A., and Ruegg, U.T. (2004). Role of superoxide as a signaling molecule. *Physiology* 19 (3): 120–123.

39 Job, C., Rajjou, L., Lovigny, Y. et al. (2005). Patterns of protein oxidation in *Arabidopsis* seeds and during germination. *Plant Physiol.* 138 (2): 790–802.

40 Leymarie, J., Vitkauskaité, G., Hoang, H.H. et al. (2012). Role of reactive oxygen species in the regulation of *Arabidopsis* seed dormancy. *Plant Cell Physiol.* 53 (1): 96–106.

41 Chen, F. and Bradford, K.J. (2000). Expression of an expansin is associated with endosperm weakening during tomato seed germination. *Plant Physiol.* 124 (3): 1265–1274.
42 Man, J., Yang, Y., Zhang, C. et al. (2013). Morphology and structural characterization of high-amylose rice starch residues hydrolyzed by porcine pancreatic α-amylase. *Food Hydrocoll.* 31 (2): 195–203.
43 Kim, J.H., Lee, Y., Kim, E.J. et al. (2014). Exposure of iron nanoparticles to *Arabidopsis thaliana* enhances root elongation by triggering cell wall loosening. *Environ. Sci. Technol.* 48 (6): 3477–3485.
44 Acharya, P., Jayaprakasha, G.K., Crosby, K.M. et al. (2020). Nanoparticle-mediated seed priming improves germination, growth, yield, and quality of watermelons (*Citrullus lanatus*) at multi-locations in Texas. *Sci. Rep.* 10 (1): 1–16.
45 Khan, I., Raza, M.A., Awan, S.A. et al. (2020). Amelioration of salt induced toxicity in pearl millet by seed priming with silver nanoparticles (AgNPs): the oxidative damage, antioxidant enzymes and ions uptake are major determinants of salt tolerant capacity. *Plant Physiol. Biochem.* 156: 221–232.
46 Abbasi Khalaki, M., Ghorbani, A., and Dadjou, F. (2019). Influence of nano-priming on *Festuca ovina* seed germination and early seedling traits under drought stress, in laboratory condition. *Ecopersia* 7 (3): 133–139.
47 Abou-Zeid, H. and Ismail, G. (2018). The role of priming with biosynthesized silver nanoparticles in the response of *Triticum aestivum* L to salt stress. *Egypt. J. Bot.* 58 (1): 73–85.
48 Latef, A.A.H.A., Zaid, A., Alhmad, M.F.A., and Abdelfattah, K.E. (2020). The impact of priming with Al_2O_3 nanoparticles on growth, pigments, osmolytes, and antioxidant enzymes of Egyptian Roselle (*Hibiscus sabdariffa* L.) cultivar. *Agronomy* 10 (5): 681.
49 Abdel-Aziz, H.M.M., Hasaneen, M.N.A., and Omer, A.M. (2019). Impact of engineered nanomaterials either alone or loaded with NPK on growth and productivity of French bean plants: seed priming vs foliar application. *S. Afr. J. Bot.* 125: 102–108.
50 Kasote, D.M., Lee, J.H., Jayaprakasha, G.K., and Patil, B.S. (2019). Seed priming with iron oxide nanoparticles modulate antioxidant potential and defense-linked hormones in watermelon seedlings. *ACS Sustain. Chem. Eng.* 7 (5): 5142–5151.
51 Sundaria, N., Singh, M., Upreti, P. et al. (2019). Seed priming with Iron oxide nanoparticles triggers Iron acquisition and biofortification in wheat (*Triticum aestivum* L.) grains. *J. Plant Growth Regul.* 38 (1): 122–131.
52 Maswada, H.F., Djanaguiraman, M., and Prasad, P.V.V. (2018). Seed treatment with nano-iron (III) oxide enhances germination, seeding growth and salinity tolerance of sorghum. *J. Agron. Crop Sci.* 204 (6): 577–587.
53 Afzal, S., Sharma, D., and Singh, N.K. (2021). Eco-friendly synthesis of phytochemical-capped iron oxide nanoparticles as nano-priming agent for boosting seed germination in rice (*Oryza sativa* L.). *Environ. Sci. Pollut. Res.*: 1–13.
54 Anand, K.V., Anugraga, A.R., Kannan, M. et al. (2020). Bio-engineered magnesium oxide nanoparticles as nano-priming agent for enhancing seed germination and seedling vigour of green gram (*Vigna radiata* L.). *Mater. Lett.* 271: 127792.

55 Ye, Y., Cota-Ruiz, K., Hernández-Viezcas, J.A. et al. (2020). Manganese nanoparticles control salinity-modulated molecular responses in *Capsicum annuum* L. through priming: a sustainable approach for agriculture. *ACS Sustain. Chem. Eng.* 8 (3): 1427–1436.

56 Ragab, G. and Saad-Allah, K. (2020). Seed priming with greenly synthesized sulfur nanoparticles enhances antioxidative defense machinery and restricts oxidative injury under manganese stress in *Helianthus annuus* (L.) seedlings. *J. Plant Growth Regul.*: 1–9.

57 Hussain, A., Rizwan, M., Ali, Q., and Ali, S. (2019). Seed priming with silicon nanoparticles improved the biomass and yield while reduced the oxidative stress and cadmium concentration in wheat grains. *Environ. Sci. Pollut. Res.* 26 (8): 7579–7588.

58 Janmohammadi, M. and Sabaghnia, N. (2015). Effect of pre-sowing seed treatments with silicon nanoparticles on germinability of sunflower (*Helianthus annuus*). *Botanica* 21 (1): 13–21.

59 Shah, T., Latif, S., Saeed, F. et al. (2021). Seed priming with titanium dioxide nanoparticles enhances seed vigor, leaf water status, and antioxidant enzyme activities in maize (*Zea mays* L.) under salinity stress. *J. King Saud University-Sci.* 33 (1): 101207.

60 Dehkourdi, E.H. and Mosavi, M. (2013). Effect of anatase nanoparticles (TiO_2) on parsley seed germination (*Petroselinum crispum*) in vitro. *Biol. Trace Elem. Res.* 155 (2): 283–286.

61 Abou-Zeid, H.M., Ismail, G.S.M., and Abdel-Latif, S.A. (2021). Influence of seed priming with ZnO nanoparticles on the salt-induced damages in wheat (*Triticum aestivum* L.) plants. *J. Plant Nutr.* 44 (5): 629–643.

62 Rai-Kalal, P. and Jajoo, A. (2021). Priming with zinc oxide nanoparticles improve germination and photosynthetic performance in wheat. *Plant Physiol. Biochem.* 160: 341–351.

63 Esper Neto, M., Britt, D.W., Lara, L.M. et al. (2020). Initial development of corn seedlings after seed priming with nanoscale synthetic zinc oxide. *Agronomy* 10 (2): 307.

64 Itroutwar, P.D., Kasivelu, G., Raguraman, V. et al. (2020). Effects of biogenic zinc oxide nanoparticles on seed germination and seedling vigor of maize (*Zea mays*). *Biocatal. Agric. Biotechnol.* 29: 101778.

65 Munir, T., Rizwan, M., Kashif, M. et al. (2018). Effect of zinc oxide nanoparticles on the growth and Zn uptake in wheat (*Triticum aestivum* L.) by seed priming method. *Dig. J. Nanomater. Biostruct. (DJNB)* 13 (1): 315–323.

66 Latef, A.A.H.A., Alhmad, M.F.A., and Abdelfattah, K.E. (2017). The possible roles of priming with ZnO nanoparticles in mitigation of salinity stress in lupine (*Lupinus termis*) plants. *J. Plant Growth Regul.* 36 (1): 60–70.

67 Rizwan, M., Ali, S., Ali, B. et al. (2019). Zinc and iron oxide nanoparticles improved the plant growth and reduced the oxidative stress and cadmium concentration in wheat. *Chemosphere* 214: 269–277.

13

Nanotechnology for Environmental Pollution Detection and Remedies

Nishant Srivastava and Gourav Mishra

Department of Biotechnology, Meerut Institute of Engineering and Technology (MIET) NH-58, Delhi Roorkee Bypass Road, Baghpat Crossing, Meerut 250005, Uttar Pradesh, India

13.1 Introduction

Environmental problems have gradually become a major concern globally and are alone accounted for more than 24% of diseases worldwide. Also, 33% of diseases found in children under five globally occur due to environmental problems [1]. The controlled–uncontrolled discharge of toxic pollutants, heavy metals, dyes, gases, hazardous chemicals in water, air, and soil causes serious contamination in the environment as depicted in Figure 13.1 [2]. Anthropogenic activities are majorly accountable for the global environmental crisis, whereas natural phenomena such as volcano eruption, forest fires, floods, storms are minor contributors. The global industrial and economic race leads to gradual environmental diminution. The increasing awareness about the environment and climate change motivates researchers worldwide to find a sustainable solution with sustainable development.

The advancement in science and technology and the emergence of highly advanced fields like nanotechnology embark on various sustainable routes to resolve the environmental crisis. The application of nanotechnology approaches for environmental remediation, monitoring, and diagnosis can revolutionize the overall process. Nanotechnology can be defined as the application of several incipient technologies at a nanometer scale of 1–100 nm by producing constituents, devices, and structures with necessarily innovative properties and functionality by regulatory shape and size of materials [3]. Nanodevices due to their small size and high surface-to-volume ratio have gained much importance to use as sensors for diagnosis, monitoring, and remediation. Nanomaterials such as CdO, Au, FeO, ZnO, SnO_2, carbon, CaO, TiO_2 have excellent properties for environmental remediation as sensors, absorbents, membranes, catalysts, and photocatalysts [3–11]. Nanomaterials have myriad applications in sensing pollutants, pesticides, chemical and biological entities as well as can be employed for their degradation and remediation, too. Nanomaterials mostly used for environmental diagnosis and remediation are nanoparticles (metallic, bimetallic, metal oxide, carbon, and silica),

Miniaturized Analytical Devices: Materials and Technology, First Edition.
Edited by Suresh Kumar Kailasa and Chaudhery Mustansar Hussain.

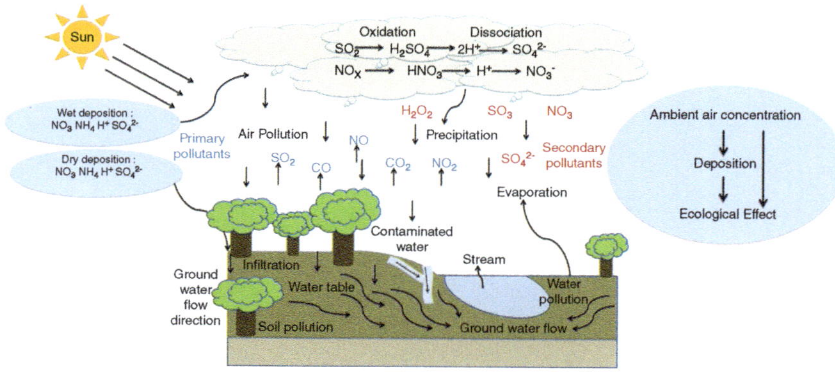

Figure 13.1 Various contributors responsible for environmental crisis.

nanotubes, and nanocomposites [12, 13]. This chapter provides a brief insight into the application of nanomaterial-based devices for environmental pollution detection, monitoring, and remediation. These nanodevices are emerging as a prodigious means for sustainable development and environmental safeguarding.

13.2 Nanotechnology for Environmental Monitoring and Diagnosis

Worldwide industrialization, urbanization, vehicle movements, and reckless deforestation have become a very serious problem for the environment by putting a severe pollution burden on the planet and the health of living organisms including humans. Scientists throughout the world have been working hard to lessen/or control the effects of pollutants, and use of nanomaterials for monitoring and diagnosis of pollutants has been emerging as the prominent technology. Various types of innumerable nanomaterials have been developed in the past few years for monitoring and diagnosis of environmental contaminants. These nanomaterials are fast, accurate, economical, small in size, lightweight, portable, and simple in use for the detection of a variety of pollutants and toxic elements in the environment. Table 13.1 summarizes some of the important nanosensors for the detection of environmental pollutants.

13.2.1 Nanosensors for Water Contamination

Water is one of the important constituents for survival on the earth. In the past few years, there is severe scarcity of fresh and pure drinking water. Anthropogenic activities such as uncontrolled and untreated industrial discharge of contaminants, urbanization, domestic and sewage discharge, high demand due to population, mismanagement of agriculture and livestock waste, and heavy metals are the major sources of severe water pollution. It has been reported that globally 1.2 billion people do not have access to clean drinking water, whereas 2.6 billion people are forced to drink untidy, adulterate, or poor quality water. Furthermore, millions of people lost

Table 13.1 Nanosensors for detection of various types of pollutants.

Environmental pollutants	Contaminants	Nanomaterials	Nanosensors strategy	Reference
Water pollutants	Pb^{II}, Cu^{II}, Hg^{II}	Gold and silver nanoparticles	Optical colorimetric	[14–17]
	Organophosphate pesticides	Reduced graphene oxide/gold nanoparticles	Electrochemical	[18]
	Benzene, toluene, ethylbenzene, and xylene	Carbon nanotubes	Chemiresistors	[19]
	Halides and pyridine	Gold nanowires	Chemiresistors	[20]
	Heavy metals	Quantum dots	Fluorescence	[21]
Air pollutants	NO_2 gas	Tin oxide	Electronic noses	[22]
	CO and CH_4	SnO_2 quantum dots	Resistive	[23]
	CO and NO_2	Au–carbon nanotubes	Resistive	[24]
	H_2S and SO_2	Germanene nanosheet	Electrical	[25]
	VOCs	ZnO–CuO/carbon nanotubes	Gravimetrics	[26]
Soil pollutants	Pesticides	Silver/copper alloy nanoparticles and graphene nanocomposites	Electrochemical	[27]
	Organophosphate pesticides	CNTs	Electrochemical	[28]
Biological/microbial	Bacteria, viruses, protozoa	Gold nanoparticles	Optical-based surface-enhanced Raman spectroscopy	[29–32]

their life due to poor water quality and associated diseases [33, 34]. One of the major causes of poor water supply is the lack of efficient and simple technologies for water quality monitoring and diagnosis. There is an immense need to monitor water continuously for efficient identification and differentiation between potable and non-potable water. Nanomaterials due to their high surface-area-to-volume ratio and small size hold excellent photoluminescence, chemiluminescence, adsorbent, and redox properties [12]. These specific and exciting properties of nanomaterials make them prime candidates for monitoring and diagnosis of water as nanosensors.

Nanosensors fall into various types including optical nanosensors, chemical nanosensors, aptamer-based nanosensors, electrochemical, thermal, and nanobiosensors. Nanosensors consist of two components [3, 35]:

(1) Receptor: to enhance the sensitivity of detection
(2) Transducer: the sensing component that works with chemical, optical, electrochemical, thermal, and biological recognition codes

The basic operating mechanism of all nanosensors involves the transfer of charge between the pollutant molecule and the receptor, which generates an electrical or optical signal depending on the type of molecule [3, 36].

Differential aptamer-based graphene nanosensors can detect water pollutants such as 17β-estradiol (E2), an endocrine-disrupting chemical specifically in complex contaminated water [37]. Nanocomposite-based voltammetric sensors are used for detection of dihydroxybenzene isomers, hydroquinone, catechol, resorcinol, bisphenol A, and nitrophenol in water. Au NPs/CNTs/graphene hybrid nanocomposite sensors are reported to detect hydroquinone, catechol, resorcinol in the river and tap water with a maximum detection limit of 0.013 μM [38]. An optical nanosensor comprising gold nanoparticles was demonstrated to detect organo-pesticides and Hg (II) very proficiently in water through fluorescence-based emission spectra [12]. Electrochemical nanosensors based on Au–ZnS hybrid nanorods with Au-mediated electron relay nanocomposites and multiwalled carbon nanotubes/MnO_2 nanocomposites were reported for detection of 4-nitrophenol in natural water samples and river water, respectively [39, 40]. Silver nanoparticle/multiwalled carbon nanotube/nanocomposite-based electrochemical sensors are reported for voltammetric detection of bisphenol A, hydroquinone, phenol, and catechol in tap water [41]. Furthermore, researchers reported the dual ligands capped $CH_3NH_3PbBr_3$ perovskite quantum dots for visual fluorescence detection of fluoride with a detection limit of 3.2 μM [42]. Similarly, the perovskite component of cesium copper bromide ($CsCuBr_3$) is reported as a nanosensor to detect dimethylformamide in water [43]. Development of many nanomaterial-based sensing devices is under way and they will soon be available for use.

13.2.2 Nanosensors for Air Pollution

Air pollution is a serious problem for human health and the environment. Air pollution can cause catastrophic damages to all fragments of the environment. As per the World Health Organization (WHO), there are six main air pollutants: nitrogen oxide, sulfur dioxide, carbon monoxide, particulate matter, ozone, and lead [44]. Air pollutants cause a wide range of moderate-to-chronic diseases such as chronic obstructive pulmonary disease, asthma, cough, respiratory diseases, cardiovascular diseases, silicosis, lung fibrosis, leading to high mortality rates. The economic and industrialization race between the nations worldwide, deforestation, and urbanization are key contributors to air pollution. Additionally, natural phenomena such as volcanic eruption, forest fire, and storms cause air pollution. These air pollutants harm the environment and cause serious effects such as acid rain, haze, climate change, greenhouse effect, ozone depletion, ground-level ozone pollution, toxic gas poisoning.

The detection of toxic pollutants in the air is very challenging and requires a precise, specific, and economical technological solution. Swift and specific development of nanosensors may provide an opportunity to detect pollutants and other hazardous elements at the molecular level. Gas and electrochemical nanosensors are of high interest due to their rapid, on-the-spot, specific, precise detection and low-cost, small, and simple design. Single-walled nanotubes (SWNTs) have chemical sensing properties for NO*x* and NH_3 gases. Electrical resistance of SWNT sensors is found to change significantly after being exposed to NO*x* and NH_3 gases [45]. Nanosensors such as carbon-based nanosensors can detect a very small quantity of ammonia gas. Optically active palladium nanoparticle–based nanosensors can detect volatile organic compounds (VOCs) in a short time with high ability [12]. Researchers reported precise detection of particulate matter using gold nanoparticles coated SiO_2. The nanosensors for air pollution detection may help environmentalists and scientists to monitor air quality more efficiently, economically, rapidly, and accurately in near future.

13.2.3 Nanosensors for Soil Contamination

Constantly rising demand for food worldwide exerts massive pressure on the agricultural sector for more productivity. The use of chemical fertilizers, pesticides, urea, and herbicides increases several folds in agriculture for high productivity to meet the demand. The use of chemical fertilizers and pesticides, industrial seepage, heavy metal and solid waste, leaching of toxic materials from industrial and other sources into land causes unembellished contamination in soil and also degrades the soil fertility gradually. Leaching of these toxic and hazardous chemicals into the soil also contaminates the groundwater, causing severe water-related problems, too. Also, the use of chemical fertilizers and pesticides in agricultural practices leads to their accumulation in food products. This accumulation of chemicals in food, directly and indirectly, affects the consumer of the product and causes various diseases including several types of cancer.

The application of nanotechnology is wide and prominent in soil pollution detection and monitoring. A range of nanosensors are available for determination of soil pollutants, pesticide concentration, contaminants, and microbial presence. The inimitable surface chemistry of nanoparticles makes them chief applicants for sensing applications. As reported by various researchers, the optical property of gold and

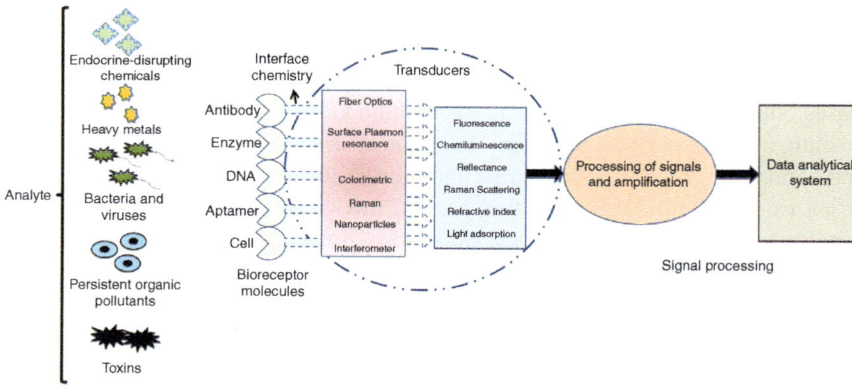

Figure 13.2 Systematic working of nanobiosensors.

silver nanoparticles helped in the detection of pesticides. Gold nanoparticle-based nanosensors have been developed to detect heavy metals and pesticides such as organophosphates, organochlorine, dichlorodiphenyltrichloroethane (DDT), and methamidophos. Also, gold nanoparticle-based surface-enhanced Raman spectroscopy-activated surfaces have been developed to detect phosmet and thiabendazole in apples. Silica-coated carbon nanotubes detect fenitrothion in orange juice. Citrate-capped silver nanoparticles are used to detect Dipterex in water [13, 46–48]. Micro-cantilever-based nanosensors are reported to determine total carbon in the soil. Multiwalled carbon nanotube-based nanosensors are well documented for their application in the detection of organophosphates and heavy metals in vegetables and other food products [12]. Nanosensing devices have the potential to transform agriculture and pollution control technologies in near future.

13.2.4 Nanobiosensors

A biosensor is a cutting-edge analytical device used to convert a biological response into an optical, electrical, or thermal signal. A typical biosensor consists of the following components:

(1) Receptor: analyte binds at the receptor site.
(2) Transducer: biosignal translates into more easily accessible forms such as optical, electrical, or thermal form.
(3) Amplifier: amplifies and processes the converted signal for clarity and understanding.
(4) Analysis or output: final processed signal available for analysis in the output form.

This process takes place in three steps: (1) binding at receptors, (2) transduction of signals, and (3) amplification and processing of signals (Figure 13.2).

Nanobiosensors are highly sensitive and economical biosensors consisting of nanomaterials. Nanobiosensors have wide applications in diagnosis, food analysis, monitoring and detection of environmental pollutants, agriculture, industrial process control, drug delivery, drug development, and biomolecule interactions [49]. The nanomaterials in biosensors increase sensitivity, selectivity to analytes and enhance response [13]. In environmental contamination control, nanobiosensors can be used for the detection of microorganisms such as total fecal coliforms, *Escherichia coli*, *Enterococcus*, viruses, and parasites for water quality monitoring [3, 50]. Researchers have reported enzyme-based platinum nanoparticles doped with surface functionalized carbon electrodes with acetylcholinesterase-immobilized nanobiosensors for detection of the pesticides carbofuran and methyl parathion. Similarly, nanobiosensors doped with gold nanorods were developed for detection of paraoxon [13, 51, 52]. In an article, the author reported various types of optical and electrochemical nanobiosensors used for rapid detection of water contaminants in few seconds to 30 minutes only. Pollutants such as Hg^{2+} and Pb^{2+} can be detected by using gold nanoparticle–based optical nanobiosensors within few seconds to 20 minutes Similarly, *E. coli* can be detected by using single-walled carbon nanotubes and/or graphene-based optical nanobiosensors in a very short time. Electrochemical nanobiosensors based on gold nanoparticles/carbon nanotubes and gold–platinum nanoparticles were reported for the detection of Hg^{2+}. Furthermore, gold nanoparticle–based nanobiosensors were reported for the detection of arsenic in water. Electrochemical nanobiosensors based on multiwalled carbon nanotubes/graphene oxide and graphene/Nafion were demonstrated to have the potential to detect Pb^{2+} in contaminated water [2]. The major advantages of using nanobiosensors for the detection of environmental pollutants are (i) fast (less time consumption), (ii) cost effective, (iii) specificity, (iv) precision (able to detect minimal quantity), (v) easy handling, (vi) portability, and (vii) continuous monitoring.

13.3 Nanotechnology for Environmental Remediation

Nanotechnological applications are not limited to detection or monitoring. Nanomaterials have been found decidedly useful for environmental remediation procedures. Nanotechnology-based remediation techniques such as photocatalytic oxidation, biodegradation, adsorption, filtration, and membrane osmosis are reported to be efficient and cost effective. Nanodevices cover the complete spectrum of environmental management processes from diagnosis to remediation with continuous monitoring.

Water and air pollution are particularly the most severe societal problem, and improper treatment or negligence can lead to serious health risks in the menacing living environment. Industrial sewage such as textile sewage containing dyes, mining sewage containing heavy metals and toxic chemicals, oil refinery waste, food industry waste, paper industry discharge, domestic waste, automobile smoke containing hazardous gases, industrial smoke containing toxic gases, fumes, construction dust, particulate matter, and many more have been severely impacting the planet environment and need to be rectified or controlled immediately. These pollutants are toxic, carcinogenic, and mutagenic. Industrial and commercial activities are certainly important for economic development, but they severely impact the

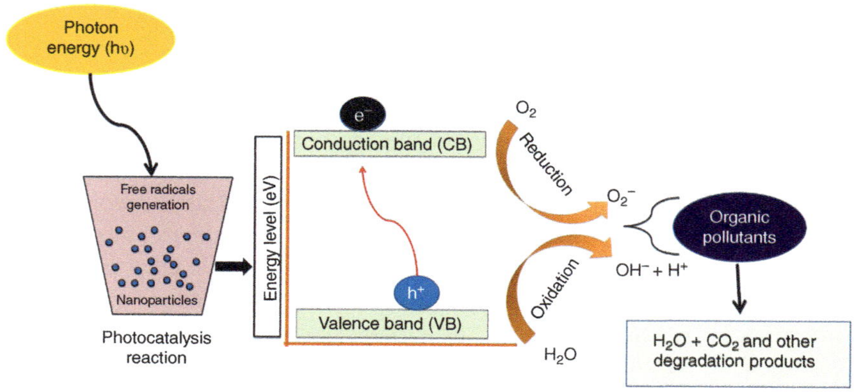

Figure 13.3 Mechanism of advanced oxidation process or photocatalysis.

entire environment, and it demands the present situation to find and opt sustainable development methods with proper waste management. Table 13.2 summarizes some of the nanomaterial-based environmental remediation methodologies.

13.3.1 Photocatalysis Or Advanced Oxidation Process for Environmental Remediation

Many organic dyes, organic chromogenic agents, and organic macromolecular compounds used in various industries such as pharmaceutical, paper and textile, ink, dyeing, leather, and many more have highly stable complex chemical structures and very difficult to degrade [8, 11, 72]. These pollutants pose serious environmental and health-related risks. A photocatalytic approach in the presence of nanocatalyst facilitates broad removal of these types of complex pollutants from the environment [11]. As depicted in Figure 13.3, photocatalysis methodology consists of activation of nanoparticles in UV radiation followed by the formation of redox atmosphere in aqueous solution, which further acts as a sensitizer for the light-tempted redox mechanism. Recently, nanooxide materials have emerged as efficient photocatalysts. Nanooxides with explicit structures, mainly with hollow structure, have received significant attention of researchers worldwide for their large specific surface area and low density [72].

The degradation of pollutants in the photocatalytic process in the presence of oxide nanoparticles is initiated by the photoexcitation process. The surface of the nanooxide material is illuminated with light energy, which leads to the generation of valance band holes and conduction band electrons. The holes were lastly trapped on the photocatalyst surface to provide highly oxidizing species, OH^- radicals. The reaction of conduction band electrons with dissolved oxygen generates superoxide radical anions O_2^-. Hydroxyl radicals HO_2 are generated due to protonation of superoxide radical anions O_2^-. Finally, due to the activity of the superoxide anions, pollutants get degraded. The oxidation–reduction of pollutants happened due to the conduction band electrons and valance band holes. The nanooxide particles have

Table 13.2 Nanomaterial-based solution for environmental remediation.

Environmental problem	Type	Nanomaterials	Type of remediation	Reference
Water	BSA and humic acid	CeO_2@HNTs	Nanopolymeric membrane	[53, 54]
	Wastewater	HNTs	Nanopolymeric membrane	[55]
	Textile dye	SnO_2, TiO_2, and Au	Advanced oxidation process	[8, 11, 56]
	Heavy metals	BiNPs@$Ti_3C_2T_x$	Electrochemical	[57]
	Carbendazim and BPA	Magnetic nanoparticles mesoporous nanosponge	Adsorption and plasmonic	[58]
	Other gases	Nanorod bimetallic organic frameworks	Fluorescence quenching	[59]
Air	Sulfur dioxide	Ni_3BTC_2/OH-CNTs	FET	[60]
	Nitrogen oxides	Pristine graphene	Raman spectra band-shift	[61]
	Carbon monoxide	Bismuth ferrite	Conductive surface phenomenon	[62]
	Particulate matter	SiO_2 coated with Au nanoparticles	Surface-enhanced Raman spectroscopy	[63]
Soil	Aromatic hydrocarbons	Nanocomposites	Fluorescence, surface-enhanced Raman spectrometry, and electrochemistry	[64]
	Pesticides	rGQDs and MWCNTs	Aptamer	[65]
	Heavy metals	$MnCo_2O_4$	Electrochemical	[66]
	Fertilizers	Polymer CuO nanoparticles	Nanofertilizers	[67]
	Pesticides	Ag and TiO_2	Fluorometric	[68, 69]
Biological/microbial	SARS-CoV-2	TiO_2 and Ag zeolite	ROS mediated	[70]
	Dengue detection	Porous silicon Microcavity (PSMC)	UV–vis spectroscopy	[71]

great potential for photocatalytic degradation of pollutants such as dyes, phenols, chromogenic agents, and other organic macromolecular compounds [8, 11]. Nanooxides such as SnO_2 are reported to be efficient photocatalysts for dye degradation [11]. SnO_2 nanofibers decorated with N-doped ZnO nanonodules were also reported as significant photocatalysts [5]. Researchers have reported successful and efficient water treatments using TiO_2-decorated functionalized halloysite nanotubes (TiO_2@ HNTs) and photocatalytic PVC (polyvinyl chloride) membranes [56]. Au nanoparticles due to their high surface-to-volume ratio were demonstrated to be an effectual photocatalyst for degradation of 2,4 dichlorophenol [8]. One study reported that Sn_3O_4–CNT hybrid nanocomposites can generate more $\cdot OH$, $\cdot O_2^-$, and H^+, which accelerates the degradation of pollutants into nontoxic materials [72, 73]. The simultaneous reduction of Cr (VI) and oxidation of 4-nitrophenols in the presence of TiO_2 nano-photocatalysts under visible light were reported effective [74].

13.3.2 Nanocomposites and Nanodevices for Environmental Remediation

Pollutants such as toxic gases, VOC, airborne particles, particulate matter in air and heavy metals, organic substances, agriculture and livestock waste, and microbial pathogens are released in the environment in various forms and cause health and environmental hazards [3]. Global warming is another serious issue occurring majorly due to greenhouse gases, and carbon dioxide contributes 75% of total greenhouse gases, others are methane, nitrous oxide, and fluorinated gases [3, 75]. Nanomaterials can play a significant role in the control and removal of these pollutants.

The extensively utilized nanomaterial for pesticide remediation is graphene oxide–based nanocomposite fabrication with metal or metal-oxide nanoparticles [13]. The FeO-based nanocomposite shows a promising response for water desalination (removal of salt) [6]. Silver nanoparticles are well known for their antimicrobial properties, and presently many studies have been conducted to derive their application in textile, cosmetics, spray disinfectants, refrigerators, and many other products [45]. In another study, nickel nanoparticles were used as a catalyst for the decomposition of methane into hydrogen and carbon nanofibers [76]. Exposure of zinc oxide and zirconium hydroxide nanoparticles for adsorption of sulfur dioxide and nitrogen dioxide reported promising, and the same was confirmed by using photoluminescence spectra of nanoparticles [7]. Researchers demonstrated the Li-LSX (low-silica type-X zeolite) mediated removal of CO_2, CH_4, and water vapors via adsorption [77]. In another study, adsorption of atmospheric lead was demonstrated on silica nanoparticles [21]. Carbon dioxide was also successfully adsorbed on nanosized NaKa zeolite and the nanomaterial showed rapid carbon dioxide adsorption kinetics [75]. Various nanomaterials are available for remediation of environmental pollution. Single and multiwalled carbon nanotubes, metal-oxide and metal nanoparticles, bimetallic nanoparticles, nanorods, and nanocomposites can be efficiently used for removal or remediation of environmental pollution. Several studies have demonstrated promising and efficient responses for pollution remediation using nanomaterials. Nanotechnology is efficient to convert pollutants into labile from non-labile phases.

13.4 Conclusion

Global industrialization and expansively rising urbanization impact the environment harshly. The gradually rising pollution contaminates air, water, and land extensively and intimidates health and life on the planet. Despite of many attempts to achieve maximum environmental remediation, none of them was very efficient. In the past few years, nanotechnology emerges as a promising tool to encounter pollution-related problems. The development of various nanodevices for diagnosis, monitoring, and remediation of environmental pollution embarks on a new hope for achieving sustainable development. Nanotechnology offers rapid, robust, advanced, and economical processes for monitoring, sensing, and treatment of pollution. The application of nanodevices in environmental operations provides an opportunity to achieve the target of zero waste discharge, conversion of toxic pollutants to nontoxic entities, safe and pure drinking water availability, and clean air. The use of the photocatalysis process, adsorption of toxic materials on nanomaterial surfaces, and pollution removal using nanocomposites show very exciting outcomes in various studies worldwide. Nanotechnology covers all the aspects of environmental problems by involving cutting-edge technologies for monitoring, diagnosis, and treatment of various toxic and hazardous contaminants to unfold pulsating and adaptable systems for sustainable development.

References

1 WHO (2006). *Almost a Quarter of All Diseases Caused by Environmental Exposure*. Geneva: World Health Organization (WHO) https://www.who.int/mediacentre/news/releases/2006/pr32/en/.

2 Hernandez-Vargas, G., Sosa-Hernández, J.E., Saldarriaga-Hernandez, S. et al. (2018). Electrochemical biosensors: a solution to pollution detection with reference to environmental contaminants. *Biosensors.* 8 (2): 29.

3 Ibrahim, R.K., Hayyan, M., AlSaadi, M.A. et al. (2016). Environmental application of nanotechnology: air, soil, and water. *Environ. Sci. Pollut. Res.* 23 (14): 13754–13788.

4 Baltrusaitis, J., Jayaweera, P.M., and Grassian, V.H. (2011). Sulfur dioxide adsorption on TiO2 nanoparticles: influence of particle size, coadsorbates, sample pretreatment, and light on surface speciation and surface coverage. *J. Phys. Chem. C* 115 (2): 492–500.

5 Lee, J.S., Kwon, O.S., and Jang, J. (2012). Facile synthesis of SnO2 nanofibers decorated with N-doped ZnO nanonodules for visible light photocatalysts using single-nozzle co-electrospinning. *J. Mater. Chem.* 22 (29): 14565–14572.

6 Nishant Srivastava, M.M. (ed.) (2014). Bacteria mediated biosynthesis of FeO nanoparticles for desalination of sea water using thin film nanocomposite membrane. In: *AIChE Annual Meeting: Sepration Division: Membranes for Water Treatment Applications II; 2014*, 2014. Atlanta, USA: AIChE.
7 Singh, J., Mukherjee, A., Sengupta, S.K. et al. (2012). Sulfur dioxide and nitrogen dioxide adsorption on zinc oxide and zirconium hydroxide nanoparticles and the effect on photoluminescence. *Appl. Surf. Sci.* 258 (15): 5778–5785.
8 Srivastava, N. and Mukhopadhyay, M. (2015). Ralstonia eutropha (*Cupriavidus* metallidurans) mediated biosynthesis of gold nanoparticles and catalytic treatment of 2, 4 dichlorophenol. *Synth. React. Inorg. Metal-Org. Nano-Metal Chem.* 45 (2): 238–247.
9 Srivastava, N. and Mukhopadhyay, M. (2016). Green synthesis and structural characterization of CdO nanoparticles. *Adv. Sci. Lett.* 22 (4): 929–934.
10 Zhao, W., Wang, Z., Shen, X. et al. (2012). Hydrogen generation via photoelectrocatalytic water splitting using a tungsten trioxide catalyst under visible light irradiation. *Int. J. Hydrog. Energy* 37 (1): 908–915.
11 Srivastava, N. and Mukhopadhyay, M. (2014). Biosynthesis of SnO2 nanoparticles using bacterium erwinia herbicola and their photocatalytic activity for degradation of dyes. *Ind. Eng. Chem. Res.* 53 (36): 13971–13979.
12 Debnath, N. and Das, S. (2020). Nanobiosensor: current trends and applications. In: *NanoBioMedicine* (eds. S.K. Saxena and S.M.P. Khurana), 389–409. Singapore: Springer Singapore.
13 Rawtani, D., Khatri, N., Tyagi, S., and Pandey, G. (2018). Nanotechnology-based recent approaches for sensing and remediation of pesticides. *J. Environ. Manag.* 206: 749–762.
14 Liu, J. and Lu, Y. (2006). Fast colorimetric sensing of adenosine and cocaine based on a general sensor design involving aptamers and nanoparticles. *Angew. Chem.* 118 (1): 96–100.
15 Beqa, L., Singh, A.K., Khan, S.A. et al. (2011). Gold nanoparticle-based simple colorimetric and ultrasensitive dynamic light scattering assay for the selective detection of Pb (II) from paints, plastics, and water samples. *ACS Appl. Mater. Interfaces* 3 (3): 668–673.
16 Zhang, Z., Chen, Z., Qu, C., and Chen, L. (2014). Highly sensitive visual detection of copper ions based on the shape-dependent LSPR spectroscopy of gold nanorods. *Langmuir* 30 (12): 3625–3630.
17 Vilela, D., González, M.C., and Escarpa, A. (2012). Sensing colorimetric approaches based on gold and silver nanoparticles aggregation: chemical creativity behind the assay. A review. *Anal. Chim. Acta* 751: 24–43.
18 Yang, Y., Asiri, A.M., Du, D., and Lin, Y. (2014). Acetylcholinesterase biosensor based on a gold nanoparticle–polypyrrole–reduced graphene oxide nanocomposite modified electrode for the amperometric detection of organophosphorus pesticides. *Analyst* 139 (12): 3055–3060.
19 Rasheed, T., Nabeel, F., Adeel, M. et al. (2019). Carbon nanotubes-based cues: a pathway to future sensing and detection of hazardous pollutants. *J. Mol. Liq.* 292: 111425.

20 Muratova, I.S., Mikhelson, K.N., Ermolenko, Y.E. et al. (2016). Chemiresistors based on ultrathin gold nanowires for sensing halides, pyridine and dopamine. *Sensors Actuators B Chem.* 232: 420–427.

21 Yang, X., Shen, Z., Zhang, B. et al. (2013). Silica nanoparticles capture atmospheric lead: Implications in the treatment of environmental heavy metal pollution. *Chemosphere* 90 (2): 653–656.

22 Sayago, I., Aleixandre, M., and Santos, J.P. (2019). Development of tin oxide-based nanosensors for electronic nose environmental applications. *Biosensors.* 9 (1): 21.

23 Sedghi, S.M., Mortazavi, Y., and Khodadadi, A. (2010). Low temperature CO and CH4 dual selective gas sensor using SnO2 quantum dots prepared by sonochemical method. *Sensors Actuators B Chem.* 145 (1): 7–12.

24 Zanolli, Z., Leghrib, R., Felten, A. et al. (2011). Gas sensing with Au-decorated carbon nanotubes. *ACS Nano* 5 (6): 4592–4599.

25 Hussain, T., Kaewmaraya, T., Chakraborty, S. et al. (2018). Defected and functionalized germanene-based nanosensors under sulfur comprising gas exposure. *ACS Sens.* 3 (4): 867–874.

26 Abraham, N., Krishnakumar, R.R., Unni, C., and Philip, D. (2019). Simulation studies on the responses of ZnO-CuO/CNT nanocomposite based SAW sensor to various volatile organic chemicals. *J. Sci.: Adv. Mater. Dev.* 4 (1): 125–131.

27 Sreedhar, N., Kumar, M.S., and Krishnaveni, K. (2015). Sensitive determination of chlorpyrifos using Ag/Cu alloy nanoparticles and graphene composite paste electrode. *Sensors Actuators B Chem.* 210: 475–482.

28 Ko, W.-K., Heo, D.N., Moon, H.-J. et al. (2015). The effect of gold nanoparticle size on osteogenic differentiation of adipose-derived stem cells. *J. Colloid Interface Sci.* 438: 68–76.

29 Ackermann, K.R., Henkel, T., and Popp, J. (2007). Quantitative online detection of low-concentrated drugs via a SERS microfluidic system. *ChemPhysChem* 8 (18): 2665–2670.

30 Hadjigeorgiou, K., Kastanos, E., and Pitris, C. (eds.) (2013). Multi-bacteria multi-antibiotic testing using surface enhanced Raman spectroscopy (SERS) for urinary tract infection (UTI) diagnosis. In: *European Conference on Biomedical Optics*. Optical Society of America.

31 Zhang, C., Wang, L., Tu, Z. et al. (2014). Organophosphorus pesticides detection using broad-specific single-stranded DNA based fluorescence polarization aptamer assay. *Biosens. Bioelectron.* 55: 216–219.

32 Halvorson, R.A. and Vikesland, P.J. (2010). *Surface-Enhanced Raman Spectroscopy (SERS) for Environmental Analyses*. ACS Publications.

33 Patel, B.R., Noroozifar, M., and Kerman, K. (2020). Review—nanocomposite-based sensors for voltammetric detection of hazardous phenolic pollutants in water. *J. Electrochem. Soc.* 167 (3): 037568.

34 WHO (2019). *Water, Sanitation, Hygiene and Health: A Primer for Health Professionals*. Geneva: Water, Sanitation, Hygiene and Health Unit WHO https://who.int/water_sanitation_health/en/.

35 Su, S., Wu, W., Gao, J. et al. (2012). Nanomaterials-based sensors for applications in environmental monitoring. *J. Mater. Chem.* 22 (35): 18101–18110.

36 Di Francia, G., Alfano, B., and La Ferrara, V. (2009). Conductometric gas nanosensors. *J. Sens.* 2009: 659275.

37 Li, Y., Zhu, Y., Wang, C. et al. (2019). Selective detection of water pollutants using a differential aptamer-based graphene biosensor. *Biosens. Bioelectron.* 126: 59–67.

38 Chen, T.-W., Yu, X.-N., and Li, S.-J. (2019). Simultaneous determination of dihydroxybenzene isomers using glass carbon electrode modified with 3D CNT-graphene decorated with au nanoparticles. *Int. J. Electrochem. Sci.* 14: 7037–7046.

39 Kim, Y., Giribabu, K., Kim, J.G. et al. (2019). Electrochemical sensors based on Au-ZnS hybrid nanorods with au-mediated efficient electron relay. *ACS Sustain. Chem. Eng.* 7 (4): 4094–4102.

40 Anbumannan, V., Dinesh, M., Rajendra Kumar, R.T., and Suresh, K. (2019). Hierarchical α-MnO_2 wrapped MWCNTs sensor for low level detection of p-nitrophenol in water. *Ceram. Int.* 45 (17, Part B): 23097–23103.

41 Goulart, L.A., Gonçalves, R., Correa, A.A. et al. (2017). Synergic effect of silver nanoparticles and carbon nanotubes on the simultaneous voltammetric determination of hydroquinone, catechol, bisphenol A and phenol. *Microchim. Acta* 185 (1): 12.

42 Lu, L.-Q., Ma, M.-Y., Tan, T. et al. (2018). Novel dual ligands capped perovskite quantum dots for fluoride detection. *Sensors Actuators B Chem.* 270: 291–297.

43 Aamir, M., Sher, M., Malik, M.A. et al. (2016). A facile approach for selective and sensitive detection of aqueous contamination in DMF by using perovskite material. *Mater. Lett.* 183: 135–138.

44 Manisalidis, I., Stavropoulou, E., Stavropoulos, A., and Bezirtzoglou, E. (2020). Environmental and health impacts of air pollution: a review. *Front. Public Health* 8: 14.

45 Yunus, I.S., Harwin, Kurniawan, A. et al. (2012). Nanotechnologies in water and air pollution treatment. *Environ. Technol. Rev.* 1 (1): 136–148.

46 Luo, H., Huang, Y., Lai, K. et al. (2016). Surface-enhanced Raman spectroscopy coupled with gold nanoparticles for rapid detection of phosmet and thiabendazole residues in apples. *Food Control* 68: 229–235.

47 Lisa, M., Chouhan, R.S., Vinayaka, A.C. et al. (2009). Gold nanoparticles based dipstick immunoassay for the rapid detection of dichlorodiphenyltrichloroethane: an organochlorine pesticide. *Biosens. Bioelectron.* 25 (1): 224–227.

48 Li, Z., Wang, Y., Ni, Y., and Kokot, S. (2014). Unmodified silver nanoparticles for rapid analysis of the organophosphorus pesticide, dipterex, often found in different waters. *Sensors Actuators B Chem.* 193: 205–211.

49 Mukherjee, A., Bhattacharya, J., and Moulick, R.G. (2020). Nanodevices: the future of medical diagnostics. In: *NanoBioMedicine* (eds. S.K. Saxena and S.M.P. Khurana), 371–388. Singapore: Springer Singapore.

50 Theron, J., Eugene Cloete, T., and de Kwaadsteniet, M. (2010). Current molecular and emerging nanobiotechnology approaches for the detection of microbial pathogens. *Crit. Rev. Microbiol.* 36 (4): 318–339.
51 Yang, L., Wang, G., and Liu, Y. (2013). An acetylcholinesterase biosensor based on platinum nanoparticles–carboxylic graphene–nafion-modified electrode for detection of pesticides. *Anal. Biochem.* 437 (2): 144–149.
52 Lang, Q., Han, L., Hou, C. et al. (2016). A sensitive acetylcholinesterase biosensor based on gold nanorods modified electrode for detection of organophosphate pesticide. *Talanta* 156–157: 34–41.
53 Mishra, G. and Mukhopadhyay, M. (2021). Well dispersed CeO2@ HNTs nanofiller in the poly (vinyl chloride) membrane matrix and enhanced its antifouling properties and rejection performance. *J. Environ. Chem. Eng.* 9 (1): 104734.
54 Mishra, G. and Mukhopadhyay, M. (2017). Flux improvement, rejection, surface energy and antibacterial properties of synthesized TiO 2-Mo. HNTs/PVC nanocomposite ultrafiltration membranes. *New J. Chem.* 41 (24): 15049–15057.
55 Mishra, G. and Mukhopadhyay, M. (2018). Enhanced antifouling performance of halloysite nanotubes (HNTs) blended poly (vinyl chloride)(PVC/HNTs) ultrafiltration membranes: For water treatment. *J. Ind. Eng. Chem.* 63: 366–379.
56 Mishra, G. and Mukhopadhyay, M. (2019). TiO 2 decorated functionalized halloysite nanotubes (TiO 2@ HNTs) and photocatalytic PVC membranes synthesis, characterization and its application in water treatment. *Sci. Rep.* 9 (1): 1–17.
57 He, Y., Ma, L., Zhou, L. et al. (2020). Preparation and application of bismuth/MXene nano-composite as electrochemical sensor for heavy metal ions detection. *Nanomaterials.* 10 (5): 866.
58 Zhang, L., Hao, R., You, H. et al. (2021). Ultra-rapid and highly efficient enrichment of organic pollutants via magnetic nanoparticles/mesoporous nanosponge compounds for ultrasensitive nanosensors. *J. Water Process Eng.* https://doi.org/10.21203/rs.3.rs-127668/v1.
59 Lian, X. and Yan, B. (2019). A dual-functional bimetallic-organic framework nanosensor for detection and decontamination of lachrymator in drinking water. *Sensors Actuators B Chem.* 281: 168–174.
60 Ingle, N., Mane, S., Sayyad, P. et al. (2020). Sulfur dioxide (SO2) detection using composite of nickel benzene carboxylic (Ni3BTC2) and OH-functionalized single walled carbon nanotubes (OH-SWNTs). *Front. Mater.* 7: 93.
61 Santos, J., Polichctti, T., Hontañón, E. et al. (eds.) (2018). Study of graphene based nanosensors for the detection of nitrogen dioxide. In: *2018 Spanish Conference on Electron Devices (CDE)*. IEEE.
62 Chakraborty, S. and Pal, M. (2018). Highly efficient novel carbon monoxide gas sensor based on bismuth ferrite nanoparticles for environmental monitoring. *New J. Chem.* 42 (9): 7188–7196.
63 Zhang, D., Peng, L., Shang, X. et al. (2020). Buoyant particulate strategy for few-to-single particle-based plasmonic enhanced nanosensors. *Nat. Commun.* 11 (1): 1–7.

64 Nsibande, S., Montaseri, H., and Forbes, P. (2019). Advances in the application of nanomaterial-based sensors for detection of polycyclic aromatic hydrocarbons in aquatic systems. *TrAC Trends Anal. Chem.* 115: 52–69.

65 Talari, F.F., Bozorg, A., Faridbod, F., and Vossoughi, M. (2021). A novel sensitive aptamer-based nanosensor using rGQDs and MWCNTs for rapid detection of diazinon pesticide. *J. Environ. Chem. Eng.* 9 (1): 104878.

66 Antunović, V., Ilić, M., Baošić, R. et al. (2019). Synthesis of MnCo2O4 nanoparticles as modifiers for simultaneous determination of Pb (II) and Cd (II). *PLoS One* 14 (2): e0210904.

67 Leonardi, M., Caruso, G.M., Carroccio, S.C. et al. (2021). Smart nanocomposites of chitosan/alginate nanoparticles loaded with copper oxide as alternative nanofertilizers. *Environ. Sci.: Nano.* 8 (1): 174–187.

68 Hsu, C.-W., Lin, Z.-Y., Chan, T.-Y. et al. (2017). Oxidized multiwalled carbon nanotubes decorated with silver nanoparticles for fluorometric detection of dimethoate. *Food Chem.* 224: 353–358.

69 Saraji, M., Jafari, M.T., and Mossaddegh, M. (2016). Halloysite nanotubes-titanium dioxide as a solid-phase microextraction coating combined with negative corona discharge-ion mobility spectrometry for the determination of parathion. *Anal. Chim. Acta* 926: 55–62.

70 Talebian, S., Wallace, G.G., Schroeder, A. et al. (2020). Nanotechnology-based disinfectants and sensors for SARS-CoV-2. *Nat. Nanotechnol.* 15 (8): 618–621.

71 Mishra, V., Patel, P., Kumari, S., and Mishra, G. (2016). Dengue NS1 detection used chemically modified porous silicon microcavity (psmc). *SILICON* 8 (3): 401–407.

72 Zhao, J., Shao, Q., Ge, S. et al. (2020). Advances in template prepared nano-oxides and their applications: polluted water treatment, energy, sensing and biomedical drug delivery. *Chem. Rec.* 20 (7): 710–729.

73 Sun, M., Yan, T., Tingting, W. et al. (2018). Self-assembled hierarchical Sn3O4multi-wall carbon nanotubes: Facile fabrication, promoted charge separation, and enhanced photocatalytic performances. *Mater. Res. Bull.* 103: 104–113.

74 Sun, B., Reddy, E.P., and Smirniotis, P.G. (2005). Visible light Cr(VI) reduction and organic chemical oxidation by TiO2 photocatalysis. *Environ. Sci. Technol.* 39 (16): 6251–6259.

75 Cheung, O., Bacsik, Z., Liu, Q. et al. (2013). Adsorption kinetics for CO2 on highly selective zeolites NaKA and nano-NaKA. *Appl. Energy* 112: 1326–1336.

76 Wang, H.Y. and Lua, A.C. (2012). Development of metallic nickel nanoparticle catalyst for the decomposition of methane into hydrogen and carbon nanofibers. *J. Phys. Chem. C* 116 (51): 26765–26775.

77 Epiepang, F.E., Li, J., Liu, Y., and Yang, R.T. (2016). Low-pressure performance evaluation of CO2, H2O, and CH4 on Li-LSX as a superior adsorbent for air pre purification. *Chem. Eng. Sci.* 147: 100–108.

Index

Miniaturized Analytical Devices: Materials and Technology, First Edition.
Edited by Suresh Kumar Kailasa and Chaudhery Mustansar Hussain.

Z

Printed and bound by CPI Group (UK) Ltd, Croydon, CR0 4YY